RESILIENCE AND SUSTAINABILITY FOR ENERGY AND WATER

New Advancements, Opportunities, and Framework

RESILIENCE AND SUSTAINABILITY FOR ENERGY AND WATER

New Advancements, Opportunities, and Framework

Edited by

Cristóbal Noé Aguilar, PhD

A. K. Haghi, PhD

Amritanshu Shukla, PhD

Ann Rose Abraham, PhD

Swati Gokul Talele, PhD, MPharm

AAP | APPLE ACADEMIC PRESS

First edition published 2025

Apple Academic Press Inc.
1265 Goldenrod Circle, NE,
Palm Bay, FL 32905 USA

760 Laurentian Drive, Unit 19,
Burlington, ON L7N 0A4, CANADA

CRC Press
2385 NW Executive Center Drive,
Suite 320, Boca Raton FL 33431

4 Park Square, Milton Park,
Abingdon, Oxon, OX14 4RN UK

Library and Archives Canada Cataloguing in Publication

Title: Resilience and sustainability for energy and water : new advancements, opportunities, and framework / edited by: Cristóbal Noé Aguilar, A.K. Haghi, Amritanshu Shukla, Ann Rose Abraham, Swati Gokul Talele.

Names: Aguilar, Cristóbal Noé, editor | Haghi, A. K., editor | Shukla, Amritanshu, editor | Abraham, Ann Rose, editor | Talele, Swati Gokul, editor

Description: First edition. | Includes bibliographical references and index.

Identifiers: Canadiana (print) 20240421566 | Canadiana (ebook) 20240421620 | ISBN 9781774916643 (hardcover) | ISBN 9781774916650 (softcover) | ISBN 9781003515845 (ebook)

Subjects: LCSH: Energy industries—Environmental aspects. | LCSH: Water-supply—Management. | LCSH: Nanotechnology. | LCSH: Sustainability.

Classification: LCC HD9502.A2 R47 2024 | DDC 333.79—dc23

Library of Congress Cataloging-in-Publication Data

CIP data on file with US Library of Congress

ISBN: 978-1-77491-664-3 (hbk)
ISBN: 978-1-77491-665-0 (pbk)
ISBN: 978-1-00351-584-5 (ebk)

About the Editors

Cristóbal Noé Aguilar, PhD
Universidad Autónoma de Coahuila, BBG-DIA—Bioprocesses & Bioproducts Research Group, Food Research Department - Facultad de Ciencias Químicas, Saltillo, Coahuila, México

Cristóbal Noé Aguilar, Ph.D., is the Director of Research and Postgraduate Programs at the Universidad Autonoma de Coahuila, Mexico. Dr. Aguilar has published more than 530 papers in indexed journals, and has 550 contributions in scientific meetings. He has also edited more than 30 books. He has been awarded several prizes and awards, the most important of which are the National Prize of Research (2010) from the Mexican Academy of Sciences; the Prize CCC-2008 from the Mexican Society of Biotechnology and Bioengineering; National Prize AgroBio-2005; and the Mexican Prize in Food Science and Technology. In 2019, he received the Outstanding Research Award from the International Association of Bioprocessing (IBA). Dr. Aguilar is a member of the Mexican Academy of Science, the Central Council of the International Bioprocessing Association, the Mexican Academy of Sciences, the Mexican Society for Biotechnology and Bioengineering, and the Mexican Association for Food Science and Biotechnology. He has developed more than 41 research projects, including 25 international exchange projects.

A. K. Haghi, PhD
Research Associate, Department of Chemistry, University of Coimbra, Portugal

A. K. Haghi, Ph.D., is a retired professor and has written, co-written, edited, or co-edited more than 1000 publications, including books, book chapters, and papers in refereed journals with over 3800 citations and h-index of 32, according to the Google Scholar database. He is currently a research associate at the University of Coimbra, Portugal. Professor Haghi has received several grants, consulted for several major corporations, and is a frequent speaker to national and international audiences. He is the Founder and former Editor-in-Chief of the *International Journal of Chemoinformatics and Chemical Engineering and Polymers Research Journal*. Professor Haghi has acted as an editorial board member of many international journals. He has served as a member of the Canadian Research and Development Center of Sciences & Cultures. He has supervised several Ph.D. and M.Sc. theses at the University of Guilan (UG) and co-supervised international doctoral projects. Professor Haghi holds a BSc degree in urban and environmental engineering from the University of North Carolina (USA) and holds two MSc degrees, one in mechanical engineering from North Carolina State University (USA) and another in applied mechanics, acoustics, and materials from the Université de Technologie de Compiègne (France). He was awarded a PhD in engineering sciences at the Université de Franche-Comté (France). He is a regular reviewer for leading international journals.

Amritanshu Shukla, PhD
Professor in Physics, University of Lucknow, Lucknow, Uttar Pradesh, India

Amritanshu Shukla, Ph.D., is a Professor of Physics at the University of Lucknow, India. Prior to this, he has worked as an associate professor at the Rajiv Gandhi Institute of Petroleum Technology (RGIPT) (set up through an Act of Parliament by the Ministry of Petroleum & Natural Gas, as an Institute of National Importance on the lines of IITs). His research interests include nuclear physics and physics of renewable energy resources. He has published more than 150 research papers in various international and national journals and conferences. He has delivered invited talks at national and international institutes. Currently, he is working on several national and international projects and has active research collaborations with institutions in India and abroad. His recent book publications include *Low Carbon Energy Supply: Trends, Technology, Management; Sustainability through Energy-Efficient Buildings; Energy Security and Sustainability;* and *Latent Heat-Based Thermal Energy Storage Systems*. Dr. Shukla earned his master's degree in Physics and completed his Ph.D. from IIT Kharagpur, India. He did his postdoctoral research work at some of the premier institutes across the globe, namely, at the Institute of Physics Bhubaneswar (Department of Atomic Energy, Government of India); University of North Carolina Chapel Hill, USA; University of Rome/Gran Sasso National Laboratory, Italy; and Physical Research Laboratory Ahmedabad (Department of Space, Government of India).

Ann Rose Abraham, PhD
Assistant Professor, Sacred Heart College (Autonomous), Thevara, Kochi, Kerala, India

Ann Rose Abraham, PhD, is an assistant professor at the Department of Physics, Sacred Heart College (Autonomous), Thevara, Kochi, Kerala, India. She has expertise in the field of condensed matter physics, nanomagnetism, multiferroics, polymeric nanocomposites, etc. She has research experience at various reputed national institutes, including Bose Institute, Kolkata, India; SAHA Institute of Nuclear Physics, Kolkata, India; UGC-DAE CSR Centre, Kolkata, India; and she has collaborated with various international laboratories. She is a recipient of a young researcher award in physics and best paper awards in 2020 and 2021. She served as an assistant professor and examiner at the Department of Basic Sciences, Amal Jyothi College of Engineering, under APJ Abdul Kalam Technological University, Kerala, India. Dr. Abraham is a frequent speaker at national and international conferences. She has a good number of publications to her credit in many peer-reviewed high-impact international journals.

Swati Gokul Talele, PhD, MPharm
Assistant Professor, Department of Pharmaceutics, Sandip Institute of Pharmaceutical Sciences, Savitribai Phule Pune University, Pune, Maharashtra, India

Swati Gokul Talele, PhD, is currently serving as an assistant professor, Department of Pharmaceutics at Sandip Institute of Pharmaceutical Sciences. She has 18 years of experience in research along with teaching. She has published more than 20 research papers in various reputed international and national journals as well as more than 30 review papers. She has also presented research work at several conferences and has received several awards. She has authored the book titled *Natural Excipients* and has many chapters and books that are in progress. She has supervised many M.Pharm. students and is currently associated with many research projects. She also worked as a college examination officer (CEO) for more than three years and is a member of the university examination committee. Dr. Talele has delivered interactive talks at continuous education programs for registered pharmacists and is a life member of the Association of Pharmacy Teachers of India (APTI) and a member of the Indian Pharmaceutical Congress Association (IPCA). Her interests are in the field of nanotechnology, natural polymers, herbal formulations, and radio-labeling-based biodistribution studies.

Contents

Contributors .... *xi*

Preface .... *xv*

1. **Introduction to Sustainable Development and Resilience** **1**
Mónica M. Rodríguez-Garza, José Sandoval Cortes, and Cristóbal N. Aguilar

2. **Feasibilities and Difficulties of Scaling-up Sophisticated Nanomaterials for Green Energy** **9**
Mamata Singh and Keshav S. Pentapati

3. **Nanotechnology for Quaternary Cu_2ZnSnS_4 (CZTS) Solar Cells** **31**
M. V. Santhosh, R. Geethu, and Anju Nair

4. **Electrical Energy Transport through Correlated Barrier Hopping Mechanism in Silver-Decorated Graphene Oxide Nanolayers** **57**
I. Dhanya, S. Heera, Anju Elizabeth Yohannan, Aswini Mariam Vinod, Meenakshy Salim, Asiya Shafeek, and Sreejith K. Pisharady

5. **Laser Device as Optical Energy Source and Its Applications** **73**
Rinsy Thomas, N. V. Unnikrishnan, and K. V. Arun Kumar

6. **Carbon-Based Materials for High-Yield Optical Device Applications** .. **105**
Jincemon Cyriac, Claris Alex, Aneetta Shaji, Sunny Mathew, V. N. Sheeja, Tina Sebastian, and Saritha K. Jose

7. **An Overview of Nanocellulose-Based Nanocomposites for Water Purification** **143**
Eapen Philip, P. H. Vaisakh, Sneha Sara Varghese, Tressia Alias Princy Paulose, Linda E. Jacob, and H. Akhina

8. **Nanocellulose: A Versatile Green Sustainable Material for Water Remediation** **161**
Rekha Mirle

9. **Reliable, Resilient, and Sustainable Water Management of Pharmaceutical Industry** **169**
Mansi P. Bandawane, Deepti N. Chaudhari, Priyanka R. Gupta, Swati G. Talele, and Laxmikant B. Borse

10. Rare Earth-Activated Phosphors for Solid-State Lighting Applications: A Case Study on Analyzing the Role of Ce^{3+} as an Efficient Sensitizer in Nanophosphors 273

S. Sisira, Ann Rose Abraham, Kamal P. Mani, Linju Ann Jacob, Cyriac Joseph, and M. Kailasnath

Index...299

Contributors

Ann Rose Abraham
Department of Physics, Sacred Heart College (Autonomous), Thevara, Ernakulam, Kerala, India

Cristóbal N. Aguilar
School of Chemistry, Universidad Autónoma de Coahuila, Saltillo, México

H. Akhina
International and Interuniversity Centre for Nanoscience and Nanotechnology, Mahatma Gandhi University, Kottayam, Kerala, India
Govt Technical High School, Krishnapuram, Kerala, India

Claris Alex
Physics Department, Marian College Kuttikkanam (Autonomous), Kuttikkanam, Peermade, Idukki, Kerala, India

Mansi P. Bandawane
Department of Pharmaceutics, Sandip Institute of Pharmaceutical Sciences, Nashik, Maharashtra, India

Laxmikant B. Borse
Department of Pharmaceutics, Sandip Institute of Pharmaceutical Sciences, Nashik, Maharashtra, India

Deepti N. Chaudhari
Department of Pharmaceutics, Sandip Institute of Pharmaceutical Sciences, Nashik, Maharashtra, India

José Sandoval Cortes
School of Chemistry, Universidad Autónoma de Coahuila, Saltillo, México

Jincemon Cyriac
Department of Physics, Christ College (Autonomous) Irinjalikkuda, Calicut University, Irinjalakuda, Kerala, India

I. Dhanya
Department of Physics, Catholicate College, Pathanamthitta, Kerala, India

R. Geethu
SCMS School of Engineering and Technology, Karukutty, Ernakulam, Kerala, India

Priyanka R. Gupta
Department of Pharmaceutics, Sandip Institute of Pharmaceutical Sciences, Nashik, Maharashtra

S. Heera
Department of Physics, Catholicate College, Pathanamthitta, Kerala, India

Linda E. Jacob
Post Graduate and Research Department of Chemistry, Bishop Moore College, Mavelikara, Kerala, India

Linju Ann Jacob
Department of Physics, Baselius College, Kottayam, Kerala, India

Saritha K. Jose
Physics Department, Deva Matha College, Kuravilangad, Kottayam, Kerala, India

Cyriac Joseph
School of Pure and Applied Physics, Mahatma Gandhi University, Kottayam, India

M. Kailasnath
Inter National School of Photonics, CUSAT, Ernakulam, Kerala, India

K. V. Arun Kumar
Department of Physics, CMS College (Autonomous), Kottayam, Kerala, India
Nanotechnology and Advanced Materials Research Centre, CMS College (Autonomous), Kottayam, Kerala, India

Sunny Mathew
Physics Department, Marian College Kuttikkanam (Autonomous), Kuttikkanam, Peermade, Idukki, Kerala

Kamal P. Mani
Inter National School of Photonics, CUSAT, Ernakulam, Kerala, India

Rekha Mirle
Assistant Professor, Department of Chemistry and Biochemistry, School of Sciences, Jain (Deemed-To-Be) University, Bangalore-27, India

Anju Nair
SCMS School of Engineering and Technology, Karukutty, Ernakulam, Kerala, India

Tressia Alias Princy Paulose
Post Graduate and Research Department of Chemistry, Bishop Moore College, Mavelikara, Kerala, India

Keshav S. Pentapati
Indus Training and Research Institute, Bengaluru, Karnataka, India

Eapen Philip
Post Graduate and Research Department of Chemistry, Bishop Moore College, Mavelikara, Kerala, India

Sreejith K. Pisharady
Department of Physics, Sanatana Dharma College, Alappuzha, Kerala, India

Mónica M. Rodríguez-Garza
School of Chemistry, Universidad Autónoma de Coahuila, Saltillo, México

Meenakshy Salim
Department of Physics, Catholicate college, Pathanamthitta, Kerala, India

M. V. Santhosh
SCMS School of Engineering and Technology, Karukutty, Ernakulam, Kerala, India

Tina Sebastian
Physics Department, Deva Matha College, Kuravilangad, Kottayam, Kerala, India

Asiya Shafeek
Department of Physics, Catholicate college, Pathanamthitta, Kerala, India

Aneetta Shaji
Physics Department, Marian College Kuttikkanam (Autonomous), Kuttikkanam, Peermade, Idukki, Kerala

V. N. Sheeja
Physics Department, Marian College Kuttikkanam (Autonomous), Kuttikkanam, Peermade, Idukki, Kerala

Mamata Singh
Department of Chemistry, GITAM University, Bengaluru, Karnataka, India
Indus Training and Research Institute, Bengaluru, Karnataka, India

S. Sisira
Department of Physics, Sree Narayana College, Alathur, Palakkad, Kerala India

Swati G. Talele
Department of Pharmaceutics, Sandip Institute of Pharmaceutical Sciences, Nashik, Maharashtra, India

Rinsy Thomas
Department of Physics, CMS College (Autonomous), Kottayam, Kerala, India
Nanotechnology and Advanced Materials Research Centre, CMS College (Autonomous), Kottayam, Kerala, India

V. Unnikrishnan
School of Pure and Applied Physics, Mahatma Gandhi University, Kottayam, Kerala, India

P. H. Vaisakh
Post Graduate and Research Department of Chemistry, Bishop Moore College, Mavelikara, Kerala, India

Sneha Sara Varghese
Post-Graduate and Research Department of Chemistry, Bishop Moore College, Mavelikara, Kerala, India

Aswini Mariam Vinod
Department of Physics, Catholicate college, Pathanamthitta, Kerala, India

Anju Elizabeth Yohannan
Department of Physics, Catholicate college, Pathanamthitta, Kerala, India

CHAPTER 1

Introduction to Sustainable Development and Resilience

MÓNICA M. RODRÍGUEZ-GARZA, JOSÉ SANDOVAL CORTES, and CRISTÓBAL N. AGUILAR

School of Chemistry, Universidad Autónoma de Coahuila, Saltillo, México

ABSTRACT

Global sustainability is going through a critical moment because since 2020 there has been a clear setback in achieving each sustainable development goal of the United Nations, so the level of uncertainty has increased significantly due to the economic and social crisis as a result of the health crisis caused by COVID-19. It is time for a greater collective effort at all levels, sectors, and regions of human society. This critical moment suggests that the possibility of achieving it becomes a utopia, so new barriers obscure the horizon of a possible social reconstruction of sustainability, reducing the perspective of opening up environmental rationality. However, the human being is an extraordinary being who, if he proposed it, would reactivate the hope of turning the unsustainability of life into the possibility of building a sustainable future from a new global understanding of the life of the human being and the relationship he has with his environment individually and collectively. It is not only about implementing policies toward sustainability in each country but also about putting into practice well-established actions in all social sectors with intelligence and priority. This chapter describes a very particular vision of the authors, who live reality as researchers and teachers in one of the institutions located in the semidesert of northern Mexico.

Resilience and Sustainability for Energy and Water: New Advancements, Opportunities, and Framework.
Cristóbal Noé Aguilar, A. K. Haghi, Amritanshu Shukla, Ann Rose Abraham, Swati Gokul Talele (Eds.)

1.1 INTRODUCTION

Sustainability implies a new stage of human life on the planet, in which each society benefits collectively and individually with permanent care and constant improvement of the environment, where all its activities are part of a new market model that meets the circular economy so that the implicit practices of sustainability allow the global population to meet the basic needs of the world, without risking the needs of the human generations of the future.

The United Nations proposed 17 Sustainable Development Goals that aim to achieve 169 goals to ensure the sustainability of human life on the planet. The planning and definition of them represent an important achievement at the global level through sustainable practices on a global scale.

The sustainable development goals have managed to become the community objectives for a better world; however, the actions, programs, and activities established to achieve them through the generation of indicators that allow measuring the progress of their fulfillment should be analyzed and possibly complemented or replaced with others that improve the progress of these objectives in order to ensure compliance by the established date. The reality is that no country has been able to meet the basic needs of its people by sustainably using its resources.

The Sustainable Development Goals are linked to potential synergies between nanotechnology, food production, access to energy, water treatment, and medical facilities, including climate change policies and renewable energy; urban ecosystems; sustainable consumption; and infrastructure development. In general, linking the goals relates to socioeconomic and environmental sustainability as raising environmental standards can be costly and beyond the capacity of smallholders.

The creation of a sustainable socioecological system is the goal of the search for the balance between the needs of the human population and its environment, and achieving it has proven to be complex and long-term due to the obstacles represented by actions such as restoration and conservation initiatives that prevent the recovery of natural systems and lead to a decrease in the size and well-being of the human population.

However, the United Nations (UN) makes continuous and considered efforts to overcome current and future sustainability challenges despite the fact that there are no efficient monitoring strategies that quantify sustainability in general nor their objectives despite the interconnection and their combined and compensatory effect to estimate the state of the global environment and human well-being.[9] Around the sustainable development goals and their interactions, there are serious uncertainties as well as their indicators.[10]

To answer whether it is correct to meet the sustainable development goals in each region of the planet in the same order as they have been established, it should be assumed that the conditions of each country and each region are the same. The reality is the opposite. At the global level, the prioritization of the Sustainable Development Goals has as its primary objectives to reduce poverty and guarantee the feeding of the population, particularly in regions such as sub-Saharan Africa, where having access to drinking water and better sanitation conditions should be prioritized.[8]

The pursuit of socioeconomic sustainability and environmental sustainability can cause detrimental effects on the expected outcome. This is because they can hinder the well-being of the human population, force people to migrate, and cause environmental degradation to spread to other regions, ultimately undermining environmental policy.

It is then a question of avoiding the collapse of nonrenewable resources associated with a continuous growth of the human population and a better quality of life. The Sustainable Development Goals are based on the need to find a balance between human population, wealth distribution, food production, supply and trade, education, health, industrial production, pollution and available resources, inequality, well-being, social learning, food trade, and investment in sustainable development so that growth is not at the expense of the environment, with economic and ecological stability in order to guarantee the ability of present and future generations to access resources.

The debate is growing around the human–environment relationship for sustainable growth, supported by technology, population planning, and socioecological indicators. It has recently been warned that the population must necessarily be reduced by ensuring human rights and the long-term well-being of all life on the planet, implying that the size of the human population and social infrastructure are fundamental to achieving sustainability.[2] This human–environment relationship must be rethought in terms of region, income level, and resource inequality.

The obstacles facing the world for the implementation and advancement of the global 2030 Agenda mark, as never before, an important challenge that is the participation of actors at different levels to find comprehensive solutions to the problems in each region that provide progress in achieving the Sustainable Development Goals (SDGs).

In Mexico, the National Agenda incorporated seven priority approaches and themes to ensure that the proposed sustainable development was aimed at reducing inequalities and increasing opportunities for all people, especially for individuals or groups in conditions of marginalization, exclusion, or vulnerability. These priority approaches and themes have as their main axis of the new objectives, social and economic inclusion, the adoption of a multidimensional approach to poverty by the international community, which includes the protection and strengthening of the rule of law as a prerequisite for development.

With the results obtained in the different countries after the COVID-19 pandemic, in which 70–80% of the countries have decreased economic growth, the most significant impact lies in the global interconnection system of which the flow of human capital and material and financial resources are part and that directly affects the global economy and supply and demand chains[11]; it becomes essential to establish a systematized framework of work and cooperation that allows reorganizing the fulfillment of the SDGs.

The different sectors involved will have to carry out a clear analysis of the expectation of progress on the 2030 Agenda and rethink and adjust strategies to adapt to the context, which has changed radically since the pandemic. The COVID-19 pandemic has exacerbated some of the biggest challenges in Mexico, including inequality, impunity, corruption, and violence.[7]

Although some indicators of sustainable development have improved, 68% of the targets evaluated by the Economic Commission for Latin America and the Caribbean (CEPAL) show an insufficient trend to complete the SDG by 2030. The review of the four goals of education, gender equality, marine biodiversity, and terrestrial biodiversity shows that the pandemic has made it more difficult to reach the milestones set by 2030. Three "inconspicuous crises" are still present: the loss of more than a year of face-to-face schooling for an entire generation of students, the increase in gender-based violence and inequality in the burden of care according to sex, and the significant increase in the destruction of biodiversity as a result of illegal activities, in many cases, in conjunction with the murder of environmental defenders.[1]

The challenges to overcome will have to consider as actors the public administration, the private sector, academia, civil society, and international cooperation. The main challenges detected in Mexico for the Federal Public Administration are as follows: maintaining priority attention to the most vulnerable population, promoting economic reactivation, the internal market and employment, maintaining a responsible use of resources, strengthening the institutional capacities of the governments of the states, municipalities, and mayors in areas as diverse as legal, financial, and technological to favor cooperation in the sustainable use of resources, improving intergovernmental coordination mechanisms at the local level as a measure to deal with resource scarcity and growing inequalities, and reforms to achieve universal health and quality education.[7]

Considering the Mexican private sector, a challenge that many companies identify is to increase investment for sustainable development, and that requires the incorporation of environmental, social, and governance (ESG) practices in the operations of companies.

The challenges for the fulfillment of the SDGs that should be addressed by the academy should be related to education for sustainable development and the design and improvement of strategies for the definition and effective analysis of relevant development and environmental indicators that allow a better analysis for decision-making and the establishment of social, economic, and environmental development policies along with designing conceptual approaches to increase effectiveness in the implementation of postpandemic strategies,aligning business practices with the SDGs,[3]strengthening the national participation of the different actors to include a multidisciplinary vision with solutions from their multidimensional perspective, and improving performance in terms of gender, environment, human rights, citizen participation, and financing for the 2030 Agenda. These effects cannot be understood without taking into account regional structural realities, such as inequality in access to the right to quality health and social protection, limited fiscal space, and high labor informality that occurs in some regions in Mexico and in Latin America in general. It is recommended that postpandemic SDG actions adopt a systematic "Classification–Coordination–Collaboration" framework[4] that increases the likelihood of success and progress on the SDGs and ensures collaboration between regions with similar conditions.

1.2 FINAL REMARKS

Finally, in the case of Mexico, in terms of challenges, the United Nations System considers that the greatest challenge to advance with the SDGs of the 2030 Agenda is inequality gaps. It also highlights collaboration as a very important aspect in the exchange of experiences and the integration of efforts in different sectors, in the same way, for the promotion of international cooperation (South–South and triangular). The importance of the financing gap and the lack of availability and access to resources as recurring challenges are highlighted. Thus, meeting the growing demands for resources that we face in the country requires that financing strategies be articulated and a dialog with counterparts, donors, and allies be maintained to identify challenges and joint solutions.

KEYWORDS

- **global sustainability**
- **implementation**
- **risks**
- **human life**
- **health**
- **social crises**
- **nanotechnology**

REFERENCES

1. CEPAL. Comisión Económica para América Latina y el Caribe, Una década de acción para un cambio de época (LC/FDS.5/3), Santiago, 2022. https://repositorio.cepal.org/bitstream/handle/11362/47745/1/S2100985_es.pdf
2. Crist, E.; Ripple, W. J.; Ehrlich, P. R.; Rees, W. E.; Wolf, C. Scientists' Warning on Population. *Sci. Total Environ.* **2022,** *845*, 157166.
3. Elavarasan, R. M.; Pugazhendhi, R.; Shafiullah, G. M. et al. Impacts of COVID-19 on Sustainable Development Goals and Effective Approaches to Maneuver Them in the Post-Pandemic Environment. *Environ. Sci. Pollut. Res.* **2022,** *29*, 33957–33987. https://doi.org/10.1007/s11356-021-17793-9
4. Fu, B.; Zhang, J.; Wang, S. et al. Classification–Coordination–Collaboration: A Systems Approach for Advancing Sustainable Development Goals. *Natl. Sci. Rev.* **2020,** *7*, 838–840. https://doi.org/10.1093/nsr/nwaa048
5. Jefatura de la Oficina de la Presidencia (JOP) de México. Estrategia Nacional para la Implementación de la Agenda 2030 en México, 2019. https://www.gob.mx/cms/uploads/attachment/file/514075/EN-A2030Mx_VF.pdf
6. Henderson, K.; Loreau, M. A Model of Sustainable Development Goals: Challenges and Opportunities in Promoting Human Well-Being and Environmental Sustainability. *Ecol. Modell.* **2023,** *475*, 11**0164**. https://doi.org/10.1016/j.ecolmodel.2022.110164
7. Ministry of Economy. Informe Nacional Voluntario **2021,** Agenda 2030 en México. Disruptivo.tv. Ciudad de México, 2021. https://agenda2030.mx/docs/doctos/InfNalVol_FPAN_DS_2021_es.pdf
8. Mugagga, F.; Nabaasa, B. B. The Centrality of Water Resources to the Realization of Sustainable Development Goals (SDG): A Review of Potentials and Constraints on the African Continent. *Int. Soil Water Conserv. Res.* **2016,** *4* (3), 215–223.

9. Costanza, R.; Daly, L.; Fioramonti, L.; Giovannini, E.; Kubiszewski, I.; Mortensen, L. F..; Pickett, K. E.; Vala Ragnarsdottir, K.; De Vogli, R.; Wilkinson, R. Modelling and Measuring Sustainable Wellbeing in Connection with the UN Sustainable Development Goals. *Ecol. Econ.* **2016,** *130*, 350–355. https://doi.org/10.1016/j.ecolecon.2016.07.009.
10. Horvath, S.-M.; Muhr, M. M.; Kirchner, M.; Toth, W.; Germann, V.; Hundscheid, L.; Vacik, H.; Scherz, M.; Kreiner, H.; Fehr, F.; Borgwardt, F.; Gühnemann, A.; Becsi, B.; Schneeberger, A.; Gratzer, G. Handling a Complex Agenda: A Review and Assessment of Methods to Analyse SDG Entity Interactions. *Environ. Sci. Policy* **2022,** *131*, 160–176. https://doi.org/10.1016/j.envsci.2022.01.021.
11. Zhao, W.; Yin, C.; Hua, T. et al. Achieving the Sustainable Development Goals in the Post-Pandemic Era. *Humanit. Soc. Sci. Commun.* **2022,** *9*, 258. https://doi.org/10.1057/s41599-022-01283-5

CHAPTER 2

Feasibilities and Difficulties of Scaling-up Sophisticated Nanomaterials for Green Energy

MAMATA SINGH[1,2] and KESHAV S. PENTAPATI[2]

[1]*Department of Chemistry, GITAM University, Bengaluru, Karnataka, India*

[2]*Indus Training & Research Institute, Bengaluru, Karnataka, India*

ABSTRACT

Modern, state-of-the-art technology significantly depends on the growth and expansion of nanotechnology, which uses nanostructures and nanomaterials to serve as the fundamental building blocks. Compared to their bulk counterparts, nanoparticles and nanostructures have different physical and chemical properties, favorable surface-to-volume ratios, and probable quantum limits. These characteristics have dramatically broadened the range of prospective uses for these materials. Low-dimensional nanomaterials are often divided into three groups: atomically thin-layered materials (0D), nanorods, nanowires, and nanobelts (1D), in addition to nanoparticles (0D). A thorough introduction to nanomaterials and nanostructures in low dimensions is given in this article. The first step is categorizing nanomaterials, and then, a thorough explanation of nanofabrication and characterization follows. Both bottom-up and top-down manufacturing techniques are covered in depth. The discussion of important low-dimensional nanomaterial uses follows, including photonics, sensors, catalysis, energy storage, different coatings, and other bioapplications. For scientists and engineers engaged in the study of nanotechnology and nanomaterials, this article would be a convenient and rapid reference.

Resilience and Sustainability for Energy and Water: New Advancements, Opportunities, and Framework.
Cristóbal Noé Aguilar, A. K. Haghi, Amritanshu Shukla, Ann Rose Abraham, Swati Gokul Talele (Eds.)

2.1 INTRODUCTION

Nanomaterials are the underlying principle of nanoscience and nanotechnology. Over the past few years, there has been an explosion in the study and advancement of the vast and diverse field of nanostructure science and technology. The methods used to create materials and goods, as well as the range and kind of functions that could be available, have the potential to be substantially altered. It already has a substantial business impact, and this influence will continue to develop.[1,2] Nanoscale materials are those having a minimum dimension of 100 nm or less. One-millionth of a millimeter, or 100,000 times smaller than the diameter of a human hair, is a nanometer. The fascinating properties of nanomaterials are in the areas of new optical, magnetic, and electrical fields. These new qualities have enormous potential for applications in electronics, medicine, and other fields.

Advanced nanomaterials are extremely small, with a minimum size of 100 nm. Nanomaterials can be one-dimensional (e.g., surface layers), two-dimensional (e.g., strands or fibers), or three-dimensional (e.g., particles). They can have irregular, spherical, or tubular geometries and can be found in single, fused, aggregated, or agglomerated forms.[3] Typical examples of nanomaterials include fullerenes, quantum dots, dendrimers, and nanotubes. Nanomaterials are utilized in nanotechnology and display different physical and chemical characteristics from regular substances (i.e., silver nanotube, carbon nanotube, fullerene, photocatalyst, and silica).

Green energy is energy that stems from renewable resources, such as the sun. Although "renewable energy" refers to energy derived from sources that are continuously replenished, such as hydro power, wind power, or solar energy, the phrase "clean energy" refers to energy sources that do not release airborne pollutants. Although there is still some disagreement, renewable energy is frequently considered to be the same thing.[4–6] Can a hydroelectric dam, for instance, be referred to as "green" if it alters local rivers and hurts the environment? A renewable, green, and clean source, such as wind energy, originates from a source that is nonpolluting, ecologically benign, and self-renewing. Nanomaterials help in the enhancement and production of green energy in many ways.

The production and usage of materials having a maximum size of up to 100 nm in one or more dimensions are part of the field of nanotechnology. A unique combination of physical and chemical features, including optical, electrical, thermal, magnetic, and adsorption traits, characterizes advanced nanomaterials. Additionally, compared to their microscale counterparts, nanoparticles have a proportionately greater surface area, which promotes the enhanced performance of the final material in the field of energy harvesting and utilization. Large-scale research over the past several decades has demonstrated that nanoparticles may be utilized in the gathering of green energy.[7] The goal of this study is to compile information that has been published on recent advancements in the primary synthesis techniques of the organic and inorganic nanoparticles that are now employed in green energy harvesting and associated technologies. This describes the modification of nanoparticles, their mode of operation, and their use in the production of green energy.

2.2 PRINCIPLE OF ADVANCED NANOMATERIALS FOR GREEN ENERGY

The term "designing of chemical products, as well as the techniques for minimizing or abolishing the use and manufacture of dangerous compounds" refers to green chemistry. With little waste produced, usable nanomaterials were created using this description and green chemistry principles (GCPs). Sugars, vitamins, botanicals, and agricultural waste can all be used as biodegradable capping and reducing agents in the production of nanomaterials. The concept of producing green nanoparticles (NPs) was initially introduced by Raveendran et al., who used starch as a capping agent and d-glucose as a reducing agent to make starch silver nanoparticles (AgNPs).[8,9] AgNPs were also made using natural polysaccharides such as heparin, gum arabic, chitosan, and gum kondagogu.

By using less toxic chemicals in the creation of nanomaterials, the GCP was created to reduce pollution before it starts at the source. The synthesis of nanoparticles using the 12 green chemistry tenets is a relatively recent development in the field of sustainability. The possibility of the area developing new, safer, more energy-efficient, and less hazardous synthesis processes has attracted a lot of interest in recent years. These routes have been linked to the thoughtful application of various materials in the creation of nanoparticles and synthetic processes.

Nanomaterials' structural properties, including size, structure, composition, and surface chemistry, are essential for their usage in a wide range of multidisciplinary domains, including catalysis, electronics, biology and biomedicine, material science, physics, and environmental remediation. Varied ecologically acceptable technologies producing NPs of various shapes and sizes appropriate for a certain application must be developed to widen the range of attributes. This article will examine the state of the art in the production of a range of nanomaterials by manipulating three parameters using biodegradable homo and block copolymers, such as polysaccharides and enzymes.[10] It is important to take into account the solvent, the reducing agent, and the stabilizing agent. Comparing gamma irradiation procedures to other approaches, there are several benefits to using traditional processes, which produce monodisperse nanometallic clusters with greater yields. Studies may be carried out at gentle pressure and temperature conditions, without the need for chemical reducing agents, and with high reproducibility, which is an additional benefit and potential of gamma irradiation synthesis.

Without the use of chemical reduction agents, high-quality NPs may be produced via experimental procedures, and particle size and shape can be controlled. This approach also has the benefit of being adaptable in terms of radiation sources because it can be used with gamma, e-beam, and X-rays without affecting the final result or needing composition changes, except by changing the radiation source.[11] One special advantage that ionizing radiation may offer is the capacity to combine sterilization with NP synthesis in a single procedure. Particularly when compared to other approaches for the production of NPs and other nanomaterials, this process occurs by having an impact on the material or ions and harmful or contaminating microorganisms simultaneously.

It produces NP in a much more rapid, straightforward, and efficient manner. There should not be any byproducts or major damage to the product if the irradiation process is effectively carried out and managed.[12] If the developed system is compatible with the finished product to be sold, this benefit will be much more important because the radiation may be applied inside the finished packaging, preventing additional contamination and requiring less handling.

It is crucial to indicate both the mean particle size and the size of the principal particles when discussing a nanomaterial. Additionally, details on the existence of aggregates and/or agglomerates should be included. Agglomerates or aggregates are present when the mean particle size differs (i.e., is greater) from the main particle size. The description of the nanomaterial and/or the product incorporating the nanomaterial should contain this information. A useful statistic to define particles in addition to size is the specific surface area. The BET approach (Brunauer et al. 1938) has the advantage of determining the specific surface area without regard to whether it is in the primary or agglomerated condition.

2.3 PROPERTIES OF ADVANCED NANOMATERIALS

Although nanomaterials themselves are presently covered, there is considerable disagreement about what exactly constitutes a "nano." The top limit for the definitions of nanoscale used by various organizations is around 100 nm. It should be noted that the majority of the currently suggested definitions start with the size of the principal particle or structure. However, when a nanomaterial is present as a particulate, the particles may exist either alone or in groups called aggregates or agglomerations.[13] The bulk of the particles may even be aggregates or agglomerates, depending on the nanomaterial. This might result in the misconception that aggregation or agglomerations of nanoparticles with exterior dimensions significantly larger than 100 nm are not regarded as nanomaterials.[14] However, they continue to have some physicochemical features that are unique to nanoparticles, perhaps as a result of their extensive specific surface area (SSA).

The uncertainty over the existence of nanomaterials (defined by size, 100 nm, or SSA >60 m^2/g when computed for 100 nm unit density spheres) in products becomes of substantial concern when the only information on their presence rests only on the information supplied by the manufacturer. When the questioned nanomaterial is incorporated into an intricate matrix of the completed product, it is impossible to frequently assess the nanomaterial composition of these items. Consumer goods, particularly those related to cosmetics and healthcare, as well as those for food and animal feed, all have this unsolved problem. All of these goods add to the exposure that the European population is currently experiencing.

To accurately describe toxicological responses in biological systems, the total surface area of nanoparticles is a promising metric according to scientific toxicological evidence. The SSA, which is independent of whether a particle is present as a primary, agglomerated, or aggregated particle, should not be confused with the total surface area.

Regarding nanoparticle safety, the major factors of relevance are as follows: Physical characteristics

- Dimensions, form, surface area, and aspect ratio
- Agglomeration/state of aggregation
- Distribution of sizes
- Morphology and surface topography
- Crystallinity and defect structure, among other aspects of structure
- Solubility

1. Dimensions, form, surface area, and aspect ratio:

Advanced nanomaterials are described as low-dimensional materials with a size of at least 1–100 nm in one of their axis dimensions. These nanoscale dimensions have an impact on the materials' capacity to achieve high surface-to-volume ratios and electronic quantum confinement. **Dendrimers**, **nanocomposites**, **metal-based**, and **carbon-based nanomaterials** are the four primary categories.[15,16]

a. **Dendrimer nanomaterials:** Dendrimers are nanoscale, radially symmetric molecules having distinct, uniform, and monodisperse structures. They have an inner shell, which generally has a symmetric core, and an outer shell.

b. **Nanocomposite nanomaterials:** One or more nanoscale phases (in zero, one, or two dimensions) embedded in a ceramic, metal, or polymer material make up the class of nanomaterials known as nanocomposites. These can be created at the molecular level using either inorganic or organic components to acquire new characteristics.

c. **Metal-based nanomaterials:** They have a variety of characteristics, including magnetic and optical capabilities, a large surface area, a low melting point, and mechanical strengths. Metallic nanoparticle catalysts are extremely active and selective, and they also have a long lifespan for a variety of chemical processes.

d. **Carbon-based nanomaterials:** Nanomaterials made of carbon include fullerenes, carbon nanotubes, graphene and its derivatives, graphene oxide, nanodiamonds, and carbon-based quantum dots.

Any substance that may contain nanoparticles and has a volume-specific surface area greater than 20 m^2/m^3 but less than 60 m^2/m^3 is regarded as a nanomaterial under the guidelines established by the European Commission for identifying nanomaterials. The aspect ratio is simply the ratio of the length to the diameter (L/D) for rodlike particles like carbon nanotubes and nanowires, but for planar fillers like graphene and graphite flakes, it is thought to be the ratio of the lateral dimension to sheet thickness.[17–19]

2. Agglomeration/state of aggregation:

   The specifics of aggregation and agglomeration in polymer nanocomposites reinforced with spherical nanoparticles are studied using a straightforward methodology. The aggregation/agglomeration level can be provided as a function of various factors using the mechanical property technique that has been presented. The current method also identifies the effective volume percentage of aggregated nanoparticles in nanocomposite samples.

   The computed findings for several samples show that all reported nanocomposites have nanoparticle aggregation or agglomeration. Additionally, the diameters of the aggregate/agglomerate and nanofiller both significantly influence the degree of aggregate/agglomeration in the samples.[20] By increasing the nanofiller quantity and decreasing the nanofiller size, the amount of aggregation/agglomeration rises. Agglomeration and aggregation also reduce the efficiency of nanoparticles in a polymer matrix, which ultimately leads to subpar sample quality.[21]

3. Distribution of sizes:

   The average size of the nanoparticles in FSP grew from 1.8 to 3.8 nm, with a bimodal particle-size distribution that includes both a sizable proportion of tiny particles and bigger particles with sizes between 5 and 15 nm. A nanoparticle is a minuscule object that ranges in size from 1–100 nm. Since they are too small to be seen by the human eye, nanoparticles may have physical and chemical properties that are very different from those of their larger material counterparts.

   Particle sizing is a vital step in characterizing the characteristics of nanoparticles since the particle size has a significant impact on those properties. Many commercially available devices may be used to determine the size and dispersion of nanoparticles' particles. The frequency distribution is referred to as monomodal if there is just one maximum and as bimodal if there are two maxima. Multimodal particle size distributions have more maxima.[22]

4. Surface topography and morphology:

   Long carbon nanotubes, for example, are employed to span an electrical circuit thanks to the morphologies of nanoparticles.

Anisotropic microcrystalline whiskers and spherical amorphous particles typically take on the shapes of their respective crystals.[23–25] The primary emphasis of topography is the quantitative dimension measurement of surface properties; morphology, on the other hand, looks at the shape, texture, and distribution of materials at a surface. Surface topography includes both surface roughness and profile shape (including the finish, asperity, and wavering). Surface topography affects both the lubrication regime and the film thickness-to-roughness ratio.

5. Crystallinity and defect structure, among other aspects of structure:

   Based on their size, solid nanomaterial flaws may be categorized into four primary groups: zero-dimensional (0D) point defects (such as vacancy, reconstruction, and doping), one-dimensional (1D) line defects (such as dislocation), two-dimensional (2D) planar defects (such as grain boundaries), and three-dimensional (3D) volume defects.

   In general, a flaw in a solid is any departure from its periodicity. Chemical impurities, surface boundaries, dislocations, twins, and stacking faults are just a few examples of the various imperfections and flaws that may be found in crystals.[26]

6. Solubility:

   Nanoparticle colloidal suspensions have reversible, temperature-dependent solubility and can function as solutions. As a result, nanoparticle aggregates that have precipitated can coexist in equilibrium with nanoparticle monomer suspensions.[27,28] The nanoparticle suspension features a phase diagram with one- and two-phase zones, similar to a solution.

2.4 CLASSIFICATION OF ADVANCED NANOMATERIALS FOR GREEN ENERGY

Four different types of nanomaterials may be distinguished, which are as follows: Nanomaterials with inorganic, carbon, organic, and composite bases. Inorganic-based nanomaterials frequently contain various metal and metal-oxide nanoparticles. Silver, gold, aluminum, cadmium, copper, iron, zinc, and lead are examples of metal-based inorganic nanomaterials, while zinc oxide, copper oxide, magnesium aluminum oxide, titanium dioxide, cerium oxide, silica, and iron oxide (Fe_3O_4) are examples of metal oxide-based inorganic nanomaterials. Carbon-based nanomaterials include graphene, fullerene, single- and multiwalled carbon nanotubes, carbon fiber, activated carbon, and carbon black.[29] Carbon-free organic molecules, such as dendrimers, cyclodextrin, liposomes, and micelles, are employed to make the organic-based nanomaterials. Any mix of metal, metal oxide, carbon, and/or organic nanoparticles makes up composite nanomaterials, and these nanomaterials have complicated topologies.

Organic-based nanomaterials are made from carbon-free organic molecules such as dendrimers, cyclodextrin, liposomes, and micelles. Composite nanomaterials, which have complex structural compositions, can be made up of any mixture of metal, metal oxide, carbon, and/or organic nanoparticles. The distance between nanometer-scale features on a peacock's surface is what gives its feathers their vivid colors.[30,31]

Artificial nanomaterials are those produced by human-made methods or goods. Examples include several forms of contaminants and fossil fuel engine exhaust. Automobile exhaust, for instance, was not created as a nanomaterial, but researchers are working to create it for use in anything from manufacturing to medicine. These are known as purposefully made nanomaterials.

Nanoparticles: Nanoscale forms of many other elements, such as gold, silicon, and titanium, may also be found in nanoparticles, such as fullerenes, which are a kind of carbon. A sort of nanoparticle known as a quantum dot is a semiconductor consisting of many elements, such as cadmium and sulfur.[32,33] Amazing fluorescence effects may be produced using quantum dots. Numerous applications, such as fabric dye and photovoltaic cells (used for solar energy), have been tested with quantum dots.[34]

The study of nanomedicine has benefited from a better understanding of nanoparticle characteristics. The use of gold nanoparticles to combat lymphoma, a kind of cancer that targets lipid cells, is one hopeful advancement in nanomedicine.[35,36] A gold nanoparticle with the appearance of a cholesterol cell has been created by researchers. The cancer is prevented from "feeding" on genuine cholesterol cells when this nanoparticle binds to a lymphoma cell, starving it to death.

A. Carbon-based Nanomaterials:

Fullerenes are purposefully created carbon-based nanostructures. They include buckyballs and carbon nanotubes. Carbon-aided vapor deposition is a common method for creating carbon nanotubes. (NASA uses this procedure to produce the "blacker than black" satellite hue.) Scientists create a substrate, or base material, where the nanotubes develop throughout this procedure. A frequent substrate is silicon. The chemical process that produces the nanotubes is then assisted by a catalyst. A frequent catalyst is iron. The procedure also calls for blowing hot gas across the substrate and catalyst. The carbon that forms nanotubes is present in the atmosphere.[37]

B. Metal-based Nanomaterials:

Quantum dots and gold nanoparticles are examples of metal-based nanomaterials. There are several ways to make quantum dots. One technique involves creating tiny crystals of two distinct elements at very high temperatures.[398,39] By altering temperature and other parameters, the size of the nanometer-scale crystals may be accurately controlled. The size affects the fluorescent color's intensity. These nanocrystals are small semiconductors termed quantum dots, which are suspended in a solution.[40]

C. Dendrimers:

Dendrimers are composed of linked, branching components. A core, an inner shell, and an outer shell make up each dendrimer. Each dendrimer also has branching terminals. A dendrimer's core, inner shell, outer shell, and branching ends can all be engineered to serve a particular chemical purpose. Either the core or the outer shell can be used to create dendrimers using a divergent technique (convergent method).

Dendrimers are nanomaterials with robust, cage-like chambers, similar to buckyballs and several other nanomaterials. Dendrimers are being tested by scientists and researchers as multipurpose medication delivery systems. For instance, a single dendrimer can deliver a medicine to a single cell while simultaneously monitoring the drug's effects on the tissue around it.[41,42]

D. Nanocomposites:

Nanomaterials are combined with other nanomaterials or with larger bulk materials to form nanocomposites. The three main types of nanocomposites are nanoceramic matrix composites (NCMCs), metal matrix composites (MMCs), and polymer matrix composites

(PMCs). The coating of packing materials is frequently done using NCMCs, also known as nanoclays. They increase the material's capacity to withstand heat and suppress flames.

Over bulk metals, MMCs are more lightweight and stronger. MMCs can be utilized to create vehicles that are light enough to be airlifted or to minimize heat in "server farms" of computers.

Several industrial polymers include PMCs. Making PMC "tissue scaffolding" is one potential field of nanomedical research.[43–45] Nanostructures known as tissue scaffolds offer a framework around which tissue, such as an organ or skin, can develop. This may completely alter how burn injuries and organ loss are treated.

2.5 APPLICATION OF ADVANCED NANOMATERIALS FOR GREEN ENERGY

Nanomaterials may be used in a variety of ways to reduce energy usage. Additionally, nanoparticle fuel additives may considerably enhance the efficiency of combustion fuels and minimize carbon emissions. The nanoparticles of cerium oxide are highly efficient catalysts for the decomposition of hydrocarbon emissions and other tiny particle emissions due to their high ratio of surface area to volume and capacity to reduce the pressure inside the combustion process of motors to improve engine effectiveness and decrease NO_x emissions. The inclusion of carbon nanoparticles has successfully increased both the jet fuel's ignition delay and burning rate.[46,47] Iron nanoparticle additions to diesel and biodiesel fuels have also shown a reduction in fuel usage and volume of hydrocarbon emission by 3–6%, as well as monoxide and nitrogen oxides.

While there are many ways by which nanomaterials might improve fuel economy, one disadvantage of their utilization is the impact of nanoparticles on the environment. Trace quantities of these hazardous particles may be released when gasoline is supplemented with cerium oxide nanoparticles; there is exhaust.[48] Lung inflammation and levels of bronchial alveolar lavage fluid in rats have both been associated with cerium oxide diesel fuel additives. This is concerning, especially in areas with a lot of traffic since these particles might accumulate and be harmful to one's health.[49] The toxicity of diesel fumes is also substantially impacted by naturally occurring nanoparticles created by the incomplete combustion of diesel fuel. If artificial nanoparticles are added to fuels, additional work is necessary to identify whether this total dangerous particle emission is a result of burning.[50]

There have been and will continue to be several beneficial economic effects for society as a result of the relatively recent move toward the use of nanotechnology for the collection, transport, and storage of energy. One of the most significant characteristics of nanotechnology is the control of materials that it provides to scientists and engineers for consumer items, which enables efficiency improvements in a variety of products. Future energy costs may be reduced as a result of more effective energy capture and storage using nanotechnology as the cost of preparing nanomaterials decreases with further research.

The production of waste heat as a byproduct of combustion is a significant problem with contemporary energy production. This frequently occurs in internal combustion engines. Around 64% of the energy in gasoline is lost as heat in internal combustion engines; thus, reducing this loss alone might have a sizable economic benefit. It has, however, turned out to be quite challenging to make this improvement to the internal combustion engine without compromising performance.[51,52] By employing molecularly tuned catalysts, polymer membranes, and enhanced fuel storage, increasing the efficiency of fuel cells through the application of nanotechnology becomes increasingly likely.

A noble-metal catalyst (often platinum, which is quite costly) is required to separate the electrons from the protons of the hydrogen atoms for a fuel cell to function, particularly the hydrogen variety. These catalysts are quite vulnerable to carbon monoxide reactions, though. Alcohols or hydrocarbon compounds are utilized to reduce the system's carbon monoxide content to fight this. Through the use of nanotechnology, catalysts may be created that restrict incomplete combustion and thus lower the amount of carbon monoxide, increasing the process' efficiency.[53.54]

2.6 FEASIBILITY OF ADVANCED NANOMATERIALS FOR GREEN ENERGY

Energy serves as the material basis for human activity and is the most priceless resource for societal production. Technology advancements that can provide effective renewable energy solutions in the future heavily rely on materials. The purpose of this special issue, which has been developed as a worldwide leading interdisciplinary forum, is to publish original, complete research papers of the highest caliber on all facets of the study of materials for the use of renewable and sustainable energy technologies.[55–57] The practical and theoretical features of materials and prototype devices for sustainable energy conversion, storage, and saving are covered in this special issue, along with the materials required for the creation of renewable energy. It brings together parties involved in the idea from universities, corporations, and government agencies.

Research on comprehensive wastewater treatment technologies, strategies for limiting gaseous pollutant releases within chemical plants, an analysis of the poisoning mechanism of FCC catalysis, clean technologies for the use of fossil fuels, new-type photocatalysis material design with controllable morphology for solar energy conversion, and other topics have already been published in this special issue. To address the crucial technologies and significant difficulties that continue to plague the global energy business, these studies present significant, fascinating, and methodical examinations of advanced materials and technologies.[58,59] Moreover, they integrate many facets of contemporary energy transportation research and technology, displaying exceptional industrial insights that might inspire and stimulate new viewpoints on the difficulties, chances, and solutions related to energy and environmental sustainability.

The two most favorable reactions to simultaneously address growing CO_2 levels and an unending need for energy are CO_2 photoreduction to value-added compounds and fuels and photocatalytic hydrogen production via water splitting. With around 65% of the total emissions, CO_2 is a major greenhouse gas (GHG) contributor and is known to have negative impacts including climate change, ocean acidification, greenhouse effects, etc. The concept of CO_2 collection and conversion to hydrocarbons can prevent CO_2 levels from rising further and assist in the production of alternative fuels that have a variety of additional uses.[60] Contrarily, hydrogen, a zero-emission fuel, is seen as a clean and sustainable source of energy with enormous potential for use in a wide range of industrial applications. The emphasis of the current review is the debate.

Considering the latest developments in developing effective photocatalytic materials for hydrogen evolution and CO_2 photoreduction (HER), the current study's focus is only on advanced nanomaterials based on TiO_2 and other materials (i.e., metal chalcogenides, MOFs, carbon nitrides, single-atom catalysts, and low-dimensional nanomaterials).

The impact of key elements that govern how well these photocatalysts function in CO_2 photoreduction and HER is explored in depth.[61] A brief description of CO_2 adsorption mechanisms on the catalyst surface and its subsequent reduction pathways/product selectivity is also given special focus in this review. The study is finished with further perspectives on future studies of potential nanomaterials and reactor design ideas for boosting photoreaction efficiency.

2.7 CHALLENGES WITH ADVANCED NANOMATERIALS FOR GREEN ENERGY

A deeper understanding of materials and their minute properties is the most important challenge with nanotechnology. Universities and organizations throughout the world are closely studying how atoms interact to create larger structures. The effects of quantum mechanics on materials at the nanoscale are currently being studied.[62]

Because elements act differently at the nanometer when compared to those in their bulk form, there is concern that some nanoscales might be toxic. The blood–brain barrier, a membrane that defends the brain from harmful chemicals in circulation, is a concern for some doctors since nanoparticles are so tiny that they may easily flow past it. We must be certain that nanoparticles will not poison us if we intend to coat everything with them, from our clothing to our roads.[63,64] The technological barrier is closely connected to the knowledge barrier. We must figure out a technique to mass-create tiny items like transistors and nanowires if we are to see the amazing claims about nanotechnology come true. While we can employ nanoparticles to manufacture items like tennis rackets and make wrinkle-free textiles, we cannot make extremely complicated microprocessor processors using nanowires yet.

There are also significant societal considerations concerning nanotechnology. We may also be able to develop more potent weapons, both deadly and nonlethal, thanks to nanotechnology. Several groups worry that once these weapons are developed, we will not have time to consider the moral ramifications of using nanotechnology in them.[65] They exhort researchers and decision-makers to carefully consider all the implications of nanotechnology before creating ever-more-powerful weapons.

Is it moral if medical nanotechnology allows us to improve our physical capabilities? Theoretically, medical nanotechnology might offer humans superhuman intelligence, physical prowess, and other powers like quick healing and night vision. Should we aim for such things? Would we still be able to identify as humans or would we evolve into transhumans, the next stage in man's evolution? Would this imply that we would have two races of people—a wealthy race of modified humans and a poorer population of unmodified individuals—since practically every technology starts off being quite expensive? Several groups are asking nanoscientists to think about these ramifications right away, before it's too late, even though we do not have solutions to these problems yet.[66,67]

Some concerns have nothing to do with modifying the human body; they have more to do with business and economics. What effects would molecular manufacturing have on the global economy if it were to become a reality?[68–70] What happens to all manufacturing employment if we can create everything we require with the push of a button? What happens to the money if you have a replicator that can make anything? Would we go to an all-electric economy? Do we even require money?[71,72]

It's debatable if we will truly need to respond to each of these queries. Several specialists believe that worries about gray goo and transhumans are at best unjustified and premature.[73–76] Even then, as we learn more about the vast potential of nanotechnology, it will undoubtedly continue to influence our lives.[77]

The first step in biological testing is often the introduction of nanoparticles into the organism using a dry powder or a suspension in water or another liquid. For preparations that more closely approximate how nanoparticles could travel throughout a biological system, a variety of ideas have been put forth.[78–81] These include the use of fatty materials present in the lining of the lungs as well as the common protein albumin. Synthetic detergents are one more option, although they make toxicity evaluations more difficult.[82,83]

2.8 CONCLUSIONS

Depending on their location, the appropriate methods for tracking nanoparticles can be chosen. A range of trustworthy sensors can detect airborne particles as tiny as a nanometer in diameter. Nanoparticles can be seen by electron microscopes on the surface of materials and in tissue slices.[84] By using light scattering and other microscopic methods, it is possible to quantify nanoparticles that are suspended in a gas or liquid in some solid media. Nanoparticles dispersed in gases and liquids are analyzed using the mass spectroscopy method, which separates charged particles based on size.[85]

These criteria should be evaluated for nanomaterials both during their creation and throughout their usage (e.g., in the formulation of any subsequent product). While measuring, it's important to consider how qualities could alter if nanomaterials are suspended in or combined with other chemicals.[86]

The rate at which nanoparticles disintegrate in whatever liquid they are likely to come into contact with must be measured carefully. As they are smaller than equivalent bulk (big) particle material, they dissolve more quickly.[87–89] Each of these methods has advantages and disadvantages; therefore, they may all be employed in tandem. The majority still require seasoned professionals to utilize them in the lab, although portable and even hand-held technology is also becoming accessible.[90]

To address the issues of soil, water, and air pollution, new and innovative tools are being developed in the rapidly expanding field of nanotechnology. Nanomaterials are being synthesized and characterized with inorganic and organic elements to make them more useful for biosensing, environmental cleanup, disease detection, and other applications.[91] Many facets of nanotechnology have been covered in this book, including synthesis, functionalization, and several kinds of nanomaterials, such as nanotubes, nanomagnets, quantum dots, and others. The volume's last chapter summarizes its information while reinforcing the ideas it has already covered.[92]

KEYWORDS

- **nanomaterials**
- **advanced nanomaterials**
- **nanotechnology**
- **green energy**

REFERENCES

1. Holkar, C. R.; Jain, S. S.; Jadhav, A. J.; Pinjari, D. V. Chapter 14—Scale-Up Technologies for Advanced Nanomaterials for Green Energy: Feasibilities and Challenges. In: *Micro and Nano Technologies, Nanomaterials for Green Energy*; Bhanvase, B. A., Pawade, V. B., Dhoble, S. J., Sonawane, S. H., Ashokkumar, M., Eds.; Elsevier, 2018; pp 433–455. https://doi.org/10.1016/B978-0-12-813731-4.00014-X
2. Alagarasi, A. Chapter—Introduction to Nanomaterials, 2011.
3. What Is Green Energy? twi-global.com/technical-knowledge/faqs/what-is-green-energy#:~:text=Green%20energy%20is%20that%20which,wind%20power%20or%20solar%20energy.

4. Madani, M.; Hosny, S.; Alshangiti, D. M.; Nady, N.; Alkhursani, S. A.; Alkhaldi, H.; Al-Gahtany, S. A.; Ghobashy, M. M.; Gaber, G. A. Green Synthesis of Nanoparticles for Varied Applications: Green Renewable Resources and Energy-Efficient Synthetic Routes. *Nanotechnol. Rev.* **2022,** *11* (1), 731–759. https://doi.org/10.1515/ntrev-2022-0034
5. Anastas, P.; Williamson, T. Green Chemistry-Designing Chemistry for the Environment. *ACS Symposium Series 626th*; American Chemical Society: Washington, DC, 1996. DOI: 10.1021/bk-1996-0626.
6. Anastas, P.; Eghbali, N. Green Chemistry: Principles and Practice. *Chem. Soc. Rev.* **2010,** *39* (1), 301–312. DOI: 10.1039/B918763B
7. Raveendran, P.; Fu, J.; Wallen, S. L. Completely "Green" Synthesis and Stabilization of Metal Nanoparticles. *J. Am. Chem. Soc.* **2003,** *125* (46), 13940–13941. DOI: 10.1021/ja029267j
8. Gils, P. S.; Ray, D.; Sahoo, P. K. Designing of Silver Nanoparticles in Gum Arabic Based Semi-IPN Hydrogel. *Int. J. Biol. Macromol.* **2010,** *46* (2), 237–244. DOI: 10.1016/j.ijbiomac.2009.12.014
9. Modrzejewska, Z.; Zarzycki, R.; Sielski, J. Synthesis of Silver Nanoparticles in a Chitosan Solution. *Prog. Chem. Appl. Chitin. Derivatives* **2010,** *15*, 63–72.
10. Kora, A. J.; Sashidhar, R. B.; Arunachalam, J. Gum Kondagogu (*Cochlospermum gossypium*): A Template for the Green Synthesis and Stabilization of Silver Nanoparticles with the Antibacterial Application. *Carbohydr. Polym.* **2010,** *82* (3), 670–679. DOI: 10.1016/j.carbpol.2010.05.034
11. Mohan, Y. M.; Raju, K. M.; Sambasivudu, K.; Singh, S.; Sreedhar, B. Preparation of Acacia-Stabilized Silver Nanoparticles: A Green Approach. *J. Appl. Polym. Sci.* **2007,** *106* (5), 3375–3381. DOI: 10.1002/app.26979
12. Kemp, M. M.; Kumar, A.; Mousa, S.; Dyskin, E.; Yalcin, M.; Ajayan, P. et al. Gold and Silver Nanoparticles Conjugated with Heparin Derivatives Possess Anti-Angiogenesis Properties. *Nanotechnology* **2009,** *20* (45), 455104. DOI: 10.1088/0957-4484/20/45/455104
13. Pandey, S.; Goswami, G. K.; Nanda, K. K. Green Synthesis of Biopolymer–Silver Nanoparticle Nanocomposite: An Optical Sensor for Ammonia Detection. *Int. J. Biol. Macromol.* **2012,** *51* (4), 583–589. DOI: 10.1016/j.ijbiomac.2012.06.033
14. Horváth, I. T.; Anastas, P. T. *Introduction: Green Chemistry*; ACS Publications: New Haven, 2007. DOI: 10.1021/cr078380v
15. Anastas, P. T.; Warner, J. C. *Green Chemistry: Theory and Practice*; Oxford University Press: New York, 1998.
16. Stubbs, D.; Gilman, P. *Nanotechnology Applications in Environmental Health: Big Plans for Little Particles*; Oak Ridge Center for Advanced Studies: Oak Ridge, 2007.
17. Restrepo, C. V.; Villa, C. C. Synthesis of Silver Nanoparticles, the Influence of Capping Agents, and Dependence on Size and Shape: A Review. *Environ. Nanotechnol. Monit. Manag.* 2021. *15*, 100428. DOI: 10.1016/j.enmm.2021.100428
18. İspirli, H.; Sagdic, O.; Dertli, E. Synthesis of Silver Nanoparticles Prepared with a Dextran-Type Exopolysaccharide from *Weissella cibaria* MED17 with Antimicrobial Functions. *Prep. Biochem. Biotechnol.* **2021,** *51* (2), 112–119. DOI: 10.1080/10826068.2020.1795673
19. Ashtiyani, M. K.; Hajipour-Verdom, B.; Satari, M.; Abdolmaleki, P.; Hosseinkhani, S. Estimating the Two Graph Dextran-Stearic Acid-Spermine Polymers Based on Iron Oxide Nanoparticles as Carrier for Gene Delivery, 2021. DOI: 10.21203/rs.3.rs-811842/v1

20. Zhao, C.-L.; Gao, Y.-Z.; Wu, M.-Y.; Zhang, H.-T.; Wu, Y.-X. Biocompatible, Hemocompatible, and Antibacterial Acylated Dextran-g-Polyisobutylene Graft Copolymers with Silver Nanoparticles. *Chin. J. Polym. Sci.* **2021,** *39*, 1550–61.10.1007/s10118-021-2621-0
21. Anghelache, M.; Turtoi, M.; Petrovici, A. R.; Fifere, A.; Pinteala, M.; Calin, M. Development of Dextran-Coated Magnetic Nanoparticles Loaded with Protocatechuic Acid for Vascular Inflammation Therapy. *Pharmaceutics* **2021,** *13* (9), 1414. DOI: 10.3390/pharmaceutics13091414
22. National Renewable Energy Laboratory Best Research-Cell Efficiencies. https://www.nrel.gov/pv/assets/pdfs/best-research-cell-efficiencies.20200406.pdf (accessed 12 May 2020).
23. Qbox code: www.qboxcode.org.
24. QD Solar: Our Technology. https://qdsolarinc.com/technology/ (accessed 24 June 2020).
25. Renewable Energy Explained. http://californiasolarcenter.org/history-pv/ (accessed 24 June 2020).
26. Abakumov, V. N.; Perel, V. I.; Yassievich, I. N. *Nonradiative Recombination in Semiconductors*; Elsevier: Amsterdam, 1992.
27. Achtstein, A. W.; Schliwa, A.; Prudnikau, A.; Hardzei, M.; Artemyev, M. V.; Thomsen, C.; Woggon, U. Electronic Structure and Exciton-Phonon Interaction in Two-Dimensional Colloidal Case Nanosheets. *Nano Lett.* **2012,** 12 (6), 3151–3157.
28. Adamo, C.; Barone, V. Toward Reliable Density Functional Methods Without Adjustable Parameters: The PBE0 Model. *J. Chem. Phys.* **1999,** *110* (13), 6158–6170.
29. Adams, D. M.; Brus, L.; Chidsey, C. E. D.; Creager, S.; Creutz, C.; Kagan, C. R.; Kamat, P. V.; Lieberman, M.; Lindsay, S.; Marcus, R. A.; Metzger, R. M.; Michel-Beyerle, M. E.; Miller, J. R.; Newton, M. D.; Rolison, D. R.; Sankey, O.; Schanze, K. S.; Yardley, J.; Zhu, X. Charge Transfer on the Nanoscale: Current Status. *J. Phys. Chem. B* **2003,** *107* (28), 6668–6697.
30. (a) Amo-Ochoa, P.; Castillo, O.; Zamora, F. Cu(I), Co(II), and Fe(II) Coordination Polymers with Pyrazine and Benzoate as Ligands: Spin Crossover, Spin Canting, and Metamagnetism Phenomena. *Dalton Trans.* **2013,** *42*, 13453–13460. (b) Bahder, T. B. Eight-Band k p Model of Strained Zinc-Blende Crystals. *Phys. Rev. B* **1990,** *41* (17), 11992.
31. Baroni, S.; de Gironcoli, S.; Dal Corso, A.; Giannozzi, P. Phonons and Related Crystal Properties from Density-Functional Perturbation Theory. *Rev. Mod. Phys.* **2001,** *73*, 515–562.
32. (a) Baroni, S.; Giannozzi, P.; Testa, A. Green's-Function Approach to Linear Response in Solids. *Phys. Rev. Lett.* **1987,** *58*, 1861. (b) Bastard, G. et al. Wave Mechanics Applied to Semiconductor Heterostructures, 1988.
33. Baumgardner, W. J.; Whitham, K.; Hanrath, T. Confined-butConnected Quantum Solids via Controlled Ligand Displacement. *Nano Lett.* **2013,** *3* (7), 3225–3231.
34. Becke, A. D. Density-Functional Exchange-Energy Approximation with Correct Asymptotic Behavior. *Phys. Rev. A* **1988,** *38* (6), 3098–3100.
35. Benchamekh, R.; Gippius, N. A.; Even, J.; Nestoklon, M. O.; Jancu, J.-M.; Ithurria, S.; Dubertret, B.; Efros, A. L.; Voisin, P. Tight-Binding Calculations of Image-Charge Effects in Colloidal Nanoscale Platelets of Case. *Phys. Rev. B* **2014,** *89* (3), 035307.
36. Benedict, L. X.; Puzder, A.; Williamson, A. J.; Grossman, J. C.; Galli, G.; Klepeis, J. E.; Raty, J.-Y.; Pankratov, O. Calculation of Optical Absorption Spectra of Hydrogenated

Si Clusters: Bethe-Salpeter Equation Versus Time-Dependent Local-Density Approximation. *Phys. Rev. B.* **2003,** *68,* 085310.
37. Benedict, L. X.; Shirley, E. L. Ab Initio Calculation of ε2(ω) Including the Electron-Hole Interaction: Application to GaN and CaF2. *Phys. Rev. B.* **1999,** *59,* 5441.
38. Bertolotti, F.; Dirin, D. N.; Ib´a˜nez, M.; Krumeich, F.; Cervellino, A.; Frison, R.; Voznyy, O.; Sargent, E. H.; Kovalenko, M. V.; Guagliardi, A.; Masciocchi, N. Crystal Symmetry Breaking and Vacancies in Colloidal Lead Chalcogenide Quantum Dots. *Nat. Mater.* **2016,** *15* (9), 987–994.
39. Bisri, S. Z.; Degoli, E.; Spallanzani, N.; Krishnan, G.; Kooi, B. J.; Ghica,C.; Yarema, M.; Heiss, W; Pulci, O.; Ossicini, S.; Loi, M. A. Determination of the Electronic Energy Levels of Colloidal Nanocrystals Using Field-Effect Transistors and Ab-Initio Calculations. *Adv. Mater.* **2014,** *26* (32), 5639–5645.
40. Blanton, S. A.; Leheny, R. L.; Hines, M. A.; Guyot-Sionnest, P. Dielectric Dispersion Measurements of CdSe Nanocrystal Colloids: Observation of a Permanent Dipole Moment. *Phys. Rev. Lett.* **1997,** *79* (5), 865–868.
41. Bloch, W. M.; Babarao, R.; Hill, M. R.; Doonan, C. J.; Sumby, C. J. Post-Synthetic Structural Processing in a Metal-Organic Framework Material as a Mechanism for Exceptional CO2/N2 Selectivity. *J. Am. Chem. Soc.* **2013,** *135* (28), 10441–10448.
42. (a) Boxer, S. G. Stark Realities. *J. Phys. Chem. B* **2009,** *113* (10), 2972–2983. (b) Braun, P. V.; Osenar, P.; Stupp, S. I. Semiconductibandlng Superlattices Are Templated by Molecular Assemblies. *Nature* **1996,** *380* (6572), 325–328.
43. Brawand, N. P.; Govoni, M.; V¨or¨os, M.; Galli, G. Performance and Self-Consistency of the Generalized Dielectric Dependent Hybrid Functional. *J. Chem. Theory Comput.* **2017,** *13,* 3318.
44. Brawand, N. P.; V¨or¨os, M.; Galli, G. Surface Dangling Bonds Are a Cause of B-Type Blinking in Si Nanoparticles. *Nanoscale* **2015,** *7* (8), 3737–3744.
45. Brawand, N. P.; V¨or¨os, M.; Govoni, M.; Galli, G. Generalization of Dielectric-Dependent Hybrid Functionals to Finite Systems. *Phys. Rev. X* **2016,** *6,* 041002.
46. Britt, D.; Furukawa, H.; Wang, B.; Glover, T. G.; Yaghi, O. M. Highly Efficient Separation of Carbon Dioxide by a Metal-Organic Framework Replete with Open Metal Sites. *PNAS* **2009,** *106* (49), 20637–20640.
47. Brown, P. R.; Kim, D.; Lunt, R. R.; Zhao, N.; Bawendi, M. G.; Grossman, J. C.; Bulovi, V. Energy Level Modification in Lead Sulfide Quantum Dot Thin Films Through Ligand Exchange. *ACS Nano* **2014,** *8* (6), 5863–5872.
48. Bussi, G.; Donadio, D.; Parrinello, M. Canonical Sampling Through Velocity Rescaling. *J. Chem. Phys.* **2007,** *126* (1), 014101.
49. (a) Califano, M.; Franceschetti, A.; Zunger, A. Lifetime and Polarization of the Radiative Decay of Excitons, Biexcitons, and Trios in CdSe Nanocrystal Quantum Dots. *Phys. Rev. B* **2007,** *75* (11), 115401. (b) Car, R.; Parrinello, M. A Unified Approach for Molecular Dynamics and Density Functional Theory. *Phys. Rev. Lett.* **1985,** *55* (22), 2471.
50. Chen, O.; Zhao, J.; Chauhan, V. P.; Cui, J.; Wong, C.; Harris, D. K.; Wei, H.; Han, H.-S.; Fukumura, D.; Jain, K. et al. Compact High-Quality CdSe-CdS Core-Shell Nanocrystals with Narrow Emission Linewidths and Suppressed Blinking. *Nat. Mater.* **2013,** *12* (5), 445–451.
51. Chistyakov, A. A.; Zvaigzne, M. A.; Nikitenko, V. R.; Tameev, A. R.; Martynov, I. L.; Prezhdo, O. V. Optoelectronic Properties of Semiconductor Quantum Dot Solids for Photovoltaic Applications. *J. Phys. Chem. Lett.* **2017,** *8,* 4129–4139.

52. Engel, G. E.; Farid, B. Generalized Plasmon-Pole Model and Plasmon Band Structures of Crystals. *Phys. Rev. B.: Condens. Matter Mater. Phys.* **1993,** *47*, 15931–15934.
53. Farha, O. K.; Ozg¨ur Yazaydn, A.; Eryazici, I.; Malliakas, C. D.; G. ¨ Hauser, B.; Kanatzidis, M. G.; Nguyen, S. T.; Snurr, R. Q.; Hupp, J. T. De Novo Synthesis of a Metal-Organic Framework Material Featuring Ultrahigh Surface Area and Gas Storage Capacities. *Nat. Chem.* **2010,** *2*, 944–948.
54. Farid, B.; Daling, R.; Lenstra, D.; van Haeringen, W. GW Approach to the Calculation of Electron Self-Energies in Semiconductors. *Phys. Rev. B* **1988,** *38* (11), 7530.
55. Forero-Martinez, N. C.; Thi Le, H.-L.; Ning, N.; Vach, H.; Weissker, H. C. Temperature Dependence of the Radiative Lifetimes in Ge and Si Nanocrystals. *Nanoscale* **2015,** *7* (11), 4942–4948.
56. Franceschetti, A. First-Principles Calculations of the Temperature Dependence of the Band Gap of Si Nanocrystals. *Phys. Rev. B* **2007,** *76* (16), 161301.
57. Gali, A.; V¨or¨os, M.; Rocca, D.; Zimanyi, G. T.; Galli, G. HighEnergy Excitations in Silicon Nanoparticles. *Nano Lett.* **2009,** *9* (11), 3780–3785.
58. Gao, Y.; Weidman, M. C.; Tisdale, W. A. CdSe Nanoplatelet Films with controlled Orientation of their Transition Dipole Moment. *Nano Lett.* **2017,** *17* (6), 3837–843.
59. Geuchies, J. J.; van Overbeek, C.; Evers, W. H.; Goris, B.; de Backer, A.; Gantapara, A. P.; Rabouw, F. T.; Hilhorst, J.; Peters, J. L.; Konovalov, O.; Petukhov, A. V.; Dijkstra, M.; Siebbeles, L. D. A.; van Aert, S.; Bals, S.; Vanmaekelbergh. In Situ Study of the Formation Mechanism of Two-Dimensional Superlattices from PbSe Nanocrystals. *Nat. Mater.* **2016,** *15* (12), 1248–1254.
60. Giannozzi, P.; Baroni, S.; Bonini, N.; Calandra, M.; Car, R.; Cavazzoni, C. et al.. Quantum Espresso: A Modular and Open-Source Software Project for Quantum Simulations of Materials. *J. Phys.: Condensed Matter* **2009,** *21* (39), 395502.
61. Giberti, F.; V¨or¨os, M.; Galli, G. Design of Heterogeneous Chalcogenide Nanostructures with Pressure-Tunable Gaps and Without Electronic Trap States. *Nano Lett.* **2017,** *17* (4), 2547–2553.
62. Godby, R. W.; Needs, R. J. Metal-Insulator Transition in Kohn-Sham Theory and Quasiparticle Theory. *Phys. Rev. Lett.* **1989,** *62*, 1169–1172.
63. Golze, D.; Dvorak, M.; Rinke, P. The GW Compendium: A Practical Guide to Theoretical Photoemission Spectroscopy. *Front. Chem.* **2019,** *7* (377), 1–66.
64. Gonze, X.; Vigneron, J.-P. Density-Functional Approach to Nonlinear-Response Coefficients of Solids. *Phys. Rev. B* **1989,** *39* (18), 13120.
65. Govoni, M.; Galli, G. Large Scale GW Calculations. *J. Chem. Theory. Comput.* **2015,** *11* (6), 2680–2696.
66. Govoni, M.; Marri, I.; Ossicini, S. Carrier Multiplication Between Interacting Nanocrystals for Fostering Silicon-Based Photovoltaics. *Nat. Photon.* **2012,** *6* (10), 672–679.
67. Green, M. *Third Generation Photovoltaics: Advanced Solar Energy Conversion*; Springer: Berlin, 2006.
68. Gurtin, M. E.; Gurtin, M. E.; Murdoch, A. I. *Surface Stress in Solids*; 1978.
69. G¨utlich, P.; Garcia, Y.; Goodwin, H. A. Spin Crossover Phenomena in Fe(II) Complexes. *Chem. Soc. Rev.* **2000,** *29*, 419–427.
70. Gygi, F. The Architecture of Qbox: A Scalable First-Principles Molecular Dynamics code. *IBM J. Res. Dev.* **2008,** *52* (1.2), 137–144.

71. Gygi, F.; Fattebert, J.-L.; Schwegler, E. Computation of Maximally Localized Wannier Functions Using a Simultaneous Diagonalization Algorithm. *Comput. Phys. Commun.* **2003,** *155* (1), 1–6.
72. Halder, G. J.; Kepert, C. J.; Moubaraki, B.; Murray, K. S.; Cashion, J. D. Guest-Dependent Spin Crossover in a Nanoporous Molecular Framework Material. *Science* **2002,** *298* (5599), 1762–1765.
73. Hamann, D. R. Optimized Norm-Conserving Vanderbilt Pseudopotentials. *Phys. Rev. B* **2013,** *88* (8), 085117.
74. Hamann, D. R.; Schluter, M.; Chiang, C. Norm-Conserving Pseudopotentials. *Phys. Rev. Lett.* **1979,** *43*, 1494–1497.
75. Hanna, M. C.; Nozik, A. J. The Solar Conversion Efficiency of Photovoltaic and Photoelectrolysis Cells with Carrier Multiplication Absorbers. *J. Appl. Phys.* **2006,** *100* (7), 074510.
76. Hazarika, A.; Fedin, I.; Hong, L.; Guo, J.; Srivastava, V.; Cho, W.; Coropceanu, I.; Portner, J.; Diroll, B. T.; Philbin, J. P.; Rabani, E.; Klie, R.; Talapin, D. V. Colloidal Atomic Layer Deposition with Stationary Reactant Phases Enables Precise Synthesis of "Digital" IIVI Nanoheterostructures with Exquisite Control of Confinement and Strain. *J. Am. Chem. Soc.* **2019,** *141* (34), 13487–13496.
77. Hedin, L. A New Method for Calculating the One-Particle Green's Function with Application to the Electron-Gas Problem. *Phys. Rev.* **1965,** *139*, A796.
78. Hill, N. E.; Vaughn, W. E.; Price, A. H.; Davies, M. *Dielectric Properties and Molecular Behavior*; Von Nostrand Reinhold: London, 1969.
79. Hohenberg, P.; Kohn, W. Inhomogeneous Electron Gas. *Phys. Rev.* **1964,** *136*, B864–B871.
80. Hybertsen, M. S.; Louie, S. G. First-Principles Theory of Quasiparticles: Calculation of Band Gaps in Semiconductors and Insulators. *Phys. Rev. Lett.* **1985,** *55* (13), 1418.
81. Hybertsen, M. S.; Louie, S. G. Electron Correlation in Semiconductors and Insulators: Band Gaps and Quasiparticle Energies. *Phys. Rev. B* **1986,** *34* (8), **5390,** 1986.
82. Ithurria, S.; Tessier, M. D.; Mahler, B.; Lobo, R. P. S. M.; Dubertret, B.; Efros, A. L. Colloidal Nanoplatelets with Two-Dimensional Electronic Structure. *Nat. Mater.* **2011,** *10* (12), 936.
83. Ithurria, S.; Talapin, D. V. Colloidal Atomic Layer Deposition (c-ALD) Using Self-Limiting Reactions at Nanocrystal Surface Coupled to Phase Transfer Between Polar and Nonpolar Media. *J. Am. Chem. Soc.* **2012,** *134* (45), 18585–18590.
84. Jang, J.; Dolzhnikov, D.; Liu, W.; Nam, S.; Shim, M.; Talapin, D. V. Solution-Processed Transistors Using Colloidal Nanocrystals with Composition-Matched Molecular Solders: Approaching Single Crystal Mobility. *Nano Lett.* **2015,** *15* (10), 6309–6317.
85. Kamisaka, H.; Kilina, S. V.; Yamashita, K.; Prezhdo, O. V. Ab Initio Study of Temperature and Pressure Dependence of Energy and Phonon-Induced Dephasing of Electronic Excitations in CdSe and PbSe Quantum Dots. *J. Phys. Chem. C* **2008,** *112* (21), 7800–7808.
86. Kim, Y. D.; Klein, M. V.; Ren, S. F.; Chang, Y. C.; Luo, H.; Samarth, N.; Furdyna, J. K. Optical Properties of Zinc-Blende CdSe and ZnxCd1−xSe Films Grown on GaAs. *Phys. Rev. B.* **1994,** *49* (11), 7262–7270.
87. King-Smith, R. D.; Vanderbilt, D. Theory of Polarization of Crystalline Solids. *Phys. Rev. B* **1993,** *47*, 1651(R).

88. Klimov, V. I. Multicarrier Interactions in Semiconductor Nanocrystals about the Phenomena of Auger Recombination and Carrier Multiplication. *Annu. Rev. Condens. Matter Phys.* **2014,** *5* (1), 285–316.
89. Klokkenburg, M.; Houtepen, A. J.; Koole, R.; De Folter, J. W. J.; Ern´e, B. H.; Van Faassen, E.; Vanmaekelbergh, D. Dipolar Structures in Colloidal Dispersions of PbSe and CdSe Quantum Dots. *Nano Lett.* **2007,** *7* (9), 2931–2936.
90. Knowles, K. M. The Biaxial Moduli of Cubic Materials Are Subjected to an Equal-Biaxial Elastic Strain. *J. Elasticity* **2016,** *124* (1), 1–25.

CHAPTER 3

Nanotechnology for Quaternary Cu_2ZnSnS_4 (CZTS) Solar Cells

M. V. SANTHOSH, R. GEETHU, and ANJU NAIR

SCMS School of Engineering and Technology, Karukutty, Ernakulam, Kerala, India

ABSTRACT

Solar cells, in this scenario, became very familiar devices to the common man owing to the increasing need for energy day by day. Such a suitably scalable device for energy conversion has not yet been introduced to the world. This is the reason why photovoltaics have reached about the fourth generation within a few decades. Replacement of conventional devices is the major target among the researchers, which in turn reduces the cost of the device. The manufacturing and installation costs of silicon are still high, which hinders the use of solar cells by the common man. Third- and fourth-generation solar cells are meant for low-cost, eco-friendly, and flexible solar cells. Incorporating nanomaterials into the device can enhance its efficiency and reduce the cost. Within a given volume, increasing surface area will enhance charge carrier collection and separation, leading to enhanced short-circuit current density and efficiency. This work briefly discusses the nanotechnology incorporated in quaternary compounds, specifically copper zinc tin sulfide chalcogenide. Various research works on developing different types of nanomaterials like quantum dots, nanostructured nanoparticles, and nanocrystals of CZTS and incorporating them into thin-film solar cells are mainly discussed here. Several works reported device efficiency near 10%, revealing the further possibility of the nanomaterials of CZTS.

Resilience and Sustainability for Energy and Water: New Advancements, Opportunities, and Framework.
Cristóbal Noé Aguilar, A. K. Haghi, Amritanshu Shukla, Ann Rose Abraham, Swati Gokul Talele (Eds.)

3.1 INTRODUCTION

The energy demands of the growing population are escalating, and hence the energy crisis is a pivotal issue to be mitigated. There has been a 2 billion population hike in just one generation, which has necessitated the quest for inexhaustible energy sources as the sustenance for a developing economy.[1] Being the perennial source of energy, the sun is the ultimate source of energy to manage the prolonged energy crisis that we are facing all around the world. In fact, Asian countries have the highest potential to harness energy owing to the maximum sunshine duration in these temperate regions.[2] Amidst other renewable sources, namely hydrothermal, wind, geothermal, etc., solar energy is the cheapest form of energy available that can be trapped without disturbing the natural balance of the ecosystem.[3]

Photovoltaic technology is the most prominent technology in this regard, which involves the conversion of solar power into electricity without the need for an interface. These devices use semiconductor material to induce electricity by activating the electrons through additional energy.[4] Conversional semiconducting materials in PV systems include silicon (monocrystalline, polycrystalline, and microcrystalline), copper indium diselenide, cadmium telluride, etc.—research is being carried out enormously to improve the output.[5] The existing PVs are classified into four generations.[6]

The first-generation PVs include a single-junction structure of monocrystalline and polycrystalline silicon with the highest PCE of 33%. This Si wafer technology is more durable, but their manufacturing process costs are higher with regard to energy consumption and work, also their performance diminishes with high temperatures. Second-generation PV cells are industrially made by thin-film PV technology, owing to which the manufacturing cost is significantly lower; however, PCEs are compromised as they consume fewer semiconductor materials. This type of generation includes cells made of amorphous silicon, CdTe (PCE 9.4–13.8%), and copper indium gallium selenide (PCE 11–18.7%).[7] Third-generation PV incorporates solar inks, nanotubes, organic dyes, and conductive plastics into second-generation PV cells. It includes dye-sensitized solar cells (DSSC), perovskite, etc., reaching an efficiency of ~ 12% and ~ 20%, respectively. Four generations of solar cells are precisely plotted by Sahu et al. in 2020 and are given in Figure 3.1.[8]

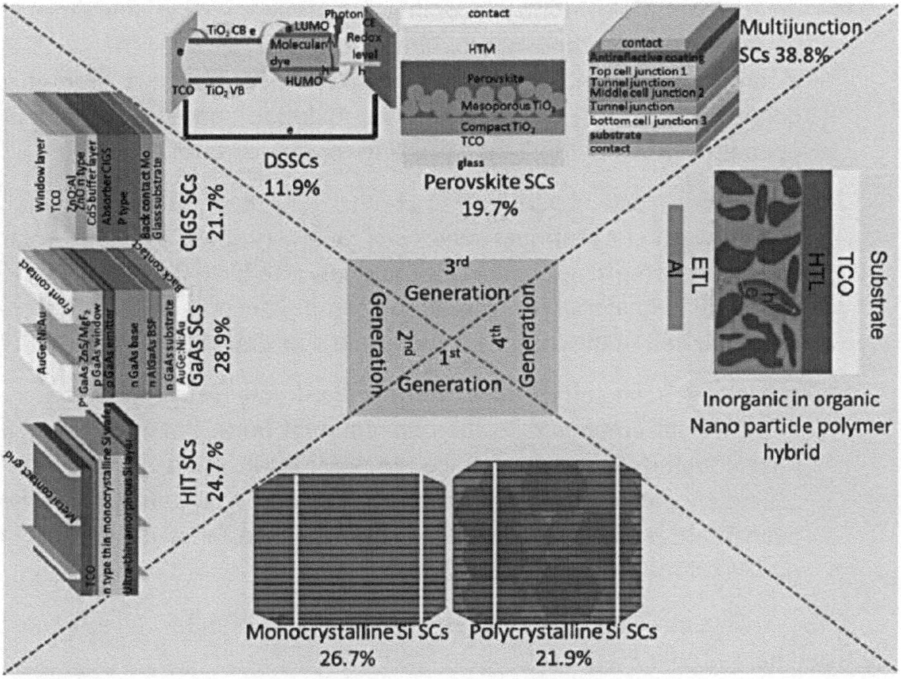

FIGURE 3.1 Diagram showing different generations of solar cells with their attained efficiencies.

Source: Reproduced with permission from Ref. [8], Elsevier.

3.2 NANOTECHNOLOGY FOR SOLAR CELLS

Efficiency and cost are the two major parameters that an industry or consumer of solar cells is looking into. Compared to conventional technology and conventional solar cells, that is, silicon-based, the most commercialized ones, nanotechnology has the potential to produce low-cost devices with better efficiency. The possibility of conducting research on silicon solar cells is declining since the practical device efficiency is very close to its theoretical value. Nanoincorporated silicon and other inorganic material (say copper indium sulfide/selenide), organic devices, etc., are developing, and a few get established and get great attention because of the low cost.

The major benefit of using nanoparticles, while fabricating solar cells, is cost reduction.

1. Fabrication cost: Conventional devices required a vacuum thermal evaporation technique for developing solar cells using crystalline silicon or other semiconducting materials. But for nanoincorporated

devices, simple solution processing techniques are available that can be done at low temperatures, eliminating the need for the vacuum technique. The deposition technique is one of the major factors that determines the cost of a device. Nanomaterial incorporated devices open up simple methods like printing for device fabrication.

2. Installation cost: Installation is another consumer of money while dealing with conventional solar cells. A few types of thin-film solar cells made from semiconducting materials also have this issue. Using nano or nanomaterial incorporation opens up the possibility of a flexible solar cell in the form of rolls, which don't need any installation.[9]

   As of now, solar cell devices developed utilizing nanotechnology are not efficient compared to the conventional ones, but lower-priced ones counterbalance this. Since the possibility of developing high-efficiency solar cells using nanomaterials, say quantum dots, in the near future is inevitable, there is a great chance for replacing heavy panels on roof top with flexible rolls.

A few solar cells utilizing nanotechnology that are under development are listed here (Table 3.1).[10]

TABLE 3.1 Nanotechnology Used in Solar Cells Under Development Stage.

Institutes	Nanotechnology used	Device type/role of incorporated nanomaterial
EPFL	Quantum dots	Perovskite
MIT	Defect-free graphene sheet	Lightweight solar cells
	Graphene coated with zinc oxide nanowires	
	Carbon nanotubes	Thermophotovoltaic
Flinders University	Phosphorene nanosheets	Perovskite solar cells
Michigan Technological University	Graphene with honeycomb structure	DSSC
Los Alamos National Lab	Quantum dots	CuInSe/S
Princeton University	30 nm thick gold nanomesh	Reduce losses due to light reflection
Duke University	Scattering nanocubes of silver on thin layer of gold	Reduce loss due to light reflection
Technical University of Munich and Ludwig Maximilians University	Layer of flexible nanoporous germanium	Produce lightweight solar cells

As said above, nanoparticles/nanostructured materials of many semiconductors are being developed by the research group all around the world owing to their wide variety of applications. Specifically, for thin-film solar cell devices, instead of thin films, nanostructured materials, like nanowires, nanorods, quantum dots, etc. of these semiconductors were synthesized and then deposited as thin film for the device fabrication. These semiconductors include binary, ternary, and quaternary compounds. Among quaternary compounds, copper zinc tin sulfide is one among the materials that has been extensively studied in the past decade. In the short term itself, it has gained huge interest among the researchers, who point out its potential to become a highly suitable part for photovoltaic devices. Comparing to copper indium gallium selenide/sulfide (CuInGaSe/S—a quaternary compound well-recognized in the field of photovoltaics), CZTS is a cost-effective and environment-friendly compound.

3.3 COPPER ZINC TIN SULFIDE

Cu_2ZnSnS_4 (CZTS) has a kesterite structure, and intense research is going on this material as a potential absorber for various solar cell technologies. The attraction of CZTS-based solar cells relies mainly on the following properties. First and foremost, they have direct bandgap materials. The optical bandgap of these materials is around 1.5 eV (ranging from 1.4 to 1.6 eV) and has a higher optical absorption coefficient ($>10^4$ cm^{-1}). Moreover, energy range is highly desired for the absorber material in various photovoltaic technologies. Comparing with well-established and proven cadmium and selenium-based solar cells, CZTS is nontoxic, and the constituent elements are abundant on earth. The theoretical efficiency limit for solar cells based on CZTS as per the Shockley–Queisser limit is 32.2%. However, solar cells based on CZTS reported the highest efficiency of 12.5%, which indicates a huge scope to improve the efficiency and other photovoltaic parameters, FF, V_{oc}, and J_{sc}.

Here we are discussing various nanotechnologies applied to CZTS-based solar cells. Nanotechnologies included in this chapter are quantum dots, nanostructures, nanoparticles, and nanocrystals of CZTS developed for fabricating different types of solar cells. The main factors analyzed in this piece of work were how device performance varied on incorporating nano into the device and how it affected devices with different configurations.

Before going into the different works on solar cells, some basic parameters of solar cells have to be explained for the understandability of this chapter. A solar cell is basically a p-n junction diode working under the phenomenon of the photovoltaic effect. It should contain at least four layers: a buffer-absorber layer and two contacts to collect charge-separated carriers. The function of each layer in the device is explained in Table 3.2. The four major parameters that must be known to analyze a device's performance are given in Table 3.3.

TABLE 3.2 Different Layers of a Solar Cell.

Order of device layer deposition	Different layers in an inorganic type solar cell	Function
4	Top contact	• Pure metal • Like silver, gold, aluminum, etc.
3	Absorber layer	• Absorption of light takes place here • Charge carrier separation as a result of photon absorption • Bandgap should be ~ 1.5 eV (matching to solar spectrum) • Absorption coefficient must be high
2	Window layer/Buffer layer	• It has similar function as room window • Collects all the light falling on the device and passes to absorber layer • Larger bandgap required, so that all the photons can be transmitted to the next layer.
1	Transparent conducting layer	• Abbreviated as TCO • Material used here should be highly transparent as plane glass plate and must have lower resistivity <10 ohm/square.

TABLE 3.3 Performance Parameters of a Photovoltaic Device.

Parameter	Abbreviation	Definition
Open circuit voltage	V_{oc}	• Voltage when circuit is under opened condition • Current is zero • Maximum voltage—from a device
Short circuit current	J_{sc}	• Current when circuit is shorted • Voltage is zero • Maximum current—from a device
Fill factor	FF	• Determines the device quality • Ratio of maximum power obtained from the device to the product of open circuit voltage and short circuit current density
Efficiency	H	• Maximum power output to power input • Ability of a device to convert incident light into electric energy

3.3.1 CZTS QUANTUM DOTS

Quantum dot solar cells (QDSC) exploit the property of tunability of size and composition. This distinct property enables the device to absorb sunlight efficiently by tuning the bandgap of QDs by changing their size. Thus, it offers tremendous future possibilities by replacing conventional materials like Si (silicon), CIGS (copper indium gallium selenide), or CdTe (cadmium telluride). Quantum dot-sensitized solar cells (QDSSCs) have arisen as an optimal choice for next-generation devices owing to their cost-effectiveness and high theoretic efficiency. By employing quantum dots (QD) as a light trapper in DSSC, it is possible to substitute the costly dyes, which offers tremendous future possibilities. A schematic diagram is shown in Figure 3.2.

Najafi et al. synthesized CZTS quantum dots (QDs) employing a noncolloidal method assisted by pulsed laser irradiations, which ensure a uniform size of 4–8 nm. They accurately controlled QDs' size by employing suitable laser energy (Nd: YAG-pulsed laser irradiation). Three solar cells were fabricated using three different TiO_2 pastes and CZTS QDs. The best efficiency obtained under this study was 1.438% for FTO/TiO_2+ZrO_2+GO/CZTS QDs/polysulfide/CuS/FTO cell structures. According to the study, small porosity of TiO_2 paste favors better open circuit voltage, and graphene oxide (GO) added to the paste of TiO_2 leads to advancement in short circuit current by expanding electron transfer on the cell.[11]

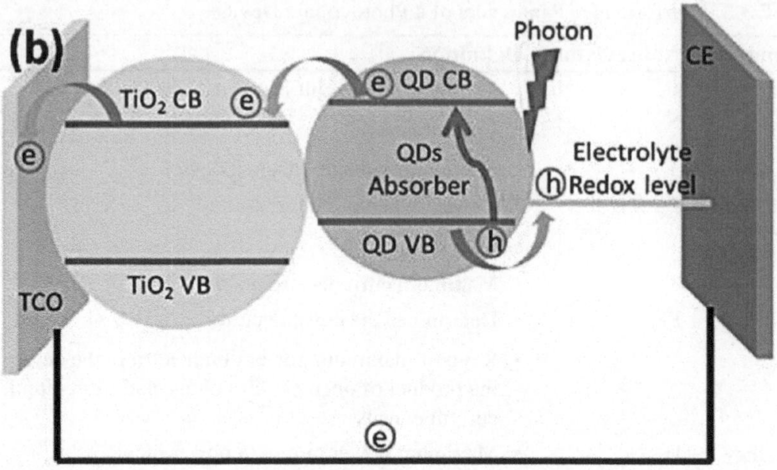

FIGURE 3.2 Schematic diagram for quantum dot-sensitized solar cells (QDSSCs).
Source: Reproduced with permission from Ref. [8], Elsevier.

Xu et al. employed a facile solvothermal route to synthesize virtually monodisperse microspheres of CZTS having diameters of around 2 μm. These are assembled from nanoflakes collected from CZTS quantum dots with diameters between 3 and 5 nm, showing a three-tiered association of the CZTS microspheres with a bandgap of 1.50 eV. CZTS microspheres are employed in QDSSCs as a productive material to be used as a counter electrode. By using these FTO coated with microspheres (SnO_2:F) as counter electrodes, the device efficiency obtained was 3.73%, illuminating under AM 1.5G (intensity of 100 mW cm^{-2}). The PCE obtained for microsphere-coated FTO counter electrodes is far better than that of bare FTO and that of FTO coated with noble Pt.[12]

Zhou et al. synthesized CZTS quantum dots using a solvothermal method by mixing easily available chemicals with an ethanol solvent. A wide bandgap nature (900 nm spectral response) of the sample was revealed from optical absorption spectrum studies and was employed as an electron acceptor for its MEH-PPV counterpart. CZTS-QDs in combination with MEH-PPV form type-II heterojunction. CZTS-QDs in the cell structure (Fig. 3.3) function as electron acceptors, and the role of donor was played by polymers. Solar cells fabricated in this study show an advantageous open-circuit voltage of 0.63 V. Researchers observed molecular capping of CZTS with 1-octadecanamine (a chemical used for synthesizing the sample) after the preparation of QDs. This seems to reduce the solar cell performance and was then removed, resulting in a better-performing device. In this device structure, CZTS possessed a suitable material, functioning as a great charge transporter from the MEH-PPV layer to the TiO_2 layer.[13]

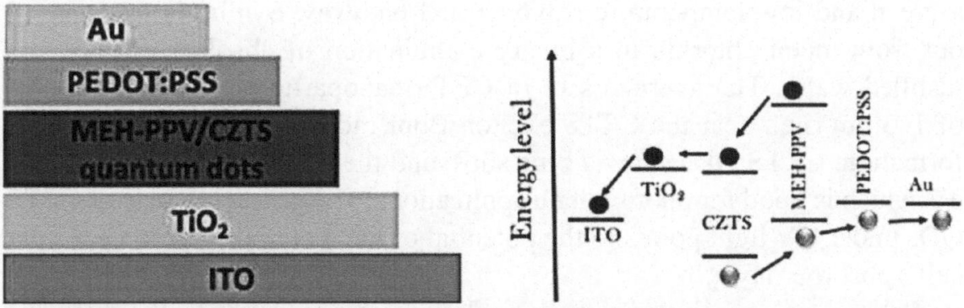

FIGURE 3.3 (a) Device structure and band alignment and (b) the flow of electrons and holes through different layers.

CZTS quantum dots offer improved charge carrier collection and stability in $CsPbBr_3$ perovskite solar cells (Zhou et al.). Here, CZTS QDs are utilized earnestly as a new inorganic hole transport material (HTM), aiming to avoid the organic HTM, which is costly and unstable. The device under this study showed an open-circuit voltage of 0.94 V, a short-circuit current density of 7.36 mA/cm^2, 70% fill factor, and 4.84% efficiency. The chapter also deals with the energy level diagram and carrier transport mechanism at the boundary between the $CsPbBr_3$ perovskite layer and the CZTS HTL (the schematic of the device is in Fig. 3.4). The electronic impedance spectroscopy (EIS) and photoluminescence (PL) demonstrate enhanced effective hole extraction and transfer properties of the CZTS QDs interface layer.[14]

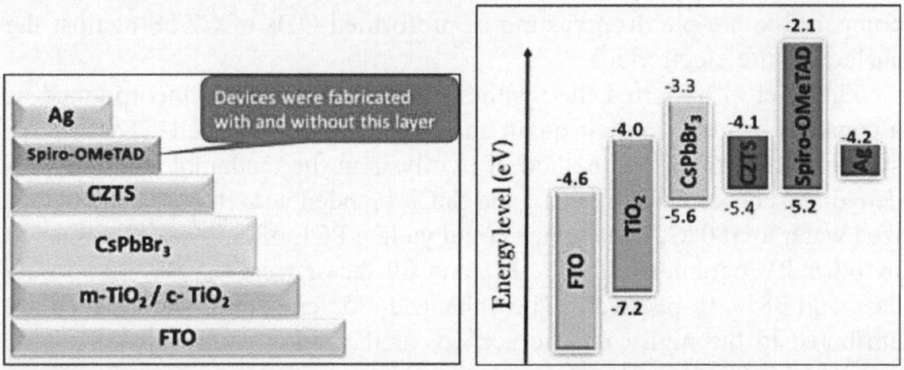

FIGURE 3.4 Schematic (left) and energy level diagram (right) of the device proposed by Zhou.

The devices were fabricated with conventional Spiro-OMeTAD, and on comparison, the device with this layer achieved an efficiency compared to that with Spiro-OMeTAD.

Das et al. reported the synthesis of spherical CTS quantum dots through a green and low-temperature solvothermal pathway. Synthesis was carried out from metal chloride in a binary combination of absolute ethanol and distilled water. The average size of CZTS nanoparticles was in the range of 1–3 nm (i.e., near the CZTS exciton Bohr radius), which confirms QDs formation. CZTS QDs show a blue shift, and the optical energy gap is 1.89 eV, which is good for photovoltaic applications. The photoresponse of CZTS QDs under UV light points to the potential of the material to become a solar cell's absorber layer.[15]

Sahoo et al. employed a numerical simulator to sequentially study the electrical properties of kesterite solar cells. The effects of each point of grain boundary defects (GB) on the electrical measurement were analyzed. Limiting factors connected with the recombination, like tail state, deep defect, and GB defect, were represented by employing Gaussian distributions and exponential tail distributions. Gaussian distributions and exponential tail distribution functions are employed to analyze the distribution of defects inside the bandgap.[16]

Sadeghnejad et al. synthesized CZTS nanocrystals in the aqueous phase through an enzymatic and ambient temperature route that ensured good compositional control. They performed biomineralization of the synthesized particles within a preformed mesoporous TiO_2 anode template where the synthesized nanocrystals attached to the TiO_2 surface, facilitating improved dispersal of the nanocrystals in the anode, which improved the performance of QDSSC. They also observed that such a uniform distribution further improves the performance of solar cells compared to simple drop casting of preformed QDs of CZTS against the surface of the electrode.[17]

Zhang et al. reported the synthesis of CZTS thin film incorporated as a counter electrode in their quantum dot-sensitized solar cells. These were synthesized through the method of pyrolysis in the methanol solution with dehydrate salts of $CuCl_2$, $ZnCl_2$, and $SnCl_2$ blended with thiourea and deionized water at 380°C. This device could yield a PCE of 2.56%, accompanied by other PV parameters—V_{OC}, J_{SC}, and fill factor were 0.52 V, 12.96 mA/cm^2, and 38%, respectively. The enhanced efficiency and performance are attributed to the high catalytic activity of the redox couple Sn^{2-}/S^{2-} over pyrolyzed CZTS thin film.[18]

Bai et al. successfully introduced quaternary CZTS QDs into QDSC via the hydrolysis approach. Low open-circuit voltage and fill factor hinder the performance of the device. To overcome these shortcomings, they employed a capping ligand, 1-dodecanethiol (DDT), for the synthesis of CZTS QDs (small-sized), which expands the inherent properties of the QDs and hence the device. The best device in this study shows a conversion efficiency of 4%. Electron transport and recombination dynamics were analyzed using intensity-modulated photovoltage spectroscopy (IMVS) and intensity-modulated photocurrent spectroscopy (IMPS) measurements. Measurements revealed that the long alkyl chain should be removed to enhance the process of electron transport. Also, this process can be improved by the type-II core/shell structure, which hinders the recombination of charge carriers. This strategy of removing ligands will be more suitable for applying CZTS QDs in quantum dot solar cells.[19]

3.3.2 NANOSTRUCTURES IN THE CZTS SOLAR CELL FOR IMPROVING PERFORMANCE OF THE DEVICE

Liu and his group, by controlling the reaction route at the interface, fabricated efficient ultrathin (~400 nm) CZTS solar cells. Developed back contact is of self-organized nanopatterns. An intermediate layer of Al_2O_3 was inserted at the Mo/CZTS interface. During the process of sulfurization, the possibility of a reaction between CZTS and Mo is high. This might lead to the formation of voids at the back contact, which was early detected by the research team. This was avoided by blocking the reaction at the interface of CZTS and Mo. The abovementioned Al_2O_3 layer inserted between molybdenum and copper zinc tin sulfide layers enhances uniformity and reduces recombination at back contact.

After completing sulfurization, transformation occurred for the Al_2O_3 intermediate layer. They organized themselves to form nanopatterns, which function as openings for charge carriers, thus empowering charge carrier transfer (Fig. 3.5). The efficiency of the device improved from 7.34 to 8.56% after introducing Al_2O_3 at the interface of Mo/CZTS. Two certified world record efficiencies (1) 9.26% (area—0.237 cm^2) and (2) 7.61% (area—1 cm^2) for CZTS solar cells with standard kesterite structure were stated. I–V characterization is depicted in Figure 3.6.[20]

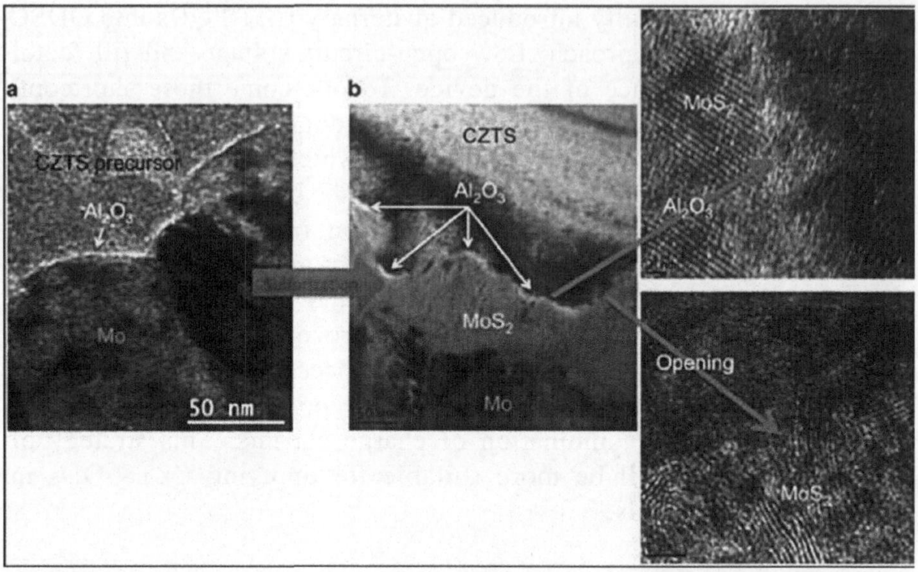

FIGURE 3.5 HRTEM image—Evolution of microstructure in the back contact of the device as a result of sulfurization (before (a) and after (b)), studied by Liu et al.

Source: Reproduced with permission from Ref. [20]. https://creativecommons.org/licenses/by/4.0/

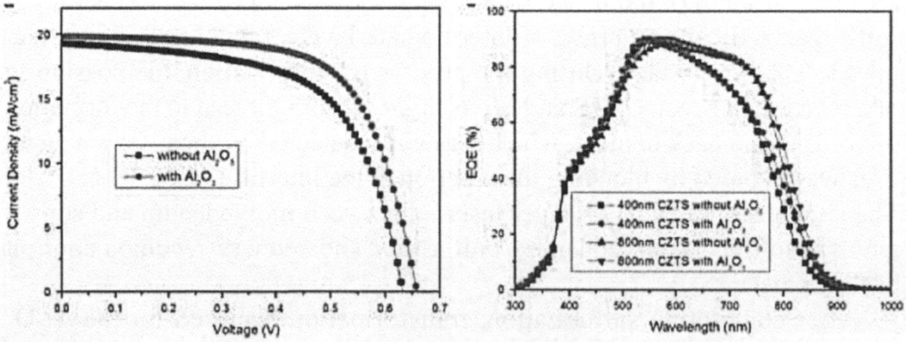

FIGURE 3.6 (a) Voltage verus current density plot and (b) EQE (external quatum efficiency) curve for ultrathin CZTS solar cells (with and without intermediate Al_2O_3 layer) studied by Liu et al.

Source: Reproduced with permission from Ref. [20]. https://creativecommons.org/licenses/by/4.0/

Lee et al. used zinc oxide nanostructures—one-dimensional (1D)—as transparent window layer for CZTS solar cells. For the device fabrication, all the layers were deposited by solution routes. The hydrothermal method and nanocrystal layer deposition (NCLD) were used to develop ZnO nanorods (NRs), which were steep and well-aligned.

The spin-coating technique was utilized to develop a light-absorbing thin film of CZTS by using cost-effective and environment-friendly molecular precursors of copper, zinc, tin, and sulfur atop nanorods of CdS/ZnO. A two-stage process of annealing was done at 250°C. A red shift in the absorption edge was observed for the CZTS sample thin film developed in this work. A detailed analysis of optical characteristics was conducted to enhance the capacity to harvest photons. The schematic diagram of the device is given in Figure 3.7. The prepared CZTS/CdS/ZnO NR device revealed a solar energy conversion efficiency of 1.2%.[21]

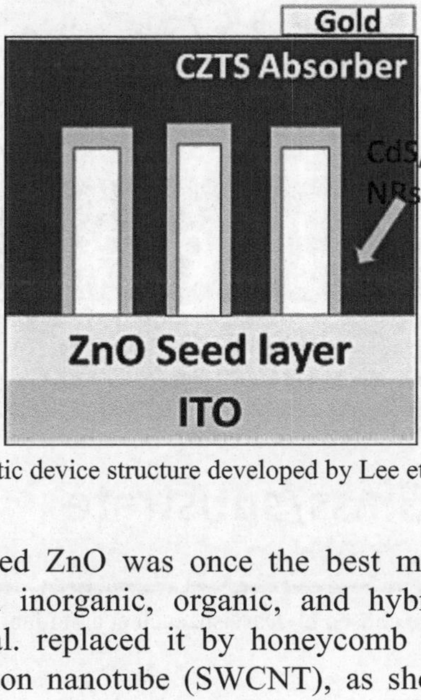

FIGURE 3.7 Schematic device structure developed by Lee et al.

Aluminum-doped ZnO was once the best material to be used as a TCO electrode in inorganic, organic, and hybrid-type devices. Here, Darvishzadeh et al. replaced it by honeycomb network-like structured single-walled carbon nanotube (SWCNT), as shown in Figure 3.8. The basic characteristics of a solar cell device (I–V), energy band diagram, and quantum efficiency calculation of the devices were done by SCAPS-1D simulation. On introducing metallic nanotubes, the device exhibited suitable optoelectronic behavior, say, wide work function, high charge transport conductivity, etc. The improvement in J_{sc} is associated with quantum efficiency spectra. On comparing with the ZnO:Al-contacted devices, those using SWCNTs exhibit enhanced quantum efficiency by 30%. They also performed Au:Cu doping in nanotube bundles, and the device fabricated with this layer exhibited a better open-circuit voltage compared with the others. Moderating degradation of devices was another highlight of these nanolayers, by surpassing ion migration under aging conditions and by covering the grain boundaries.[22]

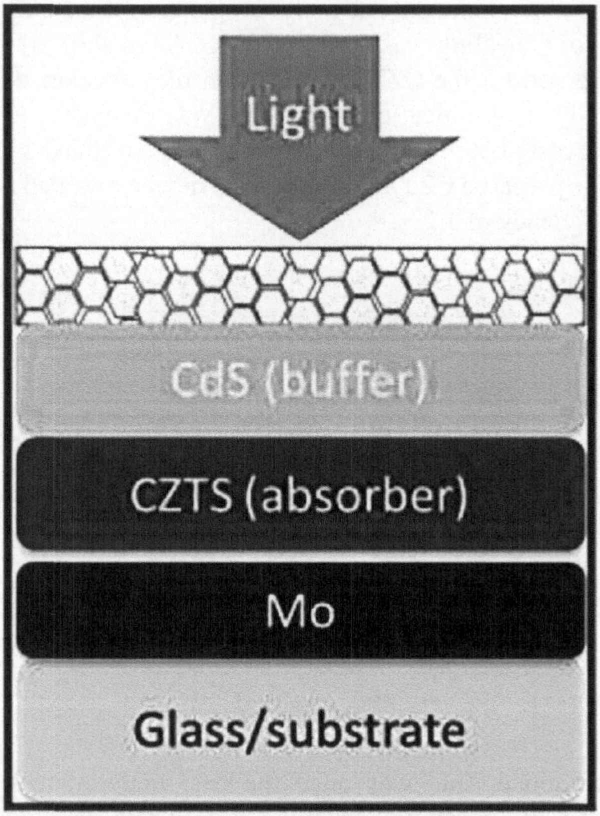

FIGURE 3.8 Device structure chosen by Darvishzadeh et al. to fabricate solar cell.

Wang et al. performed nanoengineering at the interface of the p-n junction Cu_2ZnSnS_4-TiO_2 (device structure in Fig. 3.9) by depositing an Al_2O_3 nanolayer via the atomic layer deposition (ALD) technique. The aluminum oxide layer functions as a protective layer for TiO_2 from tin and also helps to optimize the alignment of the energy levels for easy charge separation and collection. A CdS copassivation layer was also introduced near CZTS, ultimately resulting in a device as shown in Figure 3.9. Thus, the fabricated device exhibits a device efficiency of 4.2%.[23]

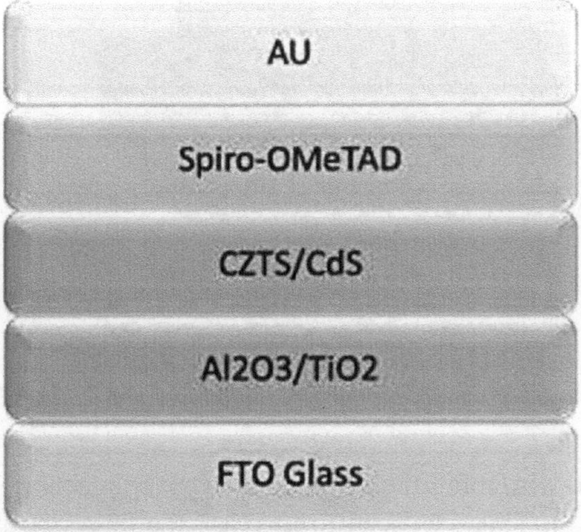

FIGURE 3.9 Device structure developed by Wang et al.

Abdelraouf et al. have done theoretical modeling work on CZTS-based solar cells. The optical modeling of solar cells was coupled with the electrical modeling and then analyzed for the possibilities of getting maximum output from a particular device structure. This research group chose different nanostructures, like nanostructured thin film, nanorods, nanotextures, and nanotubes, to fabricate solar cells based on CZTS. They make use of optical modeling to determine the best structure that possesses the highest light absorption as well as charge carrier (electron hole) generation, followed by coupling the result with the electrical modeling of the designed device to reveal the electrical data, that is, device parameters, and to determine how they performed. An efficiency of 6.59% was exhibited by the planar-structured CZTS/CdS device, whereas 7.25, 7.45, and 7.91% were exhibited by the nanorods, nanotextures, and nanotubes, respectively. Thus, the devices designed using CZTS/CdS nanotubes outrate all the other structures with a short-circuit current density of 23.03 mA/cm^2. Not only the nanotubes incorporated devices, but all the devices showed an enhancement in device performance. The work was a demonstration study by simulation that reveals the potential of nanomaterials to augment the charge carrier collection and hence the device efficiency by increasing the optical absorption of photons, and hence, increment in generation of charge carriers.[24] The basic device structure chosen for designing devices by Abdelraouf et al. is shown in Figure 3.10.

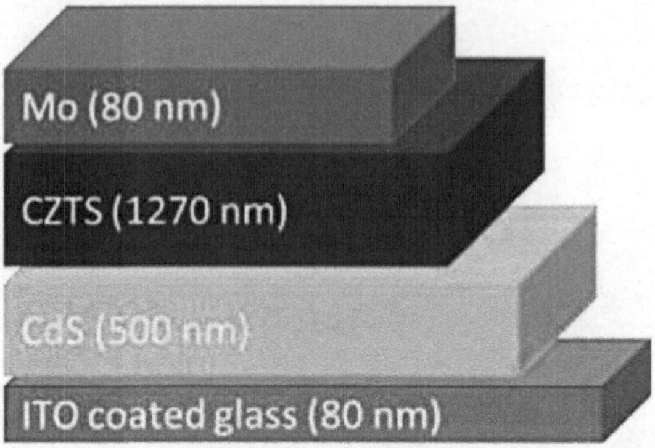

FIGURE 3.10 Device structure developed by Abdelraouf with CZTS/CdS nanostructures.

CZTS doped with lithium via ethanol solvent spin coating was fabricated by Z. Wang et al. This was seen to increase the kesterite grain growth on TiO_2 nanorods. This reduces interface charge recombination, as confirmed through EIS analysis. A thickness of ~350 nm of CZTS layer

was deposited over TiO_2, which renders a higher interaction of light inside the nanorod framework, which compensates for the thinner film as compared to conventional thin-film kesterite photovoltaic devices. The device (Fig. 3.11) displayed excellent stability in environments such as light soaking or air exposure. Device performance was improved with the replacement of Spiro-OMeTAD with a more stable inorganic HTM.[25]

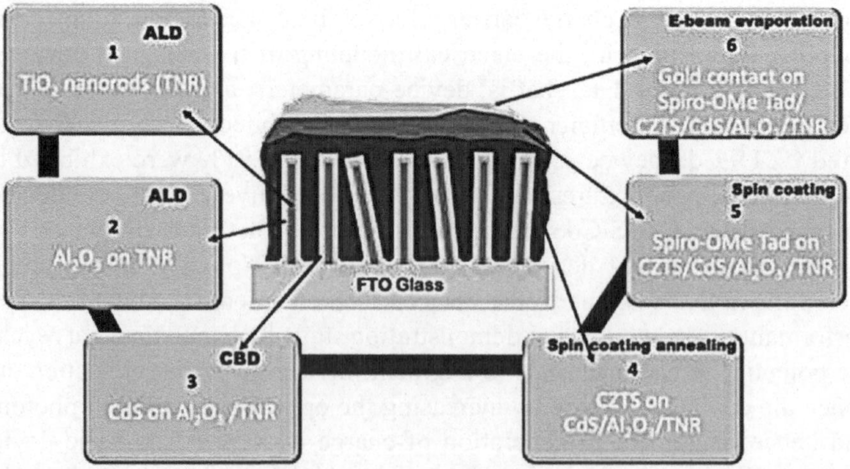

FIGURE 3.11 Schematic representation of the basic device structure put forward by Wang et al. and the step-by-step process of device fabrication.

Ghosh et al. reported CZTS solar cells with the configuration ITO/ZnO (NRs)/ZnS/CZTS/Au. The hydrothermal route was employed for depositing the vertically aligned ZnO nanorod arrays. Another solution processing technique, chemical bath deposition, was employed for ZnS and CZTS thin film deposition. In the proposed device structure (Fig. 3.12), the device roll of the window layer was played by ZnO nanorods, the buffer layer by ZnS film, and the photoactive layer by copper zinc tin sulfide. Compared to thin films of flat surface, the ZnO/ZnS layer with nanostructures boosted photon trapping. Avoiding toxic CdS was another highlight of this proposed and fabricated device structure by Ghosh and team, yielding an efficiency of ~3.6%.[26]

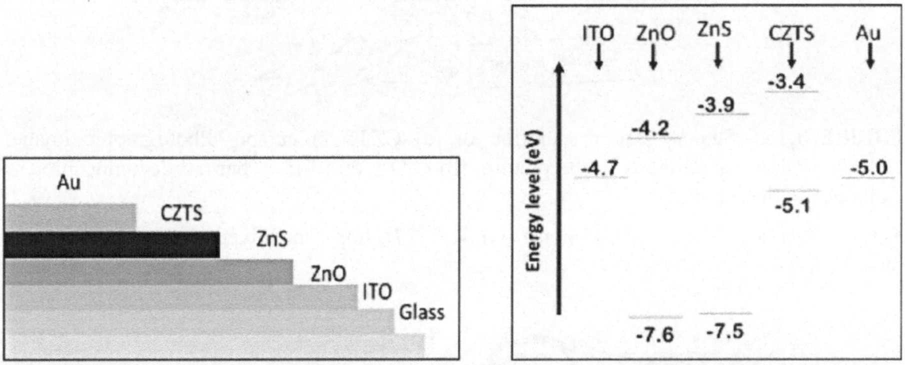

FIGURE 3.12 Schematic of the device structure (left) and energy level diagram of the device (right) proposed by Ghosh et al.

Chen et al. proposed a low-cost and eco-friendly thermal decomposition of metal-thiourea-oxygen sol-gel complexes to synthesize thin films of CZTS. The deposition procedure adopted by Chen et al. is depicted in Figure 3.13. Inverted-structure solar cells were fabricated by coupling CZTS thin films with ZnO@CdS heterojunction nanoarrays. Charge carrier collection, separation, and transfer were promoted by the vertically aligned nanowires (NWs) of ZnO, mainly due to their large interface areas. Researchers of this work initially conducted the optimization studies on CZTS absorber layer annealing temperature, CdS buffer layer thickness, and the morphology of ZnO nanowires. The device in this work yielded an efficiency of 2.27% with a good open circuit voltage of 589 mV. All the layers are deposited using a solution processing technique that permits the realization of a cost-effective CZTS solar cell.[27] In Figure 3.14, the scheme of the device, band diagram, and SEM are shown.

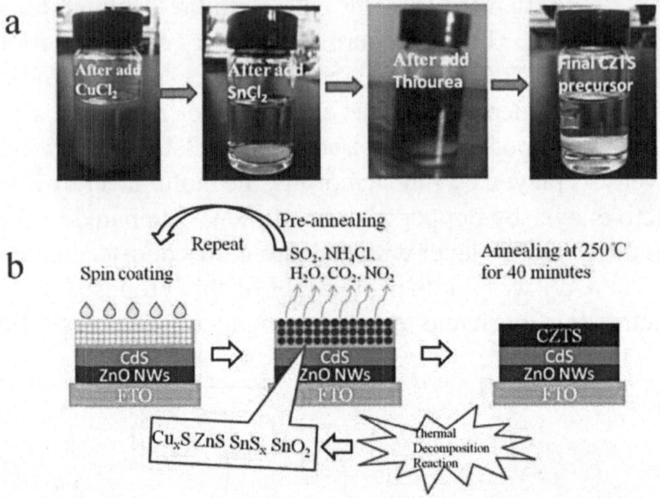

FIGURE 3.13 Step-by-step preparation of (a) CZTS precursor—Photographic image. Color variation was obvious in the picture. (b) CZTS thin film. Thermal decomposition is explained in the figure.

Source: Reproduced with permission from Ref. [27]. https://creativecommons.org/licenses/by/4.0/

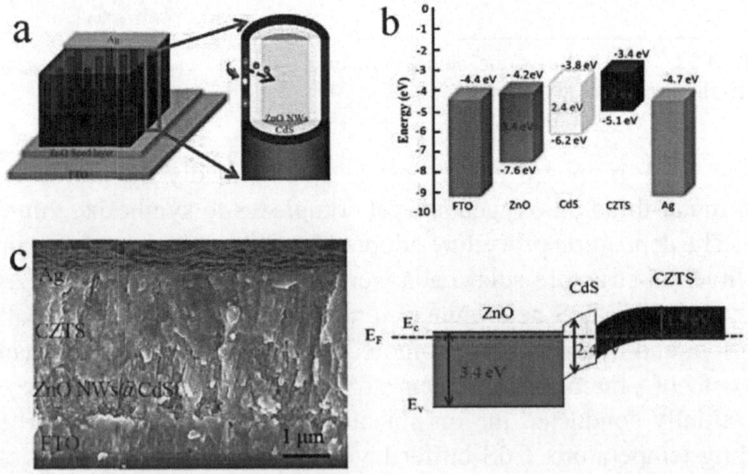

FIGURE 3.14 (a) Schematic diagram of CZTS solar cells fabricated by Chen and group, (b) energy level and band alignment of different layers in the device, and (c) SEM image of the device (cross-sectional SEM).

Source: Reproduced with permission from Solution-Processed One-Dimensional ZnO@CdS Heterojunction Toward Efficient Cu_2ZnSnS_4 Solar Cell with Inverted Structure. Scientific Reports, Publisher: Nature Portfolio 2016 (Open Access Journal).

3.3.3 CZTS SOLAR CELLS BASED ON NANOPARTICLES AND NANOCRYSTALS

Zhang et al. made use of nontoxic solvents for the synthesis of CZTS nanocrystal inks. The method chosen was wet ball milling and was tailed by filtration (Fig. 3.15). These inks were then used to develop CZTS thin film by making use of the spin-coating technique. Rapid high-pressure sulfur annealing was done on the samples with high precision and control. The annealing process was a key factor that promoted grain growth, and hence, crystallinity. It also aids in tuning the film from poor sulfur content to a sulfur-rich sample.[28]

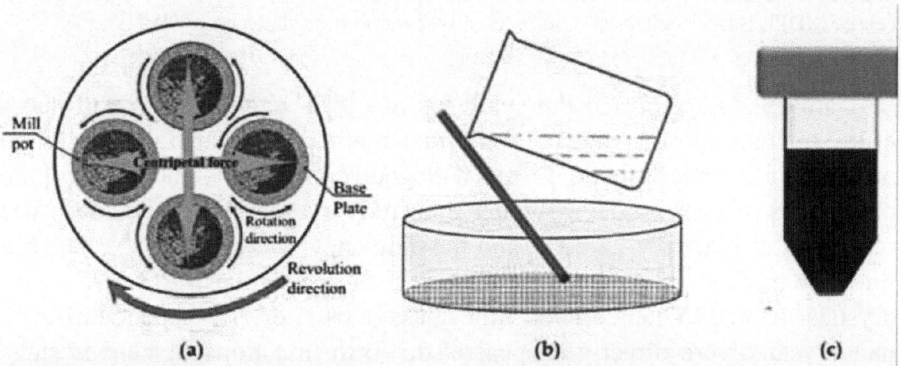

FIGURE 3.15 Step-by-step fabrication process of nanoparticles of CZTS (Cu_2ZnSnS_4) developed by Zhang et al. (a) Ball milling machine (schematic), (b) filtration process, and (c) CZTS nanoparticle ink.

Figure 3.16 shows the fabricated device configuration. The power conversion efficiency exhibited by the device was 6.2%, with a corresponding short-circuit current density of 17.6 mA/cm^2. The fill factor, which determines the quality of the device, was found to be 55.8%. Another major parameter, which is a direct result of the work function between different layers, is the open-circuit voltage, which is 633.3 mV. Further improvement in performance can be attained only through bettering the crystallinity by enhancing the grain size and reducing the defects. Also, secondary phase formation in the complex growth mechanism must be analyzed to confirm the quality of the film, which directly affects solar cell performance.[28]

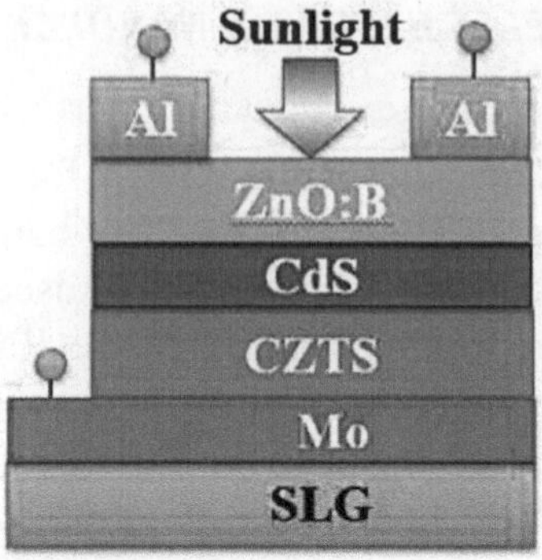

FIGURE 3.16 Device structure proposed by Zhang et al.

Source: Reproduced with permission from Ref. [28].

Camara et al. reported the synthesis of CZTS nanoparticles with kesterite structure through the hydrothermal method. Copper-rich (2.8:1:1:6) nanoparticles rendered better results with good crystallinity. The photovoltaic parameters of the best device under this study are Jsc = 23.8 mAcm^{-2}, Voc = 0.394 V, fill factor = 28.78%, and the efficiency is about 2.25%.[29] The hot injection method was utilized by Guo et al. for the synthesized CZTS nanocrystals. Atop Mo-coated soda lime glass substrate, the suspended CZTS nanocrystals were directly dropcasted to form thin films of nanocrystals. They were selenized to form CZTSSe absorbers by annealing under Se vapor in a graphite box at temperatures between 400 and 500°C. Thus, fabricated nanocrystal ink-based solar cells with CZTSSe absorber films exhibited a power conversion efficiency of 0.74%.[30]

A monocrystalline nanoparticle of CZTS was developed through a simple noninjection technique. After adding a predetermined quantity of the precursor solution, it is transferred to an autoclave and kept in the oven at around 240°C for 24 h. The as-prepared samples were well-dispersed nanoparticles. For device fabrication, these samples were mixed with the active layer, a P3HT:PCBM blend dissolved in dichlorobenzene. CZTS nanoparticles were added to this solution and made to disperse in it. This solution was then spin coated to form a thin film, a hybrid active layer. An obvious increment in the absorption of light and charge carrier collection was exhibited by the said hybrid active layer. An increment in the solar cell performance parameter was obtained with a slight quantity of CZTS NPs. A neat cross-sectional SEM image of the device is reported by the team and is included in Figure 3.17. ITO, active layer, ZnO, and nanoparticles of CZTS can be clearly identified from the figure. An enhancement in efficiency to 3.65% from 3.30%, owing to an enhancement in short-circuit current density to 10.67 mA/cm^2 from 9.90 mA/cm^2 was reported by the team. The voltage–current measurement of the device is depicted in Figure 3.18.[31]

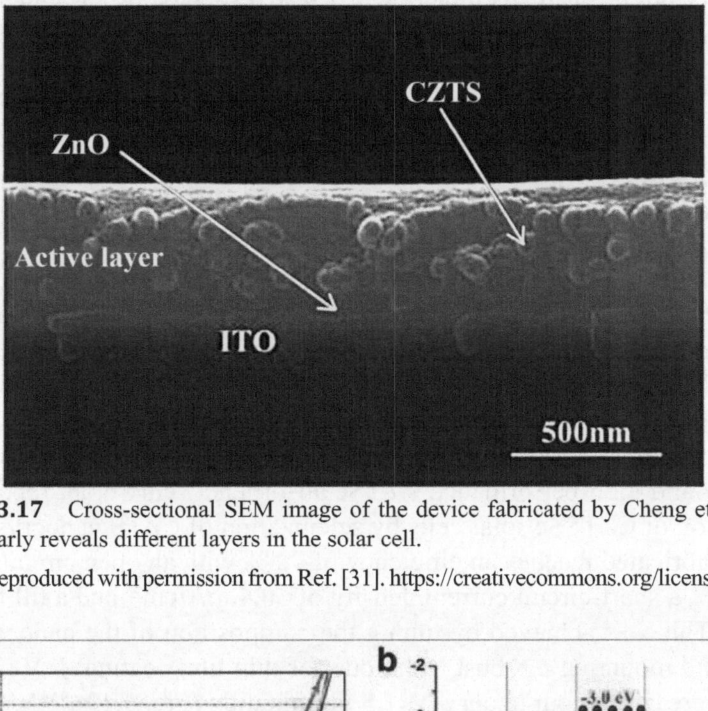

FIGURE 3.17 Cross-sectional SEM image of the device fabricated by Cheng et al.[31] The figure clearly reveals different layers in the solar cell.

Source: Reproduced with permission from Ref. [31]. https://creativecommons.org/licenses/by/4.0/

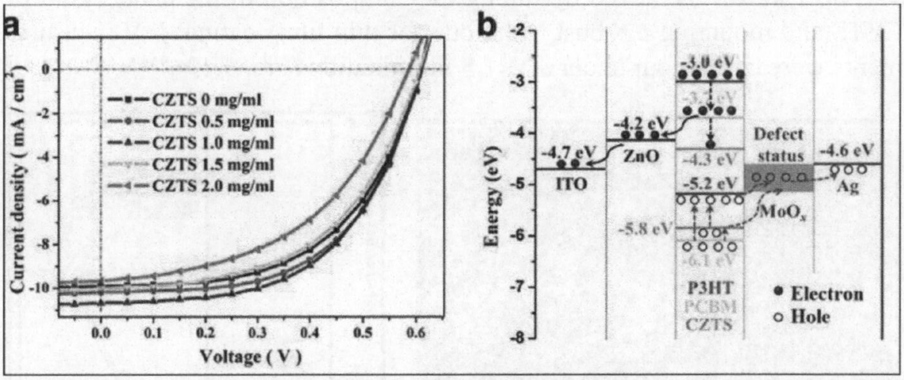

FIGURE 3.18 J-V characteristics of the device (left side) and energy level diagram of the device developed by Cheng et al.

Source: Reproduced with permission from Ref. [31]. https://creativecommons.org/licenses/by/4.0/

Mokurala et al. prepared CZTS nanoparticles using different solvents like deionized water (DW) and ethylene glycol (EG) by a wet chemical process.

It was observed that sulfurized CZTS films were excellent alternative counter electrodes for Pt in DSSCs. DSSCs fabricated with CZTS nanorods as counter electrode material exhibited an efficiency of 7.8%; nanoflowers with sheets showed an efficiency of 7.5%; and spherical nanoparticles as counter electrodes had an efficiency of 7.4%, respectively.[32] Kannan and his group explained the solvothermal route to develop nanoparticles of copper zinc tin sulfide. Using the as-prepared sample as the absorber layer, that is, p-type CZTS, heterojunction solar cells were fabricated with ZnO as the counterpart buffer layer. In this work, they varied the ratio of copper and analyzed its effect on the properties of CZTS nanoparticles, and the device performance was systematically examined. The CZTS/ZnO solar cell device under this study exhibited an efficiency of 3.2%. $V_{oc\ and}$ J_{sc} were 510 mV and 11.49 mA/cm^2. The fill factor of the device was 52.97%.[33]

Zaberca et al. synthesized surfactant-free CZTS nanoparticles via a high temperature polycondensation route as building blocks for low-cost solar cell absorbers. The films fabricated with these surfactant-free CZTS nanoparticles exhibited a low-defect concentration, crack-free, sintered nature, which can render an optimistic strategy for the manufacture of low-cost and high-performance CZTSe absorbers.[34] Guo et al. reported the fabrication of CZTSSe solar cells by selenization of CZTS nanocrystal. The device fabricated reaches an efficiency of 7.2% with an open circuit voltage of 0.42 V, a short-circuit current density of 30.4 mA/cm^2, and a fill factor of 52.7%. This was achieved by tuning the composition of the nanocrystal of CZTS and mounting a robust technique for thin film coating. I–V measurements were carried out under AM 1.5 illumination (Fig. 3.19).[35]

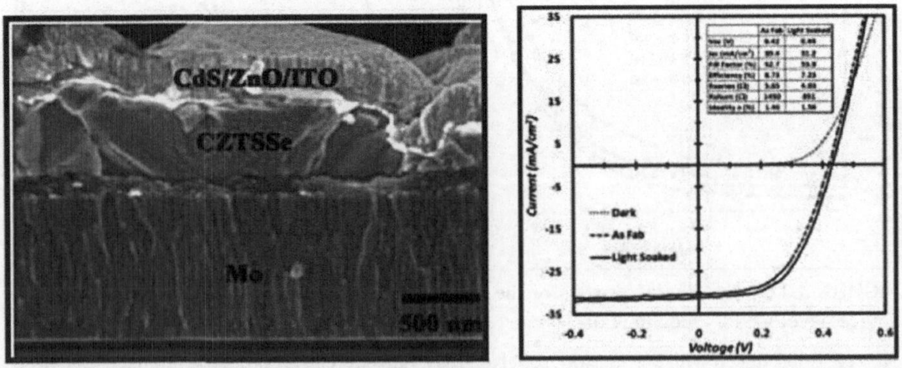

FIGURE 3.19 (a) SEM image of CZTSSe solar cell (cross-sectional image) and (b) I–V characteristics of the device showing best efficiency under the study done by Guo et al.

Source: Reproduced with permission from Ref. [35]. Copyright 2010. American Chemical Society.

SILAR—Successive ionic layer adsorption and reaction was utilized by Mahajan et al. for synthesizing the nanostructure of CZTS thin films. Both indium-doped tin oxide (ITO) and fluorine-doped tin oxide (FTO) were used as substrates. The chosen window layer was a CdS quantum dot. Voltage current measurements were taken, and hence the efficiency of the devices was calculated. The power conversion efficiency was 1.68% for the solar cells deposited over ITO substrate and 2.08% for the device deposited over FTO substrate. They also observed that sulfurization has a promising impact on the CZTS thin film. Along with that, the FTO substrate boosted the efficiency at a higher temperature.[36]

3.4 CONCLUSION

As said by famous scientist Richard P. Feynman in 1959, "There is plenty of room at the bottom," we are in the era of nanomaterials. Day by day, advancements in nanotechnology have changed not only the scientific field but the entire world. It has revised the perception of all matters around us from outer space to the storage room of a house. One among them is the field of energy, in which solar cell devices, an inevitable one. Nanotechnology opens up inclusive possibilities for solar cells. Incorporating nanomaterials (nanorods, nanowires, nanocrystals, quantum dots, or any nanostructured material) into the solar cells augments the device's performance owing to the enhancement in surface area of contact between two layers, leading to better charge carrier collection and hence the short-circuit current density. This ultimately results in an enhancement in efficiency compared to devices using micro or larger particles. "COST," an unescapable part of a device, has a considerable reduction when the usual conventional device is replaced with nanomaterials.

The above discussed work mainly focused on solar cell devices fabricated using nanomaterials of CZTS. Different synthesizing processes of nanomaterials for integration into solar cells were studied. Photovoltaic devices were fabricated using these nanomaterials with diverse device structures by different researchers. For the absorber nano-CZTS, cadmium sulfide, zinc oxide, zinc sulfide, etc. were used as the buffer layers. Even though these works are in their booming stages, working devices were reported by a large number of groups. A few reports are in the literature under this category, reaching an efficiency close to 10%. Comparing to conventional devices, efficiency may seem to be lower, but the low cost of these devices overcomes this lower efficiency. All these obviously emphasize that nanotechnology incorporating solar cells has the potential to replace silicon-based solar devices and change the face of solar cell technology.

KEYWORDS

- **CZTS**
- **nanocrystals**
- **quantum dots**
- **nanoparticles**
- **solar cells**

REFERENCES

1. Kannan, N.; Vakeesan, D. Solar Energy for Future World:—A Review. *Renewable and Sustainable Energy Reviews* **2016,** *62*, 1092–1105. https://doi.org/10.1016/j.rser.2016.05.022.
2. Löf, G. O. G.; Duffie, J. A.; Smith, C. O. World Distribution of Solar Radiation. *Solar Energy* **1966,** *10* (1), 27–37. https://doi.org/10.1016/0038-092X(66)90069-7
3. Schlamadinger, B.; Apps, M.; Bohlin, F.; Gustavsson, L.; Jungmeier, G.; Marland, G.; Pingoud, K.; Savolainen, I. Towards a Standard Methodology for Greenhouse Gas Balances of Bioenergy Systems in Comparison with Fossil Energy Systems. *Biomass Bioenergy* **1997,** *13* (6), 359–375. https://doi.org/10.1016/S0961-9534(97)10032-0
4. Green, M. A. Photovoltaic Principles, *Phys. E: Low-Dimensional Syst. Nanostruct.* **2002,** *14* (1), 11–17. https://doi.org/10.1016/S1386-9477(02)00354-5.
5. Razykov, T. M.; Ferekides, C. S.; Morel, D.; Stefanakos, E.; Ullal, H. S.; Upadhyaya, H. M. *Solar Energy* **2011,** *85* (8), 1580–1608. https://doi.org/10.1016/j.solener.2010.12.002
6. Zhang, T.; Wang, M.; Yang, H. A Review of the Energy Performance and Life-Cycle Assessment of Building-Integrated Photovoltaic (BIPV) Systems. *Energies* **2018,** *11* (11), 3157. https://doi.org/10.3390/en11113157
7. Zhao, J.; Wang, A.; Green, M. A. High-Efficiency PERL and PERT Silicon Solar Cells on FZ and MCZ Substrates. *Solar Energy Mater. Solar Cells 65* (1–4), 429–435. https://doi.org/10.1016/S0927-0248(00)00123-9
8. Sahu, A.; Garg, A.; Dixit, A. A Review on Quantum Dot Sensitized Solar Cells: Past, Present and Future Towards Carrier Multiplication with a Possibility for Higher Efficiency. *Solar Energy* **2020,** *203*, 210–239. https://doi.org/10.1016/j.solener.2020.04.044
9. https://nanografi.com/blog/future-of-solar-cells-with-nanotechnology-solar-energy/
10. How can nanotechnology improve solar cells. https://www.understandingnano.com/solarcells.html

11. Najafi, V.; Aghelifar, M.; Kimiagar, S. A Novel Synthesis of CZTS Quantum Dots Using Pulsed Laser Irradiation. *Superlattices Microstruct.* **2017,** *109*, 702–707. https://doi.org/10.1016/j.spmi.2017.05.063.
12. Xu, J.; Yang, X.; Yang, Q.-D.; Wong, T.-L.; Lee, C.-S. Cu_2ZnSnS_4 Hierarchical Microspheres as an Effective Counter Electrode Material for Quantum Dot Sensitized Solar Cells. *J. Phys. Chem. C* **2012,** *116* (37), 19718–19723.
13. Zhou, X.; Meng, W.; Dong, C.; Liu, C.; Qiu, Z.; Qi, J.; Chen, J.; Wang, M. Cu2ZnSnS4 Quantum Dots as Effective Electron Acceptors for Hybrid Solar Cells with a Broad Spectral Response. *RSC Adv.* **2015,** 5, 90217–90225.
14. Zhou, Z.-J.; Deng, Y.-Q.; Zhang, P.-P.; Kou, D.-X.; Zhou, W.-H.; Meng, Y.-N.; Yuan, S.-J.; Wu, S.-X. Cu_2ZnSnS_4 Quantum Dots as Hole Transport Material for Enhanced Charge Extraction and Stability in All-Inorganic $CsPbBr_3$ Perovskite Solar Cells. *RSC Adv.* **2015,** *5*, 90217.
15. Das, S.; Alam, I.; Raiguru, J.; Subramanyam, B. V. R. S.; Mahanandia, P. A Facile Method to Synthesize CZTS Quantum Dots for Solar Cell Applications. *Phys. E: Low-Dimensional Syst. Nanostruct.* **2019,** *105*, 19–24.
16. Sahoo, G. S.; Routray, S.; Mishra, G. P. Extensive Study on Effects of Defects in CZTS/CZTSe Quantum Dots Kesterite Solar Cells. In: *2021 IEEE 16th Nanotechnology Materials and Devices Conference (NMDC)*; 2021; pp 1–4. DOI: 10.1109/NMDC50713.2021.9677478.
17. Sadeghnejad, A.; Lu, L.; Cline, J.; Ozdemir, N. K.; Snyder, M. A.; Kiely, C. J.; McIntosh, S. In Situ Biomineralization of CuxZnySnzS4 Nanocrystals within TiO2-Based Quantum Dot Sensitized Solar Cell Anodes. ACS Appl. Mater. Interfaces **2019,** *11* (49), 45656–45664.
18. Zhang, Y.; Shi, C.; Dai, X.; Liu, F.; Fang, X.; Zhu, J. Pyrolysis Preparation of Cu_2ZnSnS_4 Thin Film and Its Application to Counter Electrode in Quantum Dot-Sensitized Solar Cells. *Electrochim. Acta* **2014,** *118*, 41–44. https://doi.org/10.1016/j.electacta.2013.11.168.
19. Bai, B.; Kou, D.; Zhou, W.; Zhou, Z.; Tian, Q.; Meng, Y.; Wu, S. Quaternary Cu_2ZnSnS_4 Quantum Dot-Sensitized Solar Cells: Synthesis, Passivation and Ligand Exchange. *J. Power Sources* **2016,** *318*, 35–40. https://doi.org/10.1016/j.jpowsour.2016.04.009.
20. Liu, F.; Huang, J.; Sun, K.; Yan, C.; Shen, Y.; Park, J.; Pu, A.; Zhou, F.; Liu, X.; Stride, J. A.; Green, M. A.; Hao, X. Beyond 8% Ultrathin Kesterite Cu_2ZnSnS_4 Solar Cells by Interface Reaction Route Controlling and Self-Organized Nanopattern at the Back Contact. *NPG Asia Mater.* **2017,** *9*, e401. DOI: 10.1038/am.2017.103
21. Lee, D.; Yong, K. Solution-Processed Cu_2ZnSnS_4 Superstrate Solar Cell Using Vertically Aligned ZnO Nanorods. *Nanotechnology* **2014,** *25*, 065401 (8pp). DOI: 10.1088/0957-4484/25/6/065401
22. Darvishzadeh, P.; Sohrabpoor, H.; Gorji, N. E. Numerical Device Simulation of Carbon Nanotube Contacted CZTS Solar Cells. *Opt. Quant. Electron.* **2016,** *48*, 480. DOI: 10.1007/s11082-016-0741-5
23. Wang, X.; Brodusch, N.; Gauvin, R.; Demopoulo, G. Nanoengineering of the Cu_2ZnSnS_4–TiO_2 Interface via Atomic Layer Deposition of Al_2O_3 for High Sensitivity Photodetectors and Solid State Solar Cells. *J. Mater. Chem.* **2018,** *6*, 11507–11520.
24. Abdelraouf, O. A. M.; Allam, N. K. Nanostructuring for Enhanced Absorption and Carrier Collection in CZTS-Based Solar Cells: Coupled Optical and Electrical Modeling. *Opt. Mater.* **2016,** *54*, 84–88. https://doi.org/10.1016/j.optmat.2016.02.021.

25. Wang, Z.; Brodusch, N.; Gauvin, R.; Demopoulos, G. P. Lithium-Doped Cu_2ZnSnS_4 Superstrate Solar Cells with 5% Efficiency—An Alternative to Thin Film Kesterite Photovoltaics. *Nano Energy* **2018,** *53*, 130–134. https://doi.org/10.1016/j.nanoen.2018.08.049.
26. Ghosh, A.; Thangavel, R.; Gupta, A. Solution-Processed Cd Free Kesterite Cu_2ZnSnS_4 Thin Film Solar Cells with Vertically Aligned ZnO Nanorod Arrays. *J. Alloys Compd* **2017,** *694*, 394–400. https://doi.org/10.1016/j.jallcom.2016.09.325.
27. Chen, R.; Fan, J.; Liu, C. et al. Solution-Processed One-Dimensional ZnO@CdS Heterojunction Toward Efficient Cu_2ZnSnS_4 Solar Cell with Inverted Structure. *Sci. Rep.* **2016,** *6*, 35300. https://doi.org/10.1038/srep35300
28. Zhang, X.; Fu, E.; Wang, Y.; Zhang, C. Fabrication of Cu_2ZnSnS_4 (CZTS) Nanoparticle Inks for Growth of CZTS Films for Solar Cells. *Nanomaterials (Basel)* **2019,** *9* (3), 336. DOI: 10.3390/nano9030336. PMID: 30832326; PMCID: PMC6473955.
29. Camara, S. M.; Wang, L.; Zhang, X. Easy Hydrothermal Preparation of Cu_2ZnSnS_4 (CZTS) Nanoparticles for Solar Cell Application. *Nanotechnology* **2013,** *24* (49), 495401. DOI: 10.1088/0957-4484/24/49/495401
30. Guo, Q.; Hillhouse, H. W.; Agrawal, R. Synthesis of Cu_2ZnSnS_4 Nanocrystal Ink and Its Use for Solar Cells. *J. Am. Chem. Soc.* **2009,** *131* (33), 11672–11673. DOI: 10.1021/ja904981r
31. Cheng, J.; Dai, Z.; Chen, B. et al. Well-Dispersed Cu_2ZnSnS_4 Nanocrystals Synthesized from Alcohols and Their Applications for Polymer Photovoltaics. *Nanoscale Res Lett* **2016,** *11*, 550. https://doi.org/10.1186/s11671-016-1761-6
32. Mokurala, K.; Kamble, A.; Bathina, C.; Bhargava, P.; Mallick, S. Effect of Solvent, Reaction Time on Morphology of Cu_2ZnSnS_4 (CZTS) Nanoparticles and Its Application in Dye Sensitized Solar Cells. *Mater. Today: Proc.* **2016,** *3* (6), 1778–1784. https://doi.org/10.1016/j.matpr.2016.04.074.
33. Kannan, A. G.; Manjulavalli, T. E.; Thambidurai, M.; Habeeba, K.; Ezhilarasi Gnana Kumari, D. V. 3.2% Efficient Cadmium Free Cu_2ZnSnS_4/ZnO Solar Cells Fabricated Using Solvothermally Synthesized Nanoparticles. In: *6th International Conference on Nanoscience and Nanotechnology (ICONN 2021), IOP Conf. Series: Materials Science and Engineering*, Vol. 1219; IOP Publishing, 2022; p 012036. DOI: 10.1088/1757-899X/1219/1/012036
34. Zaberca, O.; Oftinger, F.; Chane-Ching, J. Y.; Datas, L.; Lafond, A.; Puech, P.; Balocchi, A.; Lagarde, D.; Marie, X. Surfactant-Free CZTS Nanoparticles as Building Blocks for Low-Cost Solar Cell Absorbers. *Nanotechnology* **2012,** *23* (18), 185402. DOI: 10.1088/0957-4484/23/18/185402
35. Guo, Q.; Ford, G. M.; Yang, W.-C.; Walker, B. C.; Stach, E. A.; Hillhouse, H. W.; Agrawal, R. Fabrication of 7.2% Efficient CZTSSe Solar Cells Using CZTS Nanocrystals. *J. Am. Chem. Soc.* **2010,** *132* (49), 17384–17386. DOI: 10.1021/ja108427b
36. Mahajan, S.; Stathatos, E.; Huse, N.; Birajdar, R.; Kalarakis, A.; Sharma, R. Low Cost Nanostructure Kesterite CZTS Thin Films for Solar Cells Application. *Mater. Lett.* **2018,** *210*, 92–96. https://doi.org/10.1016/j.matlet.2017.09.001

CHAPTER 4

Electrical Energy Transport through Correlated Barrier Hopping Mechanism in Silver-Decorated Graphene Oxide Nanolayers

I. DHANYA[1], S. HEERA[1], ANJU ELIZABETH YOHANNAN[1], ASWINI MARIAM VINOD[1], MEENAKSHY SALIM[1], ASIYA SHAFEEK[1], and SREEJITH K. PISHARADY[2]

[1]*Department of Physics, Catholicate College, Pathanamthitta, Kerala, India*

[2]*Department of Physics, Sanatana Dharma College, Alappuzha, Kerala, India; Arqueoges—Archeology and Heritage Management, Guimarães, Portugal*

ABSTRACT

Graphene oxide (GO) nanosheets are prepared by modified Hammer's method. Chemically reduced silver nanocolloid is well-dispersed in graphene oxide solution, and the residue is filtered and vacuum dried to get silver-decorated graphene oxide nanocomposites. The morphology of the samples is analyzed by a scanning electron micrograph, and the chemical composition is structured through X-ray electron dispersive analysis. Structures of nanocomposites are well-analyzed, and crystalline size and crystallographic strain are calculated by X-ray diffractogram. Thermal conductivity studies are carried out over a wide range of temperatures. DC electrical conductivity studies are carried out at various temperatures between 303 and 373 K. Both intrinsic and extrinsic activation enthalpies are calculated for transferring charge between the valance band and conduction band. AC electrical

Resilience and Sustainability for Energy and Water: New Advancements, Opportunities, and Framework.
Cristóbal Noé Aguilar, A. K. Haghi, Amritanshu Shukla, Ann Rose Abraham, Swati Gokul Talele (Eds.)

conductivity studies are carried out in a frequency range of 100 Hz–1 MHz. Hopping level conduction parameters like hopping distance, hopping energy, and density of charge defect state are analyzed by the variable range hopping technique. Annealing effects on AC conductivity strongly suggest that the bipolaron-correlated barrier hopping model is the best fit for silver-doped graphene oxide two-dimensional layers.

4.1 INTRODUCTION

Recent years reflect an increase in the demand for nanocomposites, mainly in the fields of science and technology. Two-dimensional layers of materials like graphene and its oxide form can easily make nanocomposites with a variety of organic and inorganic materials.[1] The synthesis and diffusion of other materials with graphene make them excellent for the composite formation apart from other organic materials.[2] The outstanding electrical conductivity of graphene layers due to their flexible two-dimensional layered structure makes them different from other polymers in organic electronics.[3] The nanoporous nature of the surface layers makes graphene a good candidate for sensing applications.[4] Though the mentioned properties are neglected, the intrinsic charge transport property in graphene and its oxide layer and the doping of impurities like metals and rare earth ions make them engaging in the area of hopping conduction.[5] Defect level conduction can also be created on their surface by the layered surface topology and defect site formation in between surface layers.[6] The stability of graphene oxide in solid form makes it an active material for semiconductor electronics.[7] Defect-free surface layers can be achieved in the graphene oxide layers by doping them with noble metals like gold or silver.[8] Both gold and silver can create anisotropic nanoparticles and clusters which can easily migrate through the defect sites on the surface layers of graphene to form nanocomposites.[9] This feature makes them an ideal material for graphene composite fabrication. Though graphene oxide can be synthesized by various techniques, modified Hammer's method acts as the best technique for producing a bulk amount of graphene oxide with a simple reduction process.[10–12] Several studies have reported the electrical conductivity of graphene oxide and its polymer composites,[13] but very few studies are related to gold and silver-decorated graphene oxides for different applications.[14,15] An in-depth study is needed for a thorough understanding of the charge carrier transport mechanism through graphene oxide and its composites to get a better understanding of the materials to be chosen for the different applications in the semiconductor industry.[16]

The present work discusses a modest attempt made for the preparation of silver-doped graphene oxide nanolayers. A morphological study is carried out to analyze the defect sites within the topology and to analyze how they influence the electrical conductivity of the samples. The stoichiometric ratio of each component within the composites is examined by the X-ray energy dispersive analysis. The degree of crystallinity, microstrain, and size of nanoparticle's effects on conductivity is well-structured through an X-ray diffractogram. DC electrical conductivity studies are carried out to study the surface layer conductivity of composites through two probe analyses. AC electrical conductivity studies are carried out to investigate the in-depth inner layer charge carrier transport behavior of silver-doped graphene oxide nanocomposites.

4.1.1 EXPERIMENTAL STUDIES

Graphite powder and silver nitrate (purity 99.99%, Sigma Aldrich Chemical. Co.) are used as master materials for the present study. Large-scale preparation of nanosheets of graphene oxide is achieved by modified Hammer's method.[17] One gram of graphite powder and 0.5 g of sodium nitrate are added to 23 mL of 98% concentrated sulfuric acid and is mixed well using an electronic stirrer at a moderate speed of 60–80 rpm. The above mixture is kept in an ice bath (0–5°C) and is continuously stirred well for 30 min. 3 g of potassium permanganate is added slowly to the mixture, and stirring is continued for another 3 h. The solution is transferred to a 35°C water bath, followed by continued stirring for the next 30 min. 46 mL of deionized water is added to the solution and stirred at a temperature of 95°C for 30–45 min. 140 mL of deionized water is added rapidly followed by 10 mL of hydrogen peroxide solution for the complete termination of the oxidation process of potassium permanganate in the mixture. Silver nanocolloid is prepared from silver nitrate using tannic acid as the reducing agent. The chemically reduced silver colloid is added to graphene oxide solution by keeping a net uniform concentration of $Ag^0{}_X\,GO_{\,(1000-X)}$. Here, x stands for 50, 100, and 200 mL of silver colloid for 950, 900, and 800 mL of GO samples so as to maintain the composite concentration always at 1000 mL. Pure GO is labeled as sample A, while 50, 100, and 200 mL silver colloid-added samples are labeled as sample B, sample C, and sample D, respectively. The sonication process and microwave reduction method are applied to composites for better diffusion of silver with graphene oxide. Further, 30 min of microcentrifugation with a speed of 4000 rpm is also made for the better settlement of residue. The resultant residue of dark brown color is filtered using Wattman A-grade filter paper and is further cleaned by washing several times with a 5% HCl solution. The residue is vacuum-annealed to obtain the resultant powdered sample.

Various characterizations are performed on the pure graphene oxide sample as well as on the composites. Morphological analysis along with the stoichiometric ratio is examined by SEM with EDX spectroscopy using the JEOL JSM 6390 LA/ OXFORD XMX N, which has parameters like an excitation voltage of 0.5 to 30 KV with a tungsten filament and a magnification of 300000. The EDX facility is incorporated with a resolution of 136 eV, having a maximum detector area of 30 mm^2 with accurate elemental mapping. An X-ray diffractogram is taken for the powder samples with the Bruker D8 advanced X-ray diffractometer and is well-indexed in the database ISCD PDF-2 (2018). The thermal conductivity of the GO and silver-doped GO nanocomposites is analyzed using a DC electrometer set up using two probe analyses. Samples A–D in powder form are converted into pellets by applying 5 tons of uniaxial force for 5 min using a hydraulic press. Pellets thus prepared have a diameter of 10 mm and a thickness of 1.5 mm. Pellets were placed in a tube furnace for taking measurements at various temperatures. The temperature control cum monitor was used with a DPI-1100 dry temperature calibrator coupled with Keithley Electrometer Model Number 2450. AC electrical conductivity is measured using a HIOKI 3532-5 LCR meter. Silver electrodes were used to connect pellets with the instrument by applying silver paste as an ohmic contact.

4.2 THEORY

Crystalline size is calculated from an X-ray diffractogram (XRD) using Scherrer equation[18]

$$D = \frac{k\lambda}{\beta \cos\theta} \tag{4.1}$$

where D is the particle size; k is the Scherrer constant and is having a value of 0.94; λ is the wavelength of X-rays (1.541 A°); β is the full width at half maximum value and is found from the best Gaussian fit from the Origin software analysis; and Θ is the Bragg's angle. Microcrystalline strain can be formed within the graphene composite materials mainly due to nonuniform lattice distortions due to the dopant in the parent crystal system and grain surface relaxation due to layer formation.[19] The microcrystalline strain is calculated from the XRD using the equation

$$\varepsilon = \frac{\beta}{4\tan\theta} \tag{4.2}$$

Dislocation density is a measure of dislocation lines per unit volume and can be calculated using the Williamson–Smallman relation[20]

$$\delta = \frac{1}{D^2} \tag{4.3}$$

DC electrical conductivity studies are significant in graphene composites because of the bandgap energy contributions of extrinsic impurities like dopants added to graphene layers. An Arrhenius plot can be drawn from the fundamental Arrhenius equation (equation 4), and the activation energy for both the intrinsic region and that due to the dopant can be evaluated as E_{a1} and E_{a2}.[21]

$$\sigma_{DC}(T) = \sigma_0 \exp^{\left(\frac{-\Delta E}{kT}\right)} \tag{4.4}$$

where σ_0 is the conductivity of the material at infinite temperature; ΔE is the activation energy, having E_{a1} and E_{a2} values at intrinsic and extrinsic levels; k is the Boltzmann constant; and T is the temperature in kelvin. In order to explain the nature of surface level conduction mechanisms in organic materials, the composite barrier hopping theory can be implemented. Mott and Davis proposed a variable range hopping mechanism by the relation,[22]

$$\sigma_{DC}(T) = \sigma_0 \exp\left(-\frac{T_0}{T}\right)^{\frac{1}{4}} \tag{4.5}$$

Mott characteristic temperature T_0 can be determined by the relation

$$\mathrm{T_0} = \frac{\lambda\alpha^3}{\mathrm{KN(E_F)}} \tag{4.6}$$

where K is the Boltzmann constant, λ=18.1 is a dimensionless quantity, and $\alpha=10^{-9}$ m is the inverse rate of fall of the wave function, especially for organic materials that obey the VRH model.[23] Hopping distance can be calculated by the expression

$$R = \left[{}^{9}\!/_{8}\,\pi\alpha \mathrm{KTN(E_F)}\right]^{\frac{1}{4}} \tag{4.7}$$

Variable range hopping energy can be found out by

$$W=K(T_0T^3)^{\frac{1}{4}} \tag{4.8}$$

Even though most of the organic semiconductors possess stable DC electrical conductivity, AC conduction through the in-depth layers has an anomalous effect especially in high-frequency regions. The overall conductivity of a material, σ_t,can be described by Joncher's universal dynamic response (UDR) relation[24]

$$\sigma_T = \sigma_{DC} + \sigma_{AC} \tag{4.9}$$

where σ_{DC} and σ_{AC} are the DC and AC conductivity for the composite material, respectively. Jonscher's Power Law (JPL) governing AC conductivity dependence on frequency is

$$\sigma_{AC} = A\omega^s \tag{4.10}$$

where A is the temperature-dependent constant, ω is the angular frequency, and s is the frequency exponent having values between 0 and 1.[25] The density of charge carriers near bipolaron hopping centers is N_0 and is generated by the conversion of neutral defects. It can be evaluated through the expression

$$N=N_0 \exp\left[{U_{eff}}/{2kT}\right] \tag{4.11}$$

where N is the density of charged defect states and U_{eff} is the effective correlation energy. Elliott's bipolaron hopping theory suggests the relation between AC conductivity, σ_{AC}, with the hopping distance at a definite frequency, R_ω[26]

$$\sigma_{AC}(\omega) = \frac{\pi^2}{12} N^2 \varepsilon \in_0 \omega R_\omega^6 \tag{4.12}$$

Hopping distance at a fixed frequency can be calculated through the following relationship:

$$R_\omega = \frac{ne^2}{(\pi \in \varepsilon_0[W_m + kTln(\omega\tau_0)]} \tag{4.13}$$

where s is evaluated by the equation:

$$s = 1 - \frac{6KT}{w_m + kTln(\omega\tau_0)} \tag{4.14}$$

where ε is the dielectric constant for a particular frequency, k is the Boltzmann constant, τ_0 is the effective relaxation time, W_m is the maximum barrier height for the π to π* band transition, and e is the charge of electrons.

4.3 RESULT AND DISCUSSION

4.3.1 SURFACE MORPHOLOGICAL STUDIES

A scanning electron micrograph (SEM) of graphene oxide and silver-doped graphene oxide samples is shown with energy dispersive analysis (EDX) in Figure 4.1. From a morphological perspective, the agglomerated form of nanostructure in graphene oxide is transformed into a stacking of flake-like structures in a doped sample with small nanoparticle adherence to the surface. Nanolayer-by-layer formation within the morphology becomes prominent in silver-doped samples. Histograms of doped samples are shown in Figure 4.2. Doping of GO with silver restricts the height profile within the nanoregime and is fitted well with Gaussian functions. Image J software analysis found that the average particle distribution is 28.54 and 32.43 nm for samples A and D, respectively.

The stoichiometric ratio obtained from EDX analysis of maximum doped and undoped samples is shown in Table 4.1.

FIGURE 4.1 SEM and EDX spectra of GO (sample A—top) and silver-doped GO (sample D—bottom).

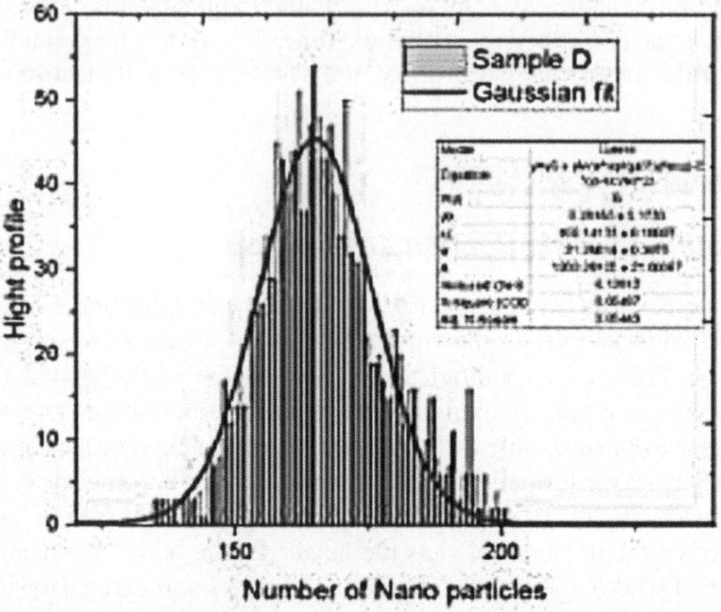

FIGURE 4.2 Variation of height profile with nanoparticle distribution in sample D.

It shows that the major constituents in the samples are contributed by carbon and oxygen. Though sulfur exists within the graphene oxide layers, its concentration is negligibly smaller compared with other major constituents.

4.3.2 STRUCTURAL STUDIES

The X-ray diffractograms (XRDs) of silver-doped GO and that of pure GO are shown in Figure 4.3. Full width at half maximum (FWHM) for all the observed peaks is analyzed through multiple curve fitting using Gaussian distribution functions in Origin software and is tabulated for samples A and D, as shown in Table 4.2. The average particle size corresponding to each FWHM, microcrystalline strain, and dislocation density is also calculated. FWHM for the prominent peak for $2\Theta = 10.61°$ is not altered much for the doped one, while the minor peaks get altered in FWHM. Particle size is found to increase with the addition of silver nanoparticles. Even with a small increase in particle size, there is an appreciable reduction in microstrain and dislocation density. Reduced microcrystalline strain makes them appropriate for electrical device fabrication.[27]

TABLE 4.1 Elemental Composition of GO (Sample A) and Silver-Doped GO (Sample D) from EDX Analysis.

Samples	Element	Line Type	Wt%	Atomic %
Sample A	C	K series	12.11	69.33
	O	K series	6.92	29.73
	S	K series	0.44	0.95
	Total:		19.47	100
Sample D	C	K series	17.98	66.43
	O	K series	11.81	32.73
	S	K series	0.08	0.11
	Ag	L series	0.53	0.73
	Total:		25.07	100

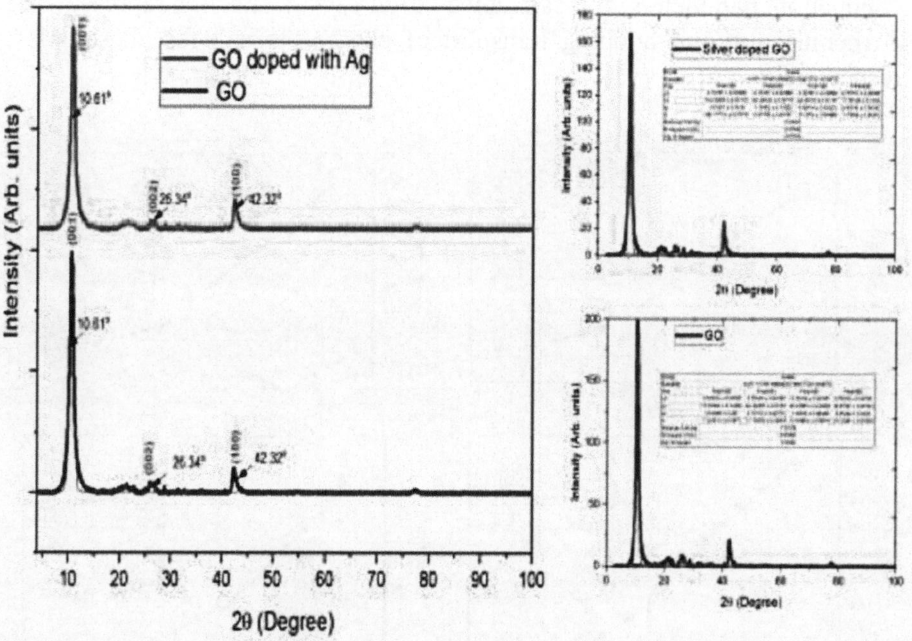

FIGURE 4.3 XRD pattern of pure GO and silver-doped GO [left] and multiple peak fit for FWHM of XRD [right].

TABLE 4.2 Structural Parameters of Sample A and Sample D from X-ray Diffraction Analysis.

Sample	FWHM β(o)	Grain Size D (nm)	Microstrain × 10−3	Dislocation density × 10−4
Sample A	1.13	7.35	53.56	185.12
	1.34	6.63	15.09	248.05
	0.87	12.25	4.72	95.14
	0.76	14.09	4.11	50.34
Sample D	1.19	7.01	55.78	203.75
	1.31	6.85	14.62	232.92
	0.94	11.28	5.13	112.04
	0.69	15.50	3.74	41.61

4.4 ELECTRICAL CONDUCTIVITY STUDIES

4.4.1 DC ELECTRICAL CONDUCTIVITY STUDIES

Thermal conductivity studies are carried out on all the samples, and hopping conduction parameters are calculated. Figures 4.4 and 4.5 represent the Arrhenius plot and hopping transport of charge carriers for all samples, respectively.

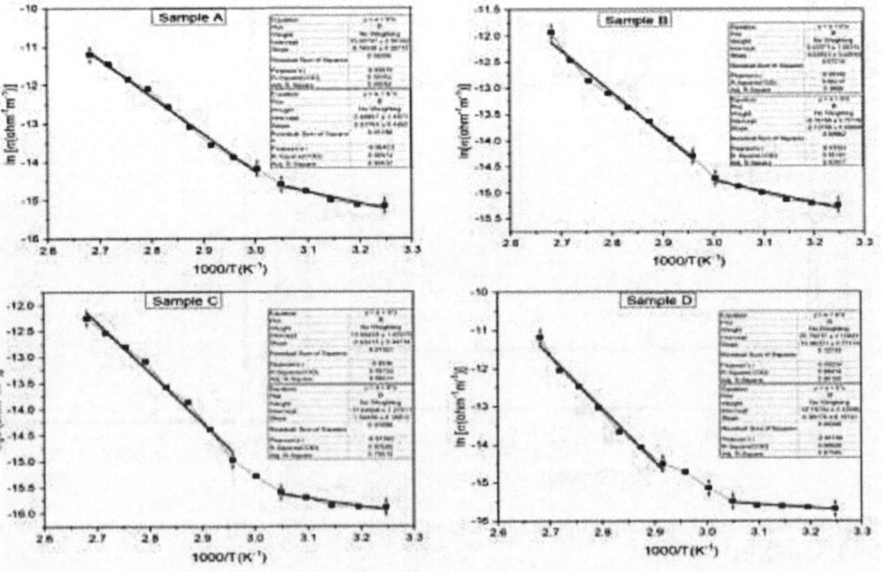

FIGURE 4.4 DC electrical conductivity (ln σ vs. 1000/T) plot for samples A, B, C, and D.

The best straight-line fit is done on all the samples using Origin software analysis. Activation enthalpy is calculated for both intrinsic and extrinsic charge carriers, and different hopping parameters are tabulated in Table 4.3. For graphene oxide, the intrinsic activation energy for crossing over the bands is nearly 1.93 eV ± 0.01 ev, while in doped samples, it increases much to 2.75 eV ± 0.01 eV. This shows the enhanced dielectric nature of doped samples with respect to pure GO. In the extrinsic region, the energy for activation decreases with dopants. It shows the creation of trap levels near the conduction band region of doped samples and the prominence of hopping conduction in them.

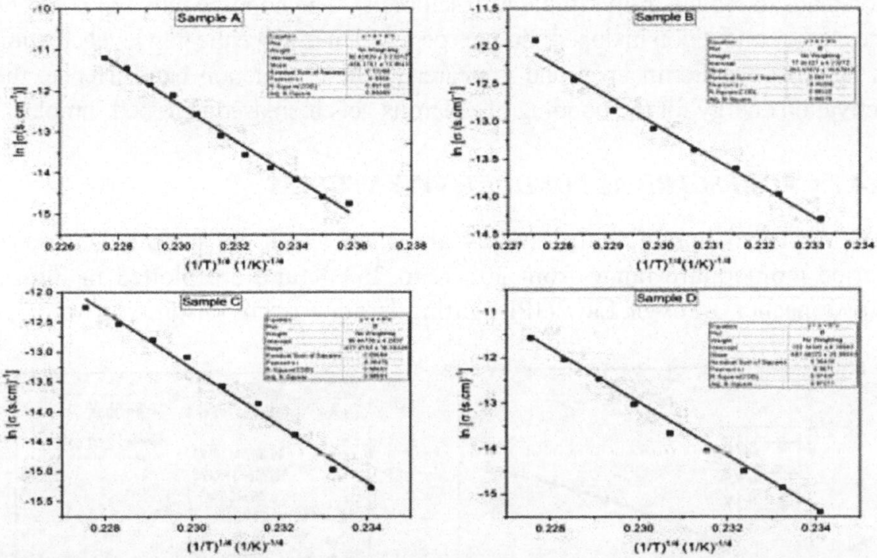

FIGURE 4.5 Variable range hopping conduction [ln σ vs. $(1/T)^{1/4}$] plot for samples A, B, C, and D.

TABLE 4.3 Activation Enthalpy and Variable Range Hopping (VRH) Parameters for Different Samples.

Samples	Activation Energy (eV)		T1	N(Ef)	W (eV)	R (Ao)
	Ea1	**Ea2**	**x 1010**	**x 1019 (eV-1)**		
Sample A	1.93	0.58	206.27	206.27	0.07	293.22
Sample B	1.59	0.424	369.563	369.563	0.062	253.44
Sample C	1.91	0.299	171.226	171.226	0.075	307.19
Sample D	2.75	0.179	74.534	74.534	0.092	378.19

Temperature-dependent DC conductivity has a strong correlation with variable range hopping conduction in layered nanocomposites.[28] VRH equations can be applied for samples annealed at temperatures above 300 K. One of the VRH parameters T_1 is the characteristic temperature, which is a measure of the degree of disorder in the system. From pure GO to maximum silver-doped GO, the value of T_1 was found to be increased, and this may be due to the creation of more defect centers within the surface, which enhances the defect level conduction within the samples. $N(E_f)$ is a measure of the density of localized states at the Fermi level. The probability of finding free charge carriers such as electrons near the Fermi level is higher in the case of pure GO, and it migrates to different hopping sites for doped samples. So, the eigen value for a particular wave function tends to reduce from sample A to sample D. The hopping range of polarons meant for conduction in doped samples gets widened since the trap level creation is far from the Fermi level and is nearer to the conduction band. Hence, the activation energy for the hopping of polarons gets increased in doped samples.

4.4.2 *AC ELECTRICAL CONDUCTIVITY STUDIES*

AC electrical conductivity studies are carried out in sample D over a varied temperature range from 303 K to 353 K and are plotted in Figure 4.6. Jonscher's Power Law (JPL) fitting is done using Origin software for

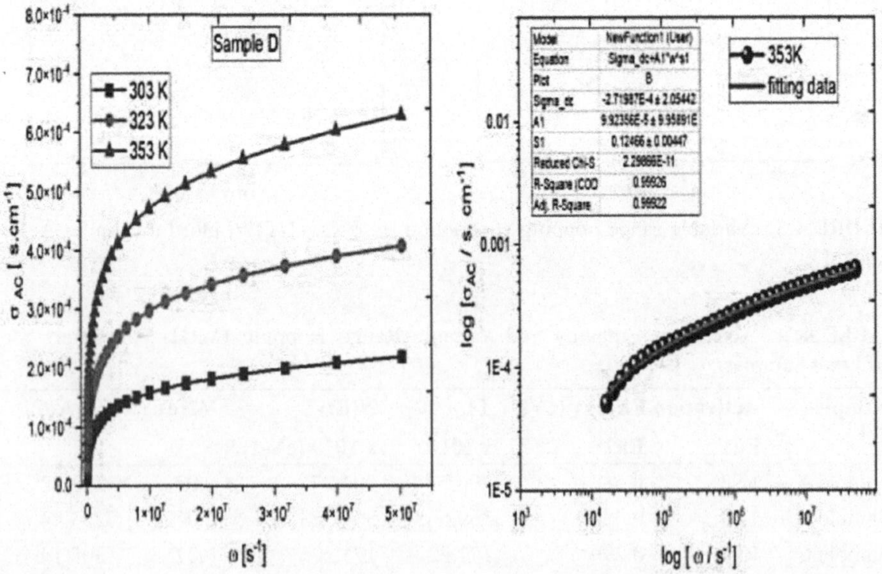

FIGURE 4.6 AC conductivity plot (left) for sample D and JPL fitting plot (right) for sample D at 353 K.

maximum annealed samples and is shown along with it. The AC conductivity of a silver-doped graphene oxide sample increases sharply at 353 K. The sharp increase in conductivity is due to the creation of more single hopping sites within the layers apart from the normal bipolaron hopping centers.[29] The correlated barrier hopping (CBH) mechanism is prominent in silver-doped graphene oxide especially at a high-frequency range and various parameters are tabulated in Table 4.4.

TABLE 4.4 Jonscher's Power Law (JPL) Fitting and Correlated Barrier Hopping (CBH) Parameters (Sample D) at $\omega=0.5 \times 10^7$ s^{-1}.

T(K)	$\lvert\sigma_{DC}\rvert \times 10^{-5}$ s.cm^{-1}	s	$A \times 10^{-5}$ s.cm^{-1}	Ω	τ_o 10^{-12}	Rω A°	$N \times 10^{19}$ (cm^{-3})
303	5.063	0.256	1.502	0.187	4.13	11.14	5.083
323	9.058	0.163	2.745	0.190	5.28	14.05	12.18
353	27.19	0.124	9.923	0.505	8.15	14.80	13.33

The AC conductivity parameters show that the temperature-dependent parameters, A and ω, increase while the value of the frequency component part, s, decreases with an increase in temperature. DC conductivity at the low-frequency range is constant, and it increases with temperature. The decrease of the s value with temperature suggests correlated barrier hopping of charge carriers over potential barriers between two charged defect states rather than quantum mechanical tunneling.[30] The best-fitted values of ω and τ_0 for each temperature are used to calculate the density of pair states for bipolaron hopping. The conversion of D^+ and D^- centers to D^0 centers increases with the rise of temperature and results in a large concentration of bipolaron centers in the material. This results in drastic conductivity changes in the material. The carrier concentration (N) also increases with temperature at large barrier heights. The density of charge carriers in defect levels creates more conductivity than at intrinsic levels.

4.5 CONCLUSIONS

A bulk amount of graphene oxide is prepared using the simple and cost-effective modified Hammer's method. Chemically reduced colloidal silver is added to graphene oxide at various concentrations to study the doping effects. Morphological studies revealed that the agglomeration of nanocrystallites in pure GO splits into sheets of flakes with nanoparticles adhering on the surface. The chemical composition of samples replicates in the EDX analysis. Structural analysis shows the polycrystalline behavior with well-defined peaks and nanocrystalline grain size. Reduced microstrain and dislocation density make the sample apt for device fabrication. Surface layer conductivity is analyzed by two-probe DC conduction analysis through the samples. With an increase in the doping concentration, the values of extrinsic activation enthalpy were found to decrease. It indicates the defect level conduction mechanism in silver-decorated graphene oxide samples. Variable range hopping mechanisms also prevail in the presence of large charge density state functions in localized defect centers rather than at the Fermi level. The presence of polaron hopping is confirmed by AC conductivity studies. A sharp increase in AC conductivity with temperature is due to bipolaron hopping especially at high-frequency regions. As frequency increases, the oppositely charged single polarons get combined to form neutral bipolarons, thereby creating an enhancement in the defect level conduction mechanism in silver-doped graphene oxide nanolayers.

AUTHOR CONTRIBUTIONS

I. Dhanya, Ph.D.: Idea of this research work, experimental data collection, coding of results, and writing the article. S. Heera, MSc.: Laboratory work, data collection, and coding of characterizations like SEM/EDX and XRD data analysis. Sreejith K. Pisharady, Ph.D.: DC and AC electrical conductivity studies and data analysis.

CONFLICTS OF INTEREST

There are no conflicts to declare.

ACKNOWLEDGMENTS

The authors acknowledge KSCSTE, Government of Kerala, and the DST, Government of India, for the research facilities provided to Sanatana Dharma College, Alappuzha, under the SARD (No. 612/2016/KSCSTE) and FIST schemes (SR/FIST/College-238/2014(c)), respectively.

KEYWORDS

- **graphene oxide**
- **conductivity studies**
- **hopping conduction**
- **electrical conductivity**
- **activation energy**

REFERENCES

1. Gao, Y.; Yang, L.; Liu, Z.; Gao, a. R.; Huc, G.; Zhao, J. *New J. Chem.* **2022,** *46*, 756–766. https://doi.org/10.1039/D1NJ03493F
2. Lai, G. H.; Chou, Y. C.; Huang, B. S.; Yang, T. I.; Tsai, M. H. *RSC Adv.* **2021,** *11*, 71–77. https://doi.org/10.1039/D0RA08535A
3. Xiao, T.; Yi, F.; Yang, M.; Liu, W.; Li, M.; Ren, M.; Zhang, X.; Zho, Z. *J. Mater. Chem. A* **2021,** *9*, 16692–16698. https://doi.org/10.1039/D1TA03608D
4. Wang, Z.; Lv, T. Y.; Shi, Z. B.; Yang, S. S.; Gu, Z. Y. *Dalton Trans.* **2021,** *50*, 13608–13619. https://doi.org/10.1039/D1DT02206G
5. Selvi, S. V.; Nataraj, N.; Chen, S. M.; Prasannan, A. *New J. Chem.* **2021,** *45*, 3640–3651. https://doi.org/10.1039/D0NJ05592A
6. Yang, L.; Zhang, L.; Jiao, X.; Qiu, Y.; Xu, W. *RSC Adv.* **2021,** *11*, 4042–4052. https://doi.org/10.1039/D0RA09684A
7. Zhao, Q.; Chen, B. *Environ. Sci.: Nano* **2021,** *8*, 1000–1009. https://doi.org/10.1039/D0EN01135E
8. Kadivara, M.; Aliakbar, A. *Anal. Methods* **2021,** *13*, 536–551. https://doi.org/10.1039/D0AY01812K
9. Zhang, Q.; Yang, Y.; Fan, H.; Feng, L.; Wen, G.; Qin, L. C. *RSC Adv.* **2021,** *11*, 15808–15816. https://doi.org/10.1039/D0RA10026A
10. Han, L.; Wang, H.; Tang, Q.; Lang, X.; Wang, X.; Zong, Y.; Zong, C. *RSC Adv*. **2021,** *11*, 38264–38272. https://doi.org/10.1039/D1RA06138K
11. Xu, J.; Zhao, X.; Liu, F.; Jina, L.; Chen, G. *New J. Chem.* **2020,** *44*, 1236–1244. https://doi.org/10.1039/C9NJ05309C
12. Nayili, A. A.; Albdiry, M. *New J. Chem.* **2021,** *45*, 10040–10048. https://doi.org/10.1039/D1NJ01658J
13. Mun, S. J.; Shim, Y. H.; Kim, G. W.; Koo, S. H.; Ahn, H.; Shin, T. J.; Kim, S. O.; Kim, S. Y. *Nanoscale* **2021,***13*, 2720–2727. https://doi.org/10.1039/D0NR07858A
14. Kumaran, R.; Kumar, A. V.; Ramaprabhu, S.; Subramanian, V. *New J. Chem.* **2021,** *45*, 16939–16948. https://doi.org/10.1039/D1NJ03536C
15. Qi, F.; Wang, C.; Peng, S.; Shuai, C.; Yanga, W.; Zhao, Z. *Mater. Chem. Front.* **2021,** *5*, 2373–2386. https://doi.org/10.1039/D0QM00958J

16. Bo, W.; Zou, Y.; Wang, J. *RSC Adv.* **2021,** *11*, 33675–33691. https://doi.org/10.1039/D1RA05902E
17. Yang, D.; Mo, W.; Zhang, S.; Li, B.; Hu, D.; Chen, S. *Dalton Trans.* **2020,** *49*, 1582–1590. https://doi.org/10.1039/C9DT03491A
18. Tanguy, N. R.; Diaye, J. N.; Arjmand, M.; Lian, K.; Yan, N. *Chem. Commun.* **2020,** *56*, 1373–1376. https://doi.org/10.1039/C9CC07613A
19. Li, Y.; Wang, S.; Xiao, Z. C.; Yang, Y.; Deng, B. W.; Yin, B.; Ke, K.; Yang, M. B. *J. Mater. Chem. C* **2020,** *8*, 4040–4048. https://doi.org/10.1039/D0TC00029A
20. Behera, A. K.; Patel, A. K.; Mehta, R.; Sarkar, S.; Das, S.; Mallik, A. *Diam. Rel. Mat.* **2020,** *109*, 108009. https://doi.org/10.1016/j.diamond.2020.108009
21. Cai, W.; Bocarsly, J. D.; Gomez, A.; Lee, R.; J. L.; A. M. Magaña, R. Seshadri and Echegoyen, L.; *Chem. Sci.* **2020,** *11*, 13129–13136. https://doi.org/10.1039/D0SC05265E
22. R. U. R Sagar, Lifang, C.; Ali, A.; Khan, M. F.; Abbas, M.; Malik, M. I.; Khan, K.; Zeng, J.; Anwarag, T.; Liang, T. *Phys. Chem. Chem. Phys.* **2020,** *22*, 25712–25719. https://doi.org/10.1039/D0CP05209D
23. Stevensa, D. L.; Gamage, G. A.; Ren, Z.; Grunlan, J. C. *RSC Adv.* **2020,** *10*, 11800–11807. https://doi.org/10.1039/D0RA00763C
24. Kubliha, M.; Maache, D.; Bosak, O.; Minarik, S.; Trnovcova, V.; Lukic-Petrovic, S.; Soltani, M. T. *Russ. J. Electrochem.* **2019,** *55*, 510–516. https://doi.org/10.1134/S1023193519060119
25. Tiwari, R.; Verma, D. K.; Kumar, D.; Yadav, S.; Kumar, K.; Krishnamoorthi, S. *Mater. Chem. Front.* **2021,** *5*, 5857–5866. https://doi.org/10.1039/D1QM00537E
26. Lua, Y.; Young, D. J. *Dalton Trans.* **2020,** *49*, 7644–7657. https://doi.org/10.1039/D0DT00872A
27. Wang, B.; Lou, W.; Wang, X.; Hao, J. *J. Mater. Chem.* **2012,** *22*, 12859–12866. https://doi.org/10.1039/C2JM31635H
28. Giménez, V. R.; Tatay, S.; Gastaldo, C. M. *Chem. Soc. Rev.* **2020,** *49*, 5601–5638. https://doi.org/10.1039/C9CS00594C
29. Oraon, R.; Adhikari, A. D.; S. K. Tiwari and Nayak, G. C.; *Dalton Trans.* **2016,** *45*, 9113–9126. https://doi.org/10.1039/C6DT00600K
30. Tarcan, R.; Boer, O. T.; Petrovai, I.; Leordean, C.; Astilean, S.; Botiz, I. *J. Mater. Chem. C* **2020,** *8*, 1198–1224. https://doi.org/10.1039/C9TC04916A

CHAPTER 5

Laser Device as Optical Energy Source and Its Applications

RINSY THOMAS[1,2], N. V. UNNIKRISHNAN[3], and K. V. ARUN KUMAR[1,2]

[1]*Department of Physics, CMS College (Autonomous), Kottayam, Kerala, India*

[2]*Nanotechnology and Advanced Materials Research Centre, CMS College (Autonomous), Kottayam, Kerala, India*

[3]*School of Pure and Applied Physics, Mahatma Gandhi University, Kottayam, Kerala, India*

ABSTRACT

One of the most interesting developments of the previous century was the advent of the laser. In this chapter, we would like to cover the basic principles, types, and applications of the laser as an optical energy source. The working principles of laser with classification based on solid, gas, semiconductor, etc. are discussed in detail. Some of the laser applications such as laser range finder, underwater laser, laser-guided antitank missiles, laser radar, ring laser gyroscope, communications, weapon firing simulators, laser micro-/nanomachining, laser cutting, laser hardening, holography, etc. are described and also used in different fields which are also mentioned. We tried to discuss laser in all the possible ways, and the chapter is fruitful to all the possible perspectives.

5.1 INTRODUCTION

The laser's origins may be traced back over a century, to 1917, when Albert Einstein proposed the idea of stimulated emission. Charles Hard Townes,

Resilience and Sustainability for Energy and Water: New Advancements, Opportunities, and Framework. Cristóbal Noé Aguilar, A. K. Haghi, Amritanshu Shukla, Ann Rose Abraham, Swati Gokul Talele (Eds.)

Nicolay Gennadiyevich Basov, and Aleksandra Mikhailovich Prokhorov have been honored with the Nobel Prize in Physics in 1964 "for fundamental work in the field of quantum electronics, which has led to the fabrication of oscillators and amplifiers based on the maser–laser concept".[1] Maser oscillation at infrared (IR) and optical wavelengths is conceivable when pumped with reasonably large amounts of incoherent light using a centimeter-sized resonant cavity with numerous resonant modes. This allows maser techniques to be extended to the IR and optical areas. Arthur Schawlow and Charles Townes, Bell Labs/Columbia University, demonstrated that a single mode may be selected by creating an angular aperture of a certain size and having only the end walls highly reflective and are commonly considered to be the "inventors" of the laser.[2] In June of 1959, the first international conference was held on the topic of "optical masers" as they were then known. During the session focused on lasers, Gould, Javan, and Schawlow presented a paper discussing the theoretical concepts of laser oscillation in gas discharges and ruby crystals.[3] Gordon Gould is often credited as the inventor of lasers as he coined the acronym LASER—Light Amplification by Stimulated Emission of Radiation).[4] In 1960, Theodore Maiman, a researcher at Hughes Research Laboratories in the United States, successfully demonstrated the first optical laser.[5] The hunt for lasers made from novel materials and able to operate at unusual frequencies has been underway in several labs across the world. Just a few short months after the first laser went into operation, Javan with his coworkers built the first gas-based laser system which is the He–Ne laser.[9] Since its discovery, laser technology has advanced significantly and has been successfully implemented in various mediums, including gaseous materials that have been ionized, liquids, solids, and semiconductors. Nearly all types of solid-state lasers require optical pumping to operate. In contrast to the semiconductor laser, which is pumped by an electric current passing through a forward-biased diode junction, a gas laser is energized by an electrical discharge.[10] So far, the laser has been put to use in a wide variety of astonishing and novel ways across all areas of scientific measurement. The unique properties of laser beams, including their high power, directionality, spectral purity, and ability to produce extremely short pulse durations, have resulted in numerous and diverse applications of laser technology. As an example, the laser's wide range of uses in surveying, remote sensing, lidar, etc. is made possible by its exceedingly tiny divergence. The ability of the laser to focus on very small objects has several uses, including in the medical field, the manufacturing industry, the recording industry, eye surgery, laser cutting, and the materialization of man-made stars in the sky. Lasers have risen to prominence in the science, technology, engineering, and mathematics (STEM) disciplines as a result of their profound effects on both fundamental scientific inquiry and a wide range of practical applications. Lasers have a wide range of applications, including consumer electronics such as music players and printers, industrial processes such as metal cutting, welding, and drilling, medical treatments, and scientific research. Some of the most innovative uses of lasers include filtering spatial frequencies, creating holograms, inducing fusion, facilitating communication through light waves, and in scientific and industrial investigations. We can find lasers in everything from consumer electronics such as music players and printers to industrial applications such as welding, metal cutting, hole drilling, and marking as well as medical and scientific uses such as interferometry, spectroscopy, and testing of the foundations of quantum mechanics. Some of the most useful applications for lasers are spatial frequency filtering, holography, laser-induced fusion, lightwave communications, and scientific and industrial research.[9]

5.2 LASER WORKING PRINCIPLES

The laser employs procedures that boost or magnify light signals that have already been produced in some other way. For a lasing process to occur, a laser essentially consists of three main components. An active medium will radiate light in the same direction and with the same characteristics as the incoming light when it is irradiated by incoming light.[6] The process of selectively introducing energy into the active medium to create population inversion is known as the pumping mechanism. In an optical resonator that contains two opposite mirrors, some of the induced emission is stored and focused into a few resonator modes.[7] The active medium is contained within a laser cavity, which is a tube polished internally and surrounded by coaxial mirrors at each end. A pumping mechanism in the form of an external energy source is used to maintain the active medium. The pumping process serves as an external energy source that stimulates the active medium, which is typically a light source, such as a flashlight or an arc lamp, but it can also take the form of an electromagnetic coil or a diode laser unit.[8] The active medium absorbs energy from this main source, which causes the emission of laser light. A coherent optical amplifier, which is an active medium, is utilized by the laser as an optical oscillator. The purpose of an optical resonator is to provide selective feedback of light emitted by the excited molecules in the active medium. To create a coherent light source or a singular oscillator system, it is necessary to implement feedback. One way to achieve this is by utilizing mirrors or reflective surfaces to construct an optical resonator, which allows light to continuously pass through the amplifying medium. This resonator is typically composed of two parallel mirrors, either planar or spherical, that enable the back-and-forth transmission of light.[6]

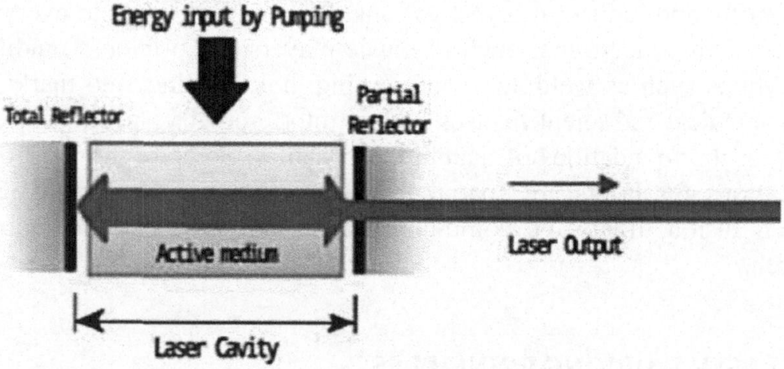

FIGURE 5.1 Basic Components of Laser System

5.2.1 LIGHT AMPLIFICATION BY STIMULATED EMISSION

Electromagnetic fields can interact with atoms via three different mechanisms.

i. Spontaneous emission—When an atom undergoes a transition between two different energy levels with an energy difference ΔE, it emits a photon with a frequency of $\upsilon = \Delta E/h$ through a process known as spontaneous emission. This transition can be thought of as an oscillator with a resonant frequency, where h represents Planck's constant. The problem lies in the fact that while these atoms have the same resonant frequency, they emit photons individually in arbitrary directions, and lack a coherent phase. This type of light has optical noise. It lacks both temporal and spatial coherence.

ii. Stimulated emission—Absorption—The atoms, initially in a state of low energy, are excited to higher energy levels. by absorbing light. Then these excited atoms come back to the ground state spontaneously either by emitting light or by using another nonradiative method.

iii. Stimulated emission—When an atom is in an excited state, it has an equal probability of being stimulated by incoming photons and emitting a photon, leading to an increase in the photon flux.[6]

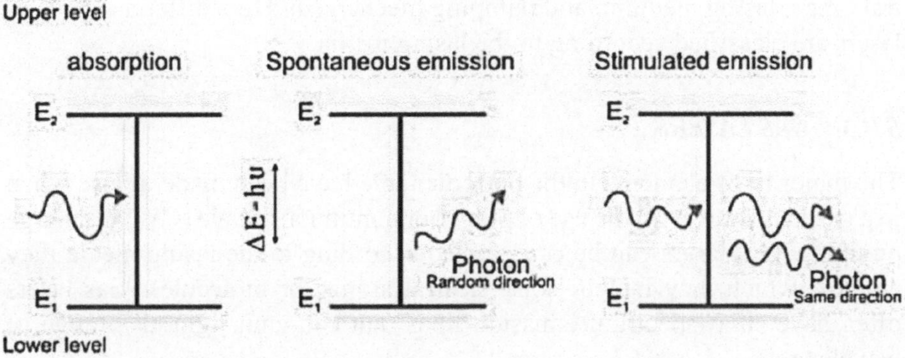

FIGURE 5.2 Mechanism of interaction between an atom and a photon

These three processes such as stimulated absorption, stimulated emission, and spontaneous emission control how photons interact with atoms. Both the production of laser light and the law of radiation in thermal equilibrium are founded upon this principle. For a medium with an excess of excited atoms, stimulated emission dominates absorption and there is a net gain. The population inversion, a nonequilibrium state, can be reached by using a technique called pumping. In a laser system, the laser light is confined within the laser cavity through the reflection of a pair of mirrors. When a sufficient amount of laser light is built up within the cavity, the shutter can be opened to release it as a laser beam. The light emitted in this manner is highly coherent, which means that it is comprised of waves that are precisely in phase with one another. Additionally, laser light is monochromatic, meaning that it consists of light of a single wavelength, and it exhibits very low divergence, meaning that the light spreads very little as it travels away from the laser source. These characteristics make the laser beam highly useful in a wide range of applications.[11] Laser emission is tuned by the laws of quantum physics, which restrict atoms and molecules possessing finite amounts of stored energy that rely on the nature of the atom or molecule.

5.3 CLASSIFICATION OF LASERS

The coherence of lasers sets them apart from other light sources. Various techniques can be used to classify different types of lasers. Some common methods include categorizing them based on their temporal coherence, spectral range, lasing medium, and pumping mechanism. Here different kinds of lasers are classified according to the lasing media.

5.3.1 GAS LASERS

The majority of elements in the periodic table have been made to lase when in a gaseous state, with the use of pulsed or continuous-wave (CW) discharge pumping. Gas lasers can be categorized according to the lasing media they employ, which may include ions, neutral atoms, or molecules. Gas lasers often have short-spectrum emission lines that can emit light at a range of wavelengths.

5.3.1.1 NEUTRAL ATOM LASERS

A broad category of lasers called neutral atom gas lasers includes, in particular, the majority of noble gases. The first CW laser invented was the helium–neon (He–Ne) laser. This belongs to the "neutral atoms" category. The relevant energy levels of the He–Ne system, which are essential for the occurrence of laser action, are illustrated in Figure 5.3.

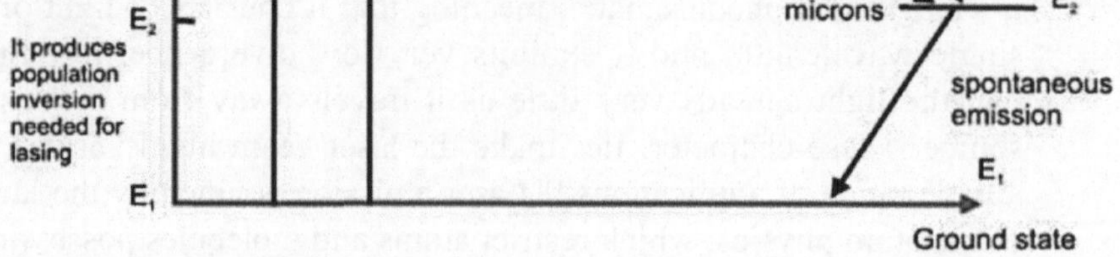

FIGURE 5.3 He-Ne laser

The original laser transitions occurred in the near-IR spectrum. The red laser, which has a wavelength of 632.8 nm, is the most widely utilized transition. It has a lifetime of 10,000 h for certain commercial units. Laser systems are capable of generating a wide range of power outputs, from less than 1 mW to over 100 mW hence their uses are restricted to low-power activities. He–Ne lasers have extremely low gain and efficiency. A He–Ne laser's active medium is a combination of helium and neon gases. Helium atoms, which are more prevalent, are preferentially stimulated to higher electronic levels because they absorb extra energy from collisions with electrons created by the electric discharge compared to neon atoms. Helium metastable atoms are formed as a result of electrons interacting with helium atoms in the excitation process. The energy of the metastable atoms is then transferred to neon laser levels. Once a population inversion is achieved, the neon atoms in an excited state return to their ground state, and produce stimulated emission.[12]

5.3.1.2 *ION LASERS*

The most significant "ion lasers" are powered by noble gas ions in different states of ionization, such as Ar, Kr, Ne, or Xe. The emission wavelength of the ion lasers ranges from near ultraviolet (UV) through the visible part of the spectrum. During the electrical discharge excitation process, neutral atoms collide with electrons in their ground states. Transferring sufficient energy to ionize them throughout this process leaves the ions in a variety of potential excited states. Maiman's ruby laser, invented in 1960, technically qualifies as the first ion laser because it uses chromium, Cr, which has been triple ionized, although in an aluminum-oxide crystal host, to provide the

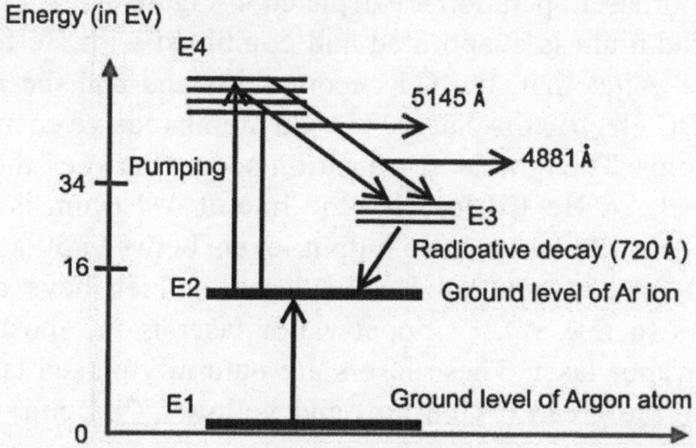

FIGURE 5.4 Argon laser

energy levels needed.[13] Ion gas lasers use the majority of noble gases and of which the Ar laser is the most famous.

Figure 5.4 depicts a condensed scheme for the pertinent energy levels in an argon laser. When an electric discharge is applied to the tube, the Argon atoms undergo a two-step collision with electrons, causing them to be excited to the energy level E_4. The first stage involves ionizing neutral argon atoms and raising them to energy level E_2. This is known as the ground state of argon ions and it is achieved by removing one electron from the 3*p* outer shell of Ar. The ion is stimulated to energy level E_4 in the second step from its ground state. The excited Argon ion at energy level E_4 either emits spontaneously or undergoes induced emission to return to energy level E_3. The ions situated in the energy level E_3 experience a spontaneous decay toward the ground state of the argon ion. During this process, a photon with a wavelength of 72 nm is emitted.[14] Argon lasers feature blue and green emission lines in the visible spectrum, with the brightest ones at 488 and 515 nm, respectively.

5.3.1.3 METAL VAPOR LASERS

Gas lasers known as "metal vapor" lasers use a metal vapor as the laser gain medium. This class includes the helium–cadmium (He–Cd) laser, a significant metal vapor laser. This laser's design is similar to that of the He–Ne laser, however, the gas used here is a combination of helium and ionized cadmium. He–Cd lasers have been employed for numerous flow cytometry applications that call for UV excitation due to their lower cost and operating requirements than water-cooled argon lasers.[15] The relevant He–Cd system energy levels for laser operation are depicted in Figure 5.5.

Metallic cadmium is evaporated and combined with He for the excitation process. After that, the Cd becomes ionized and the He ions are excited by a DC electric discharge. The Cd atoms receive energy from the excited He atoms. Then the laser transition occurs between the Cd atom's electronic levels. A He–Cd laser's blue line at 441.6 nm is its primary emission line. He–Cd lasers have output levels between 50 and 100 mW, which puts them between red He–Ne lasers (which have a few mW) and Ar^+ lasers (a few W).[14] Copper-vapor laser is the most significant neutral atom vapor laser. These lasers are naturally pulsed laser systems with output in the green (510.6 nm) and yellow (578.2 nm) wavelength ranges.[16]

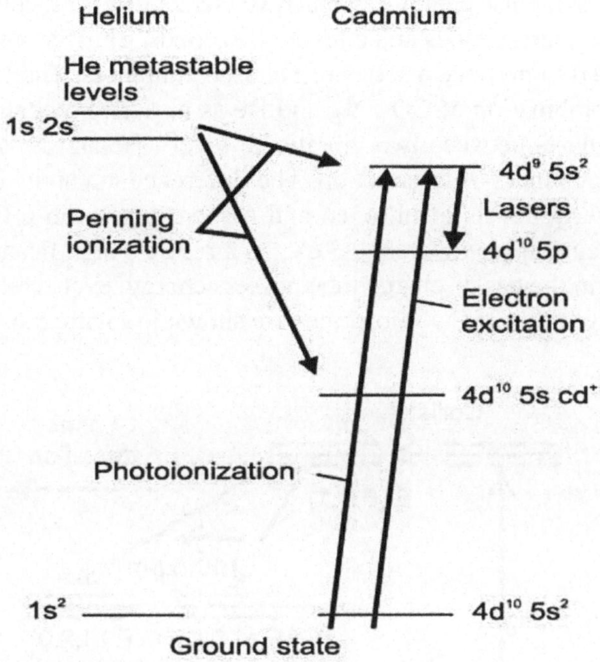

FIGURE 5.5 He-Cd laser

5.3.1.4 MOLECULAR LASERS

Molecules rather than individual atoms or ions serve as the laser-active components in molecular lasers, which are gas lasers. Transitions between a molecule's energy levels are used by these lasers. Molecular gas lasers can be classified into one of the following three groups, depending on the type of transition involved: (1) Vibrational rotational lasers, (2) vibronic lasers, and (3) pure rotational lasers. The CO_2 laser is the most significant example of a vibrational rotational laser. They have mid-IR emissions at 10.6 and 9.6 μm. CO laser and HF chemical laser also belong to this category.[14] In the invisible IR waveband, the CO_2 laser functions. Numerous temporal beam modes are available for the CO_2 laser. The most basic mode is CW, which emits the laser beam, runs it for a predetermined amount of time, and then turns it off. To produce a succession of square waveforms, a CW beam can be manually or electronically gated or chopped. Quasi-CW CO_2 lasers are more contemporary CO_2 lasers that have trains of very brief, high peak power pulses with very long interpulse intervals.[17] When operating in the pulsed or

Q-switched mode, pulse energy can reach 10,000 J/pulse with pulse durations between a few microseconds and a few milliseconds. These systems can also be mode-locked to produce a series of nanosecond pulses. The laser uses an appropriate combination of CO_2, N_2, and He as its active medium. The CO_2 molecule oscillates between two vibrational levels. He and N_2 significantly boost the effectiveness of laser action. The difference in energy between the vibrational energy levels of nitrogen and carbon dioxide in a CO_2 laser is quite slight, being approximately 0.3 eV. As a result, a significant number of electrons within the gas discharge tube possess energy levels greater than 0.3 eV. The energy levels of CO_2 molecules are shown in Figure 5.6.

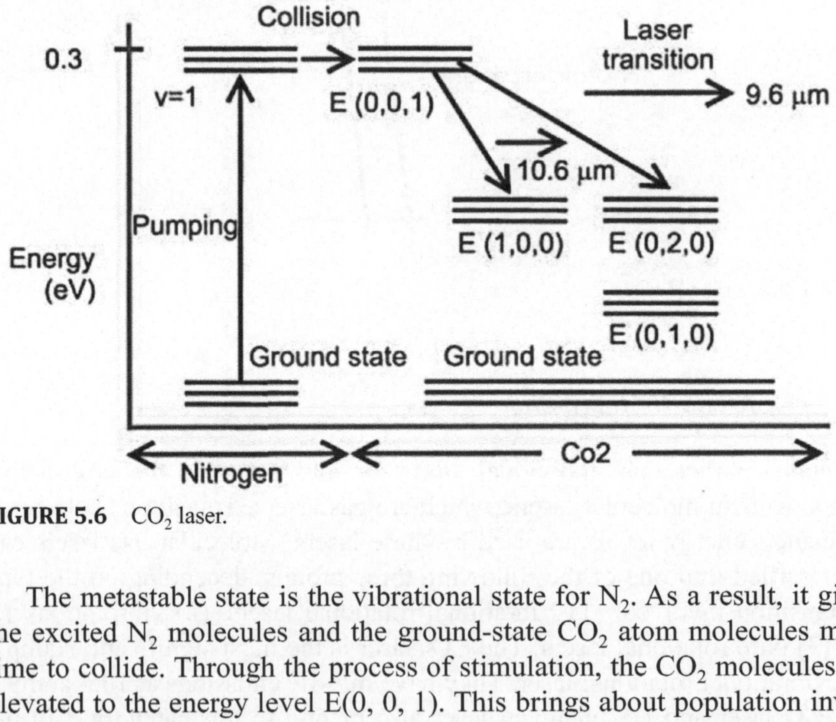

FIGURE 5.6 CO_2 laser.

The metastable state is the vibrational state for N_2. As a result, it gives the excited N_2 molecules and the ground-state CO_2 atom molecules more time to collide. Through the process of stimulation, the CO_2 molecules are elevated to the energy level E(0, 0, 1). This brings about population inversion because the quantity of CO_2 molecules within the E(0, 0, 1) state is substantially greater than those present in the vibrational states E(1, 0, 0), E(0, 0, 2), and E(0, 1, 0). The transition from energy level E(0, 0, 1) to E(1, 0, 0), and E(0, 2, 0) through the mechanism of stimulated emission gives rise to IR photons with wavelengths of 10 and 9.6 µm, respectively. This laser produces a continuous stream of light. It is quite effective. The output power is high.[14]

5.3.1.5 *EXCIMER LASERS*

Excimer lasers are a fascinating and significant type of molecular lasers that function by transitioning between distinct electronic states of unique molecules known as excimers. The most significant lasers for industrial and medical use in the near-UV to vacuum-UV range are based on rare-gas halide excimers such as XeF at 351 nm; XeCl at 308 nm; and KrF at 248 nm; F2 at 153 nm; and ArF at 193 nm. Electric discharges, electron or proton beams, or optical excitation can all be used to pump them. Excited dimer, a class of molecules created by the fusion of two identical constituents in the excited state, is where the name excimer originates. The excimer laser's energy levels are depicted in Figure 5.7 as a function of the distance between the atoms in the molecule. Halogen is represented by H, and the noble gas atom is represented by R.

A crucial characteristic is that the excimers quickly dissolve following induced or spontaneous emission, preventing the reabsorption of the laser energy produced. This enables the achievement of a fairly high gain even with a moderate excimer concentration.[18]

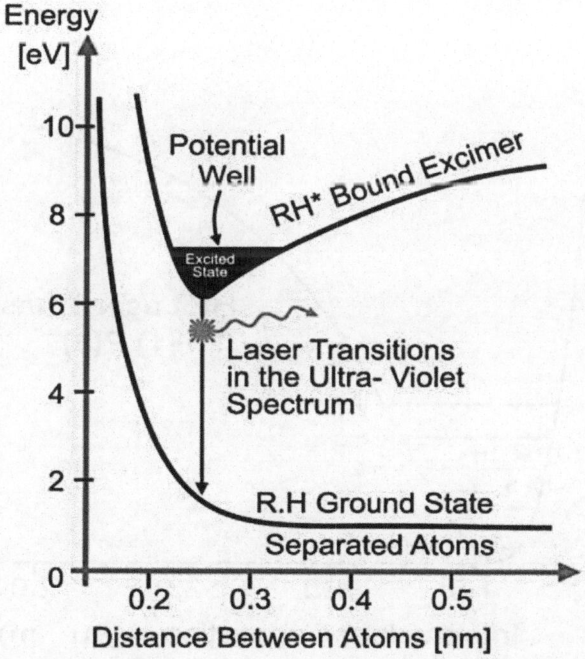

FIGURE 5.7 Excime lasers.

5.3.1.6 *CHEMICAL LASERS*

The population inversion is typically thought of as being "directly" produced by an exothermic chemical reaction in a chemical laser. With a chemical laser, it is possible to operate both pulsed and CW lasers. Chemical reactions between gaseous components are often either associative or dissociative in chemical lasers.[14] This form of pumping may cause a population inversion with regard to one of the lower states. Between the molecule's rotational or vibrational states, laser transitions can take place. An excellent illustration is the production of vibrationally excited hydrogen fluoride resulting from the reaction between atomic fluorine and hydrogen molecules.

$$F + H_2 \rightarrow HF\ (v) + H \rightarrow \quad (5.1)$$

The HF laser can perform at a number of distinct wavelengths that correspond to various rotational and vibrational transitions. These transitions are identified by their rotational (J) and vibration (v) quantum numbers. Figure 5.8 depicts the (2 → 1) P(3) transition as lasing from excited level v = 2, J = 2 to lower level v = 1, J = 3.

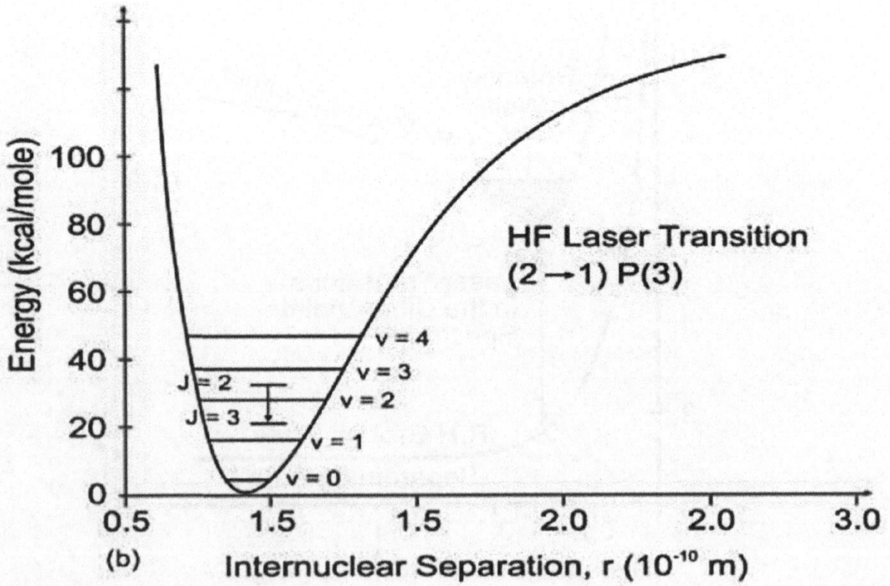

FIGURE 5.8 HF laser transition.

However, it is possible to get the deuterium fluoride (DF) laser to laser at 3.6–4.2 μm by switching out hydrogen for deuterium in the reaction shown in equation (5.1). The DF laser is preferred for applications needing air propagation because the atmosphere presents a window with good transmission in this range.[19]

5.3.2 SOLID-STATE LASERS

Typically, a minority ion in a solid-state host serves as the active medium in solid-state lasers. These can be categorized according to the kind of laser-active centers as follows: Rare-earth ions, transition metal ions, and color centers. Excitation is accomplished by optical pumping from either a different laser or light. Most solid-state lasers can be operated using conventional Q-switching and mode-locking methods.[20]

5.3.2.1 TRANSITION METAL ION LASERS

The outer 4*s* electrons and maybe some 3*d* electrons are lost by the transition metals (iron group), which follow the element Ca (40), in the fourth period of the periodic table. As a result, their electronic configurations are $1s^2\ 2s^2\ 2p^6\ 3s^2\ 3p^6\ 3d^n$ where $n < 10$. There are several low-lying energy levels in ions with incomplete 3*d* shells between which optical transitions can take place. Absorption and emission band structure dominate the spectroscopic characteristics of the transition metal ions in a laser material.[21] The positively charged transition metal ions are Cr^{3+}, Cr^{4+}, Co^{2+}, V^{2+}, Ni^{2+}, and Ti^{3+}. Optically active electrons in empty $3d^n$ electron configurations are present in the laser ions of this type.[20] Cr^{3+} is the transition metal laser ion with the best performance. The outer shell electrons which are optically active are sensitive to the local crystal field environment. The most frequent strong field host is sapphire, and the first laser was developed using $Al_2O_3{:}Cr^{3+}$, also known as ruby. Figure 5.9 depicts the energy level diagram of ruby laser.

Ruby is a laser system with three energy levels. Some Cr^{3+} ions at ground energy level E_1 are stimulated to higher energy level E_3 after absorbing photons of wavelength 5500 from xenon flash lamp. They are unstable at this energy level and fall to the metastable energy level E_2, which has a substantially longer lifetime, by losing some of their energy to the crystal lattice (about 10–3 s). Because of the pumping effect of the flashlight, Cr^{3+} ions continue to increase in the E_2 state while decreasing in the ground state,

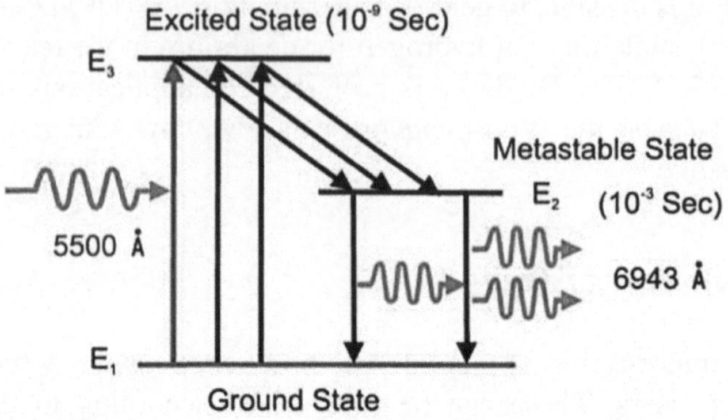

FIGURE 5.9 Ruby laser.

and soon the population inversion between the E_2 and E_1 states is reached. Currently, some Cr^{3+} ions will spontaneously decay to the ground state E_1 by producing photons with a wavelength of 695 nm. It produces visible red light. Ruby lasers are exclusively utilized in pulse mode.[22]

5.3.2.2 RARE-EARTH ION LASERS

Rare earth has been demonstrated to display laser fluorescence in liquid, crystal, glass, and, more recently, even in gaseous media. Till now, only crystal and glass lasers have been shown to be useful. All of the trivalent rare-earth lasers now in use are 4-level lasers. The terminal state is either a spin-orbit component above the ground level (Yb^{3+}, Tm^{3+}, Er^{3+}, $H0^{3+}$, and Pr^{3+}) or a crystal-field split component of the ground level (Yb^{3+}, Tm^{3+}, Eu^{3+}, Nd^{3+}, and Pr^{3+}).[23] Unfilled $4f^n$ electronic configurations are what make these ions unique. The environment of the local host material has a minimal impact on the energy levels because the inner shell *4f* electrons are shielded by outer shell electrons. Glasses and crystals can both serve as host materials for rare-earth ion lasers. Garnets such as $Y_3Al_5O_{12}$ (Known as YAG) are typical hosts for oxide crystals. $YLiF_3$ is a typical fluoride crystal host. Nd^{3+} is an active ion used to lase in a large number of host materials. One of the most successful commercially available lasers is $Y_3Al_5O_{12}$:Nd^{3+}, generally known as Nd:YAG. Figure 5.10 depicts the energy level diagram of the Nd:YAG laser.

To achieve population inversion in the Nd:YAG laser, the neodymium ions' lower energy state electrons are excited to the higher-energy state. The absorption of light radiation having wavelengths of 0.73 and 0.8 μm, specifically in the pump bands, leads to the elevation of neodymium (Nd^{3+}) atoms from the lower energy levels E_0 to the upper energy levels E_3 and E_4. Through a nonradiative transition, the atoms of neodymium ions go from these energy levels to E_2, a metastable state. Neodymium ions are gathered at level E_2, and between E_2 and E_1 the population inversion is accomplished. A photon is released when an ion switches spontaneously from E_2 to E_1. Between E_2 and E_1, this photon's emission will start a chain of stimulated photons. Despite the fact that the mean beam power is low, the beam intensity can be considerable due to shorter pulse time and improved focusing behavior. Even when falling on a reflecting surface, Nd:YAG laser is heavily absorbed due to its shorter wavelength. Nd:YAG lasers are a particularly valuable tool in the field of micromachining thanks to the development of short pulse lasers employing diode-pumped (DP) and Q-switching techniques (for frequency doubling and tripling).[24]

5.3.2.3 COLOR CENTER LASERS

To create optically pumped lasers that are widely tunable over a variety of bands in the UV, the visible, and most critically, in the 0.8–4 μm range of the IR, color centers can be used. Color centers are imperfections in insulating crystals that trap electrons (or holes). The electronic states of the center are intimately connected to the crystal phonons because the immediately surrounding crystal often provides the majority of the potential perceived by the trapped particle or particles. For the development of effective, optically pumped, broadly tunable lasers, multiple color centers are almost ideal. For instance, a typical color center in alkali halide host crystals like NaCl consists of an electron trapped at a halide ion vacancy. The three types of useful laser-active color centers in alkali halides are as follows: FA(II) centers, F^{2+} centers, $Tl^0(1)$ centers.[25] Radiation exposure or thermal processing can create color centers. Because of the high ion and electron mobility, these centers frequently only remain stable at low temperatures. Impurity ions can sometimes act to stabilize the defect centers. An illustration of this is a neutral Tl atom at a cation site adjacent to an anion vacancy in KCl. Ar, Kr, or Nd:YAG lasers are used to optically pump color center absorption, which typically occurs in the visible spectrum. A mode-locked output with pulses between 5 and 15 pcs is produced when a color center laser and a mode-locked Nd:YAG laser are pumped in a synchronous way.[20] Color center lasers can achieve higher frequency definition than dye lasers since the gain medium is static.

5.3.3 SEMICONDUCTOR LASERS

The active area of semiconductor lasers is where the radiative recombination of charge carrier pairs, such as electrons and holes, takes place, resulting in the optical emission of the laser. During this process, the energy difference is simultaneously transferred to the light field through radiative recombination, where the conduction-band electron fills the valence-band hole. Consequently, the frequency of the laser light is largely dependent on the bandgap of the semiconductor laser material. Lasing requires carrier inversion, or enough electrons in the conduction band so that the probability of light emission is greater than that of light absorption.[27] Semiconductor laser emission lines are most frequently found in the near-IR or IR. Typically, these are composed of III-V compounds, such as GaAs in the red and near IR and Pb salts in the mid-IR. The first compound semiconductor laser was a diode laser made of gallium arsenide (GaAs). It is capable of producing laser radiation at wavelengths of 750–870 nm. Gallium aluminum arsenide mixed compounds (Ga1-xAlx-As) can now be used to create a variety of laser structures, especially with the introduction of strained-layer technology. The active layer is sandwiched between two layers with bigger bandgaps and lower refractive indices in a typical GaAlAs laser. Electrical confinement is produced by the first feature and optical confinement by the second. This increases effectiveness and permits CW laser operation. The creation of quantum well lasers was made possible by methods like molecular beam epitaxy. Typically, they are based on material systems such as GaAs/GaAlAs

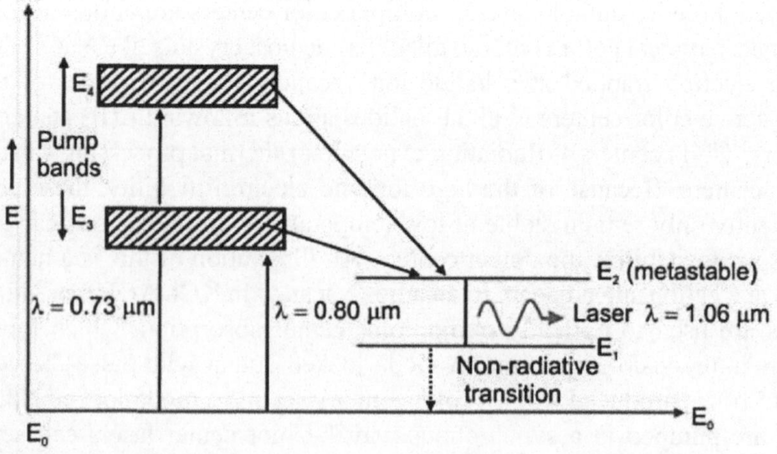

FIGURE 5.10 Nd:YAG laser.

or InP/InGaAsP. The creation of novel device configurations, including vertical cavity surface emitting lasers (VCSELs), is one of the most crucial fields of semiconducting laser research. These produce less round-trip gain.[20]

5.3.4 LIQUID LASERS

The most popular liquid lasers use organic solvents as their basis and organic dye molecules as their active laser species. Additional two varieties of liquid lasers exist. Rare-earth chelates in organic solvents are used in one instance whereas rare-earth ions are used in an inorganic solvent for the lasing system. Flashlamps or other lasers are used to optically pump these systems.[20] Dye lasers are tunable. It has been demonstrated that dye lasers oscillate between the CW regime and the femtosecond pulse domain (dye lasers). Dyes have strong visual absorption and broad luminescence that frequently covers a large spectrum of wavelengths. Dye molecules' energy level diagram consists of vibrational states. Each vibrational state consists of a group of rotational levels. The singlet-to-triplet state transitions are generally unfavorable and less frequent than the singlet-to-singlet state transitions among these states. Light energy is utilized to excite the molecules and move the population from the singlet ground state to one of the singlet excited states to produce the stimulated emission. The excited molecule in the rotational states swiftly decays to the lowest level in each vibrational band. Depending on how the laser's cavity is configured, dye lasers may be straightforward to use. The main limitations of dye lasers are that the dyes are messy and that occasionally air bubbles or dirt will enter the dye-recirculating system and cause the flow jets to spatter (Song et.al.). Rare-earth chelate lasers use an organic solvent-containing complex of rare-earth as the active lasing center. These chelates are complex groups consisting of organic molecules, also known as ligands, which are attached to an ion, specifically a rare-earth metal, through claw-like chelate linkages. Carboxylate ions, organic phosphates, or β-diketone are commonly used as chelate ligands. The coordination nonsaturation of the compound, the number of vacant orbitals on the rare-earth ion, steric considerations, and the electron-donor characteristics of the ligands are factors that influence the number of ligands attached to the ion. The complexes of rare-earth ions and Flerovium (fi) diketones are the most intriguing in the diverse family of chelates utilized for laser oscillation. Europium -benzoylacetonate [Eu(BA)] is a prevalent example. Chelates exhibit distinct spectral properties which are determined by the transfer of energy from the organic component of the molecule to the rare-earth ion. Figure 5.11 depicts how this energy transfer works.

The organic component of the molecule absorbs energy, causing it to transition from the ground state 1 to an excited singlet state 2. After a very brief time, molecules transition to the triplet state 3 occurs. If the triplet state of the ligand is higher than the upper excited state of the rare-earth ion 4, the rare-earth ion can absorb energy and generate radiation that is distinctive to rare earth. This energy transfer can be very effective in particular circumstances, resulting in high quantum efficiency of chelate luminescence. In cases where the energy-transfer efficiency is poor, the following two opposing mechanisms occur simultaneously: Emission from the triplet state of the ligand and energy transfer to the rare-earth ion. In this instance, rare-earth luminescence decays faster than ligand luminescence does. The second scenario illustrates that energy is transferred either from the triplet state with vibrational energy or from a higher-lying triplet state 3′. Eu(BA)] is predominantly used in liquid lasers. The active lasing center in inorganic rare-earth liquid lasers is a rare-earth ion combination of oxyhalides or heavy metal halides. The process of optical pumping is applied directly to the rare-earth ion, and an all-liquid phase design may be used to build a high-power output, compact, and scalable laser. This design incorporates rare-earth ions (such as Nd^{3+} or Yb^{3+}) embedded in inorganic aprotic solvents that contain a Lewis acid. These ions serve as both a coolant and a gain medium in the all-liquid phase setup. Aprotic solvent makes sure that no O–H bond can enter and extinguish the excited rare-earth ions. As a result, it prevents the liquid medium's gain from decreasing.[29]

5.4 LASER APPLICATIONS

5.4.1 COMMUNICATIONS

One of the most useful and interesting applications of laser is in the field of telecommunications. For communication purposes, properties such as narrow beam width and wide bandwidth advantages over other long distances transmission. Laser wavelength exists in a wide range of wavelengths in the electromagnetic spectrum. The laser beams in the IR regions have a specific application for military purposes, detecting them in the IR range is more difficult. Another important application of semiconductor lasers is signal transmission purpose. The advantage of such a laser is the direct excitation by electric current in the invisible infrared region. In the case of military purposes laser beams with narrow beam width have more advantages over ordinary radiowaves for the strict secrecy of the communication. Another advantage of a laser communication system is protection from interference by spurious radionoise. The channel capacity of communication is proportional to the frequency bandwidth so the optical laser has a great potential application in long-distance communication. One of the important facts about laser beams is that it is quite powerful, sharply directed, and highly coherent; so, they are ideally suited for communications purposes.

The optical frequency waves suffer a significant disadvantage over microwave frequencies like considerable attenuation by snow fog, and rain while transmitting through the atmosphere. So, laser communication through an atmospheric medium is effective only if there are no obstacles interrupt between the transmitting and the receiving stations and clear weather conditions.

The multiplexing technique is more effective in long-distance communication systems for simultaneous transmission of different massages over the same pathway. Multiplexing process. effective in improving efficiency and economy in communication. In the modulation process composite single wave was created by modulating a large number of voice signals with a single-frequency carrier wave. The laser beam was modulated by the signal for effective communication purposes. At the receiving point, the modulated beam is detached (demodulated) from the carrier beam (laser). The intensity of the signal is varied according to the output current, and then amplified and fed to the speaker.

The main mechanism involved in an optical modulator is based on the refractive index variations of the substance according to the signal wave. The optical modulator works on the principle of continuous laser passing through a KDP crystal which is placed on the ring electrode. The electric field generated proportional to the signal wave is applied to the crystal. The variations in the refractive index of the crystal, there is a change in polarization of the light beam. As a result, the intensity of the light coming out of the analyzer also changes, according to the signal. Accordingly, a semiconductor laser beam is directly modulated by varying the current through it, according to the signal. The demodulation mechanism involves two ways, namely, direct photodetectors and photomixers. In the visible and IR regions Photomultiplier detectors are good and Silicon photodiodes, have a peak response at about a wide range. Open atmosphere communication through laser is possible only if the line of sight between the transmitter and the receiver is clear in good weather conditions. Under these circumstances, communication through an optical fiber medium communication is one of the most effective methods in the future.[31–33]

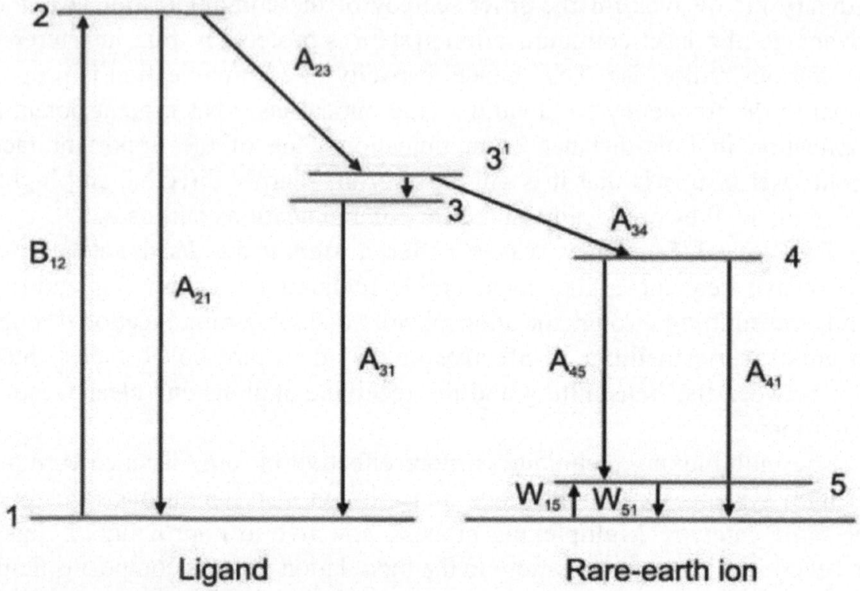

FIGURE 5.11 Rare-earth chelate laser.

5.4.2 MEDICAL APPLICATIONS

The first-time lasers were applied in medicine in 1961 to cure detached retinas and skin discolorations. Currently, lasers are employed extensively in a variety of medical applications. These include urology, surgery, ophthalmology, dermatology, angioplasty, cancer treatment, and esthetic procedures such as liposuction, laser hair removal, and tattoo removal. Here, a few of these crucial applications are briefly addressed. First, let us consider the laser application in ophthalmology. Different laser types are used in ophthalmology nowadays for either therapeutic or diagnostic objectives. Confocal laser microscopy, which enables the early detection of retinal abnormalities, is a key diagnostic tool. This increases the likelihood that glaucoma and retinal detachment will be diagnosed in time for effective treatment. Targets for all therapeutic laser eye treatments fall into two categories: front and rear segments. The front part of the eye is composed of the cornea, sclera, trabecula, iris, and lens, while the rear part consists of the vitreous body and retina. Although lasers were first used in dentistry as a medical application, it has mostly remained an area of research. Regular mechanical drills still outperform most lasers, especially CW or long-pulse lasers, especially in caries therapy, the most common dental surgery. The only thing that could replace mechanical drills would be laser systems that could deliver ultra-short pulses. Mechanical drills have advanced over many years to reach the current level, and it takes time to build appropriate application units for laser radiation as well. Laser soft tissue treatment and laser welding of dental bridges and dentures are two more dental areas of interest. Laser applications are highly significant in both gynecology and ophthalmology fields. This is mostly because using the CO_2 laser to treat cervical intraepithelial neoplasia (CIN), a rare growth of new cervical tissue, has a success rate of between 93 and 97%. In gynecology, the CO_2 laser is the norm. Vaginal intraepithelial neoplasia (VAIN) and vulvar intraepithelial neoplasia (VIN) are also treated with it. Lasers are also used in urology for a variety of purposes. The main-stay lasers used in urology are CO_2, argon ion, Nd:YAG, and dye lasers. The most accurate lasers for tissue cutting are CO_2 lasers. For the coagulation of highly vascularized tumors or malformations, two types of lasers are used: argon ion lasers and Nd:YAG lasers.[34–36]

5.4.3 BIOMEDICAL IMAGING AND SUPER RESOLUTION

Lasers enable advancements in the analysis of biological material. In biology and medicine, spectral imaging is being used in more innovative and fascinating ways. Confocal microscopy is the industry standard for producing high-resolution pictures. Ultrasound, CT, and MRI provide many chances for undertaking functional assessments of many body systems due to quick technology advancements that result in great spatial and temporal resolution. The discovery and advancement of functional magnetic resonance imaging (MRI) and radionuclide imaging [single photon emission computed tomography/positron emission tomography (SPECT/PET)] have completely changed the study of the body systems and their functions. Functional magnetic resonance (MR) includes a variety of methods that show physiological and molecular changes before they become apparent morphologically in traditional imaging. Utilizing integrated functional MRI and SPECT data, intraoperative MR guidance for neurosurgery enhances tumor resection prediction. One approach to molecular imaging is MR spectroscopy. The ability to analyze the chemical makeup of a lesion helps us to understand the metabolic process of normal and sick tissues. For the pretherapeutic diagnosis, assessment of therapy response, and therapeutic monitoring of brain, breast, and prostate cancer, MR spectroscopy has become clinically standard practice. Additionally, the assessment of tumor volume, extracapsular extension, and post-radiotherapy recurrence have all benefited from the use of MR spectroscopy. Powerful methods such as PET and SPECT allow for in vivo, real-time imaging of metabolic processes. A diagnostic technique called molecular imaging makes use of novel diagnostic and therapeutic markers as well as several imaging modalities to observe cells and molecular structures in vivo. Scintigraphy is at the cutting edge of molecular imaging because it can see molecular events that are happening at subnanomolar to millimolar levels in real time. Molecular imaging makes numerous contributions to biomedical research and development.[34–36]

5.4.4 HOLOGRAPHY

One of the other important applications of the laser beam is holography, which is the production of true three-dimensional pictures in space. The mechanism of recording the three-dimensional image of an object on a film is called a hologram. It works on the phenomenon of interfering two beams of monochromatic light waves under certain conditions. The pattern is produced due to interference by mixing two beams of laser light. One of the beams is called the reference beam which is directed toward the photographic plate and the other beam is directed toward the object to record the image. Both the beams together form the interference pattern on the photographic plate. To ensure high-quality images highly coherent sources are used. He–Ne gas laser is generally considered one of the highly coherent and continuous beams, which is used for the fabrication of the hologram. Solid-state lasers such as ruby lasers are sometimes used for this purpose. The exposure time for continuous gas lasers and giant pulsed lasers are different it may range from a few minutes to a few nanoseconds. Good-resolution special photographic plates are used for recording holograms. The developed plates are fixed in the usual way. A hologram is one kind of magic; it does not give any picture to the naked eye but when negatively exposed it shows random patterns on the surface. When it is viewed in a monochromatic light beam random patterns are transformed into sharp and detailed three-dimensional images of the object. The image feels like a solid object hanging in three-dimensional space, by changing the angle of view it feels like moving its head from side to side or up and down. Holography is considered as one of the nondestructive techniques used for information storage, display devices, and pattern matching used for tasks like credit card and identity card verification.

The holographic method used for secret communication of information is mainly used for secret documents, maps, and objects, and restructuring the images possible only at the receiver end. For the use of high-precision components with respect to their shape, dimensional accuracy standard holograms are used. In the future, target recognition from air to ground holographic technique may be used and also for viewers holographic movies and television giving better feeling of the projected objects.[31–32,37]

5.4.5 LIDAR (LASER RADAR)

The laser that used for radar applications is called lidar. Lidar has more advantages over microwave radars, that a laser beam can be focused by lenses and mirrors easily but in the case of microwaves, requires a huge antenna for focusing. Another advantage over microwave radar is higher accuracy obtained in the dimension and the distance of the target. The most commonly used lidar are carbon dioxide, GaAs semiconductors, or Q-switched neodymium, type lasers. High power output with good spectral purity is the counted advantage of carbon dioxide lasers over other lasers in lidar applications. Despite the fact that the carbon dioxide laser beam has a frequency that is several thousand times greater than that of the X-band radar and at a sharp beam width of a few microradians, coherent carbon dioxide laser tips radar performs essentially similar to a coherent microwave radar. Even from slow-moving objects, the high frequency of the carbon dioxide laser results in substantial Doppler shift. The carbon dioxide laser has unmatched imaging capabilities because of its narrow beam and high Doppler shift. This radar device tracks low-flying aircraft and slowly moving objects by monitoring radial velocities. Rain, fog, and snow greatly attenuate the laser's beam; hence, they have a negative impact on lidar performance.[31,38,39]

5.4.6 LASER RANGE FINDER

For defense purposes, destroying an enemy tank, it is necessary to range it very accurately. Because of its high intensity and extremely low divergence even after traveling a considerable distance, the laser is perfectly suited for this purpose. For artillery and tanks, typical equipment now includes laser range finders that use neodymium and carbon dioxide lasers. Compared to conventional range finders, these laser range finders are more reliable and accurate over a wider range. The laser range finder operates similarly to a radar. It uses the monochromaticity, high intensity, coherency, and directionality of the laser beam, which are its distinguishing qualities. When a collimated laser pulse is pointed at a target, the light reflected from the target is picked up by an optical system and identified. It measures how long it takes the laser beam to move back and forth between the transmitter and the target. The range, or target's distance, is determined by multiplying the time in half that has been thusly recorded by the speed of light.

The four functional components of a standard laser range finder are (i) the transmitter, (ii) the receiver, (iii) the display and printout, and (iv) the sighting telescope. An older design of a laser range finder, the transmitter emits a single, collimated, and short pulse of laser radiation at the target using a Q-switched Nd:YAG laser. A photodetector receives a small portion of light from the transmitter pulse through a scattering wire grid, which is then amplified and sent to the counter. The counter is initiated by this pulse of light. The reflected pulse is received by the telescope and, to remove any extraneous radiation, it is routed through an interference filter. After that, it is focused on a different photodetector. The counter is then fed the resulting signal. The time interval is converted into distance using a digital approach. The output shows the range, which the counter has so calculated. The illumination telescope enables the user to see the target and read the range at the same time. Range gating and special circuits have been utilized to reduce spurious signals and facilitate the usage of laser range finders. Under any weather where the targets can be viewed visually through the sighting telescope, it is possible. The most recent models of laser range finders Use a carbon dioxide laser or a high repetition pulsed Nd:YAG laser with a range gating mechanism. These systems can easily range a target that is around 10 km away with an accuracy of 5 m. The usage of portable laser range finders in field artillery fire control systems and tank laser range finders for artillery, armored vehicles, and trucks are two defense applications for medium-range (up to 10 km) laser range finders. Airborne laser range finder, pod-mounted and servo-positioned for the Air Force; Laser walkie–talkie range finder, a small, portable device, weighing less than 4 kg, useful to range objects at distances less than 5 km. These are intended for field application in conjunction with artillery fire control systems.[31,40,41]

5.4.7 UNDERWATER LASER

It is also possible to send information underwater using lasers. The ideal laser for this purpose is one that emits light in the blue-green spectrum since seawater transmits light most effectively in this range. Underwater transmission is attenuated as a result of (i) materials in the water-absorbing light, (ii) suspended particles scattering light, and (iii) fluctuations in optical density along the light's path. The use of blue-green lasers has grown significantly in systems with a focus on naval applications. The submarines now rely on sonar to locate enemy vessels and steer clear of underwater obstacles. This has significant restrictions. False signals are sent by whales, dolphins, and other marine animals. Because seawater broadens or scatters the sonar beam, a normal sonar cannot produce a clear image. The target may appear to be where it is not due to differences in the salinity of the water that might cause the sonar beam to bend. The position of the ship from which sonar is transmitted is revealed to the enemy, which is another issue with sonar use. Lasers are an effective tool for detecting and ranging underwater objects. A frequency-doubled Nd:YAG laser, an argon ion gas laser, or a xenon chloride laser with a Raman shift are all employed for this purpose. An underwater system comprises of a laser transmitter that beam splits and diffuses high-power laser pulses of around 10 ns in length at a rate of 30–50 per second to the target. The time interval counter is started by directing a small amount of the laser light that is reflected by the beam splitter onto the photodiode in the range and display circuit. Telescopic optics are used to gather the target's reflected light after an interference filter has removed any stray radiation. Unwanted echoes are reduced using a range gating circuit. The image intensifier amplifies the target's reflected pulse, and the output is supplied to the image orthicon, which displays the item. Both the target's range and image are acquired in this way. Underwater ranging up to 500 m is feasible in clear water with high power releases of several megawatts. Additionally, lasers can be employed in torpedo and other autonomous underwater vehicle navigation systems, as well as for communication between submarines while maintaining complete confidentiality. Recent underwater laser communication has been developed using satellite technology, going from the ground through a satellite before reaching an underwater station.[31,42,43]

5.4.8 LASER-GUIDED ANTITANK MISSILE

An IR laser beam with minimal divergence can be used to guide and control a missile. There are four ways to achieve this: (i) After the target tank has been illuminated by the laser beam and turned into a source of backscattered radiation, the antitank missile zeroes in on the target. (ii) The laser beam serves as the command link and provides guidance instructions to the missile. (iii) The missile travels along the laser beam, which is kept pointed in the direction of the target along the collision path. (iv) The missile itself is equipped with a laser seeker and scanner for active target tracking. A pulsed Nd:YAG laser is used as the laser target designator in the first situation. The laser beam is modified in such a way that the missile's four-quadrant detector receiver can detect if the missile's trajectory deviates from the beam axis and correct it by changing the missile's fins. Both optical and electronic components are found in the guidance unit. This makes it possible for the gunner to direct the missiles' IR guidance beam. The system with the missile acting as a beam rider intended to follow the laser pointing at the target is more appealing. Four detectors that point backward can be mounted on a missile's wing tips. The missile's flight path is maintained along the detectors' determination of the laser beam's center axis. The sustainer motor plume should absorb as little of the laser's wavelength as possible. Thus, in a laser designator, the laser lights a selected target due to its narrow beam. When a laser beam is scattered, a receiver in a bomb or missile homes in on the target it has located.[31,44]

5.4.9 RING LASER GYROSCOPE

An incredibly helpful tool for detecting and measuring minuscule rotational angles of moving objects is the ring laser gyroscope. It has now taken the place of the mechanical gyroscopes that were once commonplace in both military and civilian aircraft as well as long-range guided missiles. In comparison to a mechanical gyroscope, the ring laser gyroscope has the following three key advantages: (I) No moving parts; (II) high *g* capabilities; and (III) improved reliability. Additionally, the laser gyroscope has a wide dynamic range and quick response times needed for missile guiding.

The basic structure of a ring laser gyroscope is a cavity around which two laser light beams move in opposition to one another. The so-called Sagnac effect, which uses an interferometric approach to measure an object's rotation, is the basis for how ring laser gyroscopes work. Inertial navigation is the primary use of the ring laser gyroscope. It is utilized for gunfire directing, flight control, and inertial guidance of airplanes, ships, and missiles.[31,45]

5.4.10 LASER DRILLING

The face-heating phenomenon is the basis for the laser drilling of metals. Within the laser's penetration depth, the absorbed intensity is converted to heat. Additionally, material removal begins when the heated lighted area on the surface exceeds boiling point because of the vaporization and melt ejection processes. The drilling of a diamond die with a laser only takes a few minutes as compared to 20 h using traditional techniques. There is no waste in the procedure, and the money saved on diamond dust costs aids in recouping the cost of such a drilling machine. The main use of laser light energy is to create tiny apertures in diamonds and rubies. It is now possible to drill filament canals in refractory materials without heating the entire machined unit. One benefit of laser drilling is that it does not harm diamonds or other processed materials in any way. Nd:YAG, alexandrite, or pulsed carbon dioxide lasers are typically used for laser drilling. It is challenging to do special procedures, such as drilling holes with a diameter of less than 0.5 mm, using traditional methods. For such procedures, however, both metals and nonmetals respond well to pulsed laser microdrilling. The alexandrite laser is set to 755 nm, whereas the Nd:YAG laser emits at 1.06 m. The alexandrite laser consistently produces a cleaner entrance hole with better roundness and finer edges than the Nd:YAG laser does for various metals. The shorter wavelength and constant output spikes of the alexandrite laser are what make it perform better. With shorter wavelengths, most materials absorb more energy, providing alexandrite lasers an advantage over Nd:YAG lasers in many material processing applications.[31,46–48]

5.4.11 LASER MICRO-/NANOMACHINING

Recently, lasers have also found use in the field of microelectronics, where the laser beam directly interacts with semiconductor chips and circuit boards that include memory and logic circuits. The development of lasers with greater stability, high repetition rates, wavelengths extending into the UV range, and brief pulse durations has enabled these laser processing applications. Due to these characteristics, it is possible to regularly and dependably heat certain micrometer-sized regions to very high temperatures. Laser heating can happen when it is applied to specific areas of a silicon chip without harming nearby components or circuitry. Using a laser, ceramic substrates can be directly machined to create intricate designs for electronic circuits in a single step. A computer-assisted laser is used in the procedure, and it is configured to specify the type of circuit layout that needs to be produced. The electrical circuits are being trimmed using contemporary lasers and the automated control equipment that goes along with them. The methods for switching and moving the laser beam are automatically regulated, and each action is planned to save time. For instance, arrangements can be made to test certain circuit components as the beam position is altered. Utilizing a laser with higher power will also speed up trimming procedures.[31,49].

5.4.12 LASER CUTTING

Carbon dioxide gas is attractive to CW lasers. Lasers are widely used to cut a variety of materials, including ceramics, sapphire, graphite, diamond, tungsten, carbide, and all metallic foils. The majority of the time, assist gases such as oxygen, carbon dioxide, or air are used during continuous cutting, which intensifies the heat effects through both mechanical and chemical action. This gas-assisted cutting method works with metals with thicknesses up to 5 mm and cut widths as small as 30 μm. The cutting of thin steel and nonmetallic materials is the most promising use of laser cutting. Recently, wealthy nations have begun using laser cutters in the clothing business, which is a brand-new and highly practical application of lasers. With the help of computers, lasers can cut clothing far more quickly than traditional tailors could. Using a laser cutter, it is now able to quickly and reliably cut through many layers of thick cloth. The laser system also includes a computer that has been programmed with cutting guidelines and designs for various types of clothing. When a laser beam is directed at a substance, the fabric is sliced through, leaving flawlessly smooth edges.[31,50,51]

5.4.13 LASER HARDENING

The short-term laser irradiation of a material's surface is the basis of the laser hardening process. A thin layer is rapidly heated because of heat conduction into a metal. A significant temperature difference is created in the surface zone during the heating phase, which is quickly quenched by the cool subsurface material. It takes between one and two seconds to complete the hardening cycle. The traditional surface hardening methods, such as flame hardening and induction hardening, frequently result in distortion and waste labor or results in significant additional expenses. Plain steels with a carbon content of more than 0.3% are often good candidates for laser hardening. Their carbon concentration affects how hard they can become to a maximum. Hardenability is increased by the amount of alloying elements like chromium, manganese, or molybdenum. The possibility of replacing pricey alloy steels with cheaper steels is a benefit of laser hardening of steels. The usage of lasers to harden many other industrial items has increased due to the invention of new approaches. Typically, a laser output of more than 1 kW is regarded as appropriate for hardening a track width of a few millimeters. The world has recently seen a large number of laboratories working on laser applications and research. A carbon dioxide laser's CW beam is typically utilized for heat treatment and hardening.[31,52]

5.5 CONCLUSIONS

In this chapter, we discussed the basic principles, types, and applications of the laser. The working of laser in different media is also covered in detail. Applications of laser in different fields such as laser range finders, underwater laser, laser-guided antitank missiles, laser radar, ring laser gyroscopes, communications, weapon firing simulators, laser micro-/nanomachining, laser cutting, laser hardening, holography, etc. also discussed.

KEYWORDS

- **laser applications in nanotechnology**
- **laser micro-/nanomachining**
- **solid-state laser**
- **gas laser**
- **semiconductor laser**
- **optical communication**
- **medical applications**

REFERENCE

1. Nobel Prize. https://www.nobelprize.org/prizes/physics/1964/summary/
2. Schawlow, A. L.; Townes, C. H. Infrared and Optical Masers. *Phys. Rev.* **1958,** *112* (6), 1940–1949. https://doi.org/10.1103/physrev.112.1940.
3. Major, F. G. Optical Frequency Oscillators: Lasers. In: *The Quantum Beat*; Springer, New York, 1998. https://doi.org/10.1007/978-1-4757-2923-8_14
4. Gould, R. G. *The Ann Arbor Conference on Optical Pumping*; 1959.

5. Maiman, T. H. Addendum 10: Reprint of T. H. Maiman, 'Stimulated Optical Radiation in Ruby,' Nature, [7, 493–494 (August 6, 1960). *Laser Inventor* **2017,** pp. 299–301. https://doi.org/10.1007/978-3-319-61940-8_35.
6. Al-Amri, M. D.; El-Gomati, M.; Zubairy, M. S. *Optics in Our Time*; Springer International Publishing, 2018.
7. Demtröder, W. Fundamental Principles of Lasers. In: *Laser Spectroscopy: Springer Series in Chemical Physics*, Vol, 5; Springer: Berlin, Heidelberg, 1981.
8. Parker, S. Introduction, History of Lasers and Laser Light Production. *Br. Dental J.* **2007,** *202* (1), 21–31.
9. Thyagarajan, K.; Ghatak, A. *Lasers: Fundamentals and Applications*, 2nd ed.; Springer: New York, 2011.
10. Maini, A. K. *Lasers and Optoelectronics: Fundamentals, Devices and Applications*; John Wiley & Sons, Incorporated, 2013. *ProQuest Ebook Central*. https://ebookcentral.proquest.com/lib/inflibnet-ebooks/detail.action?docID=1343443.
11. Shokrollahi, K.; Raymond, E.; Murison, M. S. C. Lasers: Principles and Surgical Applications. *J. Surg.* **2004,** *2* (1), 28–34.
12. Abramczyk, H. *Introduction to Laser Spectroscopy*; Amsterdam: Elsevier, 2005.
13. Bridges, W. B. Ion Lasers-the Early Years. *IEEE J. Select. Topics Quant. Electron.* **2000,** *6* (6), 885–898.
14. Svelto, O.; Svelto, O. *Principles of Lasers*; Rheine: Heyden: London, 1976.
15. Telford, W. G. *Methods Cell Biol.* **2011,** *102*, 373–409.
16. Nakhe, S. V.; Rajanikanth, B. S.; Bhatnagar, R. Energy Deposition Studies in a Copper Vapour Laser Under Different Pulse Excitation Schemes. *Measure. Sci. Technol.* **2003,** *14* (5), 607–613.
17. Omi, T.; Numano, K. The Role of the CO_2 Laser and Fractional CO_2 Laser in Dermatology. *Laser Therapy* **2014,** *23* (1), 49–60.
18. Basting, D.; Marowsky, G. *Excimer Laser Technology*; Springer/Praxis: Berlin, 2005.
19. Perram, G. P. Lasers, Chemical. *Encyclopedia Phys. Sci. Technol.* **2002,** 283–295.
20. Powell, R. Types of Lasers. In: *Springer Handbook of Atomic, Molecular, and Optical Physics*; Drake, G., Ed.; Springer: New York, 2006.
21. Avram, N.; Brik, M. et al. Energy Levels of Transition Metal Ions in Laser Crystals: Crystal Field Analysis and Electron-Phonon Interaction. *Rom. J. Phys.* **2006.**
22. Potter, K. S.; Simmons, J. H. Optical Gain and Lasers. In: *Optical Materials*, 2nd edn.; 2021.
23. Reisfeld, R.; Klixbüll, J. C. *Lasers and Excited States of Rare Earths*; Springer-Verlag: Berlin, 1977.
24. Dubey, A. K.; Yadava, V. Experimental Study of nd:YAG Laser Beam Machining—An Overview. *J. Mater. Process. Technol.* **2008,** *195* (1–3), 15–26.
25. Mollenauer, L. F.; White, J. C.; Pollock, C. R. Color Centre Lasers. In: *Tunable Lasers*; Springer-Verlag: Berlin, 1992.
26. Byer, R. L.; Gustafson, E. K.; Trebino, R. *Tunable Solid State Lasers for Remote Sensing Proceedings of the NASA Conference Stanford University, Stanford, USA, October 1–3, 1984*; Springer: Berlin, 2013.
27. Koch, S. W.; Hofmann, M. R. Semiconductor Lasers. In: *Encyclopedia of Modern Optics*, 2nd ed.; 2018.

28. Song, J. et al. Study of Interpulse Delay Effects on Orthogonal Dual-Pulse Laser-Induced Breakdown Spectroscopy in Bulk Seawater. *J. Analyt. Atom. Spectrometry* **2020,** *35* (10), 2351–2357. https://doi.org/10.1039/d0ja00241k.
29. Varshney, A. K. et al. "Optical and Thermal Studies of ND^{3+} Doped Inorganic Liquid Medium for Scalable Laser Source. *Opt. Laser Technol.* **2022,** *148*, 107740. https://doi.org/10.1016/j.optlastec.2021.107740.
30. Abramczyk, H. Lasers. Introduction to Laser Spectroscopy **2005,** *4*, 59–106. https://doi.org/10.1016/B978-044451662-6/50005-8.
31. https://www.fisica.net/optica/Laser-and-its-Applications.pdf
32. Thyagarajan, K.;·Ghatak, A. Lasers Fundamentals and Application. DOI: 10.1007/978-1-4419-6442-7
33. Kaushal, H.; Kaddoum, G. Optical Communication in Space: Challenges and Mitigation Techniques. 10.1109/COMST.2016.2603518
34. Verdaasdonk, R. M. Medical Lasers: Fundamentals and Applications. *Adv. Lasers App.* **2021,** 181–226.
35. Buckley, S. Advances in the Applications of Lasers for Bioimaging: Light Sheet Microscopy. *Spectroscopy*, *34* (10), 33–32, 50.
36. Sun, Z.; Ng, K. H.; Ramli, N. Biomedical Imaging Research: A Fast-Emerging Area for Interdisciplinary Collaboration. *Biomed. Imaging Interv. J.* **2011,** *7* (3), e21.
37. Gabor, D.; Kock, W. E.; Stroke, G. W. Holography: The Fundamentals, Properties, and Applications of Holograms Are Reviewed. **1971,** *173*, , 11–23.
38. Muhadi, N. A.; Abdullah, A. F.; Bejo, S. K.; Mahadi, M. R.; Mijic, A. The Use of LiDAR-Derived DEM in Flood Applications: A Review. *Remote Sens.* **2020,** *12*, 2308.
39. Wang, Z.; Menenti, M. Challenges and Opportunities in Lidar Remote Sensing. *Front. Remote Sens.* **2021,** *2*, 641723.
40. Muzal, M.; Zygmunt, M.; Knysak, P.; Drozd, T.; Jakubaszek, M.. Methods of Precise Distance Measurements for Laser Rangefinders with Digital Acquisition of Signals. *Sensors* **2021,** *21*, 6426.
41. Forrester, P. A.; Hulme, K. F. Laser Rangefinders. *Opt. Quant. Electron.* **1981,** *13*, 259–293.
42. Jaffe, J. S.; Moore, K. D.; McLean, J.; Strand, M. P. Underwater Optical Imaging: Status and Prospects. *Oceanography* **2001,** *14* (3), 64–75.
43. Jaff, J. S. Underwater Optical Imaging: The Past, the Present, and the Prospects. *EEE J. Ocean. Eng.* **2015,** *40* , 683–700.
44. Iyer, N. Recent Advances in Antitank Guided Missile Systems. *Defence Sci. J.* **1995,** *45*, 187–197.
45. Passaro, V. M. N.; Cuccovillo, A.; Vaiani, L.; De Carlo, M.; Campanella, C. E. Gyroscope Technology and Applications: A Review in the Industrial Perspective. *Sensors* **2017,** *17* (10), 2284.
46. Wang, O.; Lin, H.; Wang, C.; Zheng, L.; Hu, X. Laser Drilling of Structural Ceramics—A Review. *J. Eur. Ceram. Soc.* **2017,** *37*, 1157–1173.
47. Schulz, W. Review on Laser Drilling I. Fundamentals, Modeling, and Simulation. *J. Laser App.* **2013,** *25*, 012006.
48. Wang, C. et al. The Development of Laser Drilling: A Review. *IOP Conf. Ser.: Mater. Sci. Eng. 782*, 022067.
49. Dongre, S.; Gujrathi, A.; Nandan, H. A Review of Laser Micromachining. *Int. J. Eng. Sci. Comput.* **2016,** *6*, 2416.

50. Naresh, P. K. Laser Cutting Technique: A Literature Review. *Mater. Today: Proc.* **2022,** *56*, 2484–2489.
51. Choi, J.; Kim, R.; Song, D.; Cho, D.-W.; Suh, J.; Kim, S.; Ahn, S.-H. Analysis of Laser Cutting Process for Different Diagonal Material Shapes. *Processes* **2022,** *10* (12), 2743.
52. Dinesh, P.; Balasubramanian, K. R.; Naidu, B. G. Laser Surface Hardening: A Review. *Int. J. Surface Sci. Eng.* **2011,** *5*, 131–151.

CHAPTER 6

Carbon-Based Materials for High-Yield Optical Device Applications

JINCEMON CYRIAC[1], CLARIS ALEX[2], ANEETTA SHAJI[2], SUNNY MATHEW[2], V. N. SHEEJA[2], TINA SEBASTIAN[3], and SARITHA K. JOSE[3]

[1]*Department of Physics, Christ College (Autonomous) Irinjalikkuda, Calicut University, Irinjalakuda, Kerala*

[2]*Physics department, Marian College Kuttikkanam (Autonomous), Kuttikkanam, Peermade, Idukki, Kerala, India*

[3]*Physics Department, Deva Matha College, Kuravilangad, Kottayam, Kerala, India*

ABSTRACT

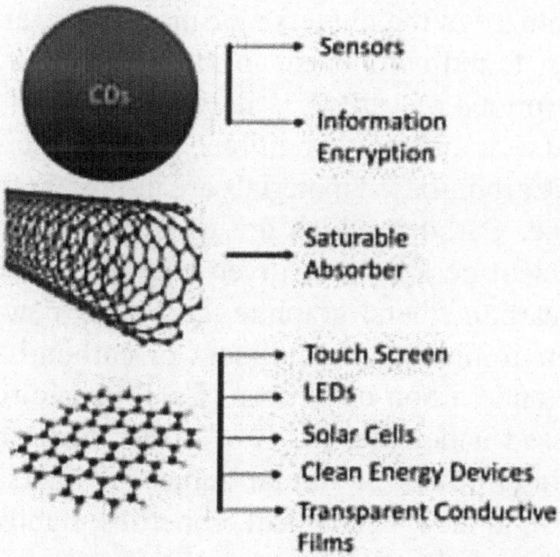

Resilience and Sustainability for Energy and Water: New Advancements, Opportunities, and Framework.
Cristóbal Noé Aguilar, A. K. Haghi, Amritanshu Shukla, Ann Rose Abraham, Swati Gokul Talele (Eds.)

For decades, carbon has been used in a variety of ways in both technology and human life. Charcoal, graphite, carbon black, and other carbon materials have been used for writing purposes since prehistoric times. For the past 25 years, conjugated carbon nanomaterials, including fullerenes, carbon nanotubes (CNTs), graphite, and activated carbon, have been utilized as a medium of energy because of their distinctive characteristics. Carbon-based materials have found application in a wide range of fields due to their superior chemical, mechanical, electrical, and thermal properties. The optical applications of carbon-based materials are one of the interesting and developing fields in the modern era.

Carbon-based materials serve as a type of smart material with different types of lasers and saturable absorbers. The optical properties of these materials facilitate their use for developing new technologies. These materials, including CNTs, carbon dots (CDs), graphene, graphite, etc. have convertible and distinctive optical properties. This makes them more appealing for different fields ranging from electronics, health, and the environment sector.

6.1 INTRODUCTION

Carbon is a one-of-a-kind and essential element that exists in the world. Carbon is the fourth most abundant substance in the solar system, the seventeenth most abundant component in the Earth's crust, and the sixth most abundant element in the universe.[1]. Surprisingly, after oxygen, carbon is the second most found component in the human body. One of the most outstanding descriptions of carbon is its wide range of metastable phases that can be formed near ambient conditions, as well as its expansive fields of kinetic stability.[2]. Carbon-based materials are important for the advancement of material science. Carbons which are used in industries that have been used from the ancient period, have given way to the contemporary period carbons such as carbon fibers, graphite, and other new forms of carbon.[3] Because of the environmental friendliness of carbon-based nanomaterials such as graphene and carbon nanotubes (CNTs) basic research and execution of carbon-based nanomaterials is of great importance in the areas of material science, and others in similar subjects.[4] CNTs, graphene, and its derivatives are good in their mechanical properties, stability, flexibility, good conductors of heat and electricity, and optoelectronic properties as well as ease of functionalization or hybridization with other soft materials.[5] So, they are ideal for photoresponsive soft actuators. Other low-cost carbon materials such as graphite and amorphous carbon have appreciable photoresponsive actuation,[7] energy conserving, and wide band light absorption capacity and are also good conductors of heat and electricity.[6] Importantly, because of the synergistic effects of the individual components, materials of carbon can be easily altered using other functional soft species to form fast responsive composites with huge sensitivity and high actuation speed.

Quantum dots (QDs) have a great importance in the present situation because of its tunable photoluminescence (PL), high density and are cheaper than other carbon materials. High Quantum yield (QY), small size, less toxicity, good biocompatibility, and sufficient low-cost sources make them useful in a variety of domains, including catalysis, biomedicine, anticounterfeiting and optoelectronic devices.[8]

CNT, referred to as the material of the 21st century, are estimated in a variety of methods to implement the Antenna Solar Energy Conversion (ASEC) concept. The interaction of CNT and CNT arrays with the optical spectrum for current and prospective future CNT technologies, as well as CNT thermomechanical and electrical properties in relation to ASEC, are of utmost importance in the modern era.[9] Single-walled carbon nanotubes (SWCNT) are used in various fields such as health care, energy, molecular biology, and catalysis due to their versatile nature. The intrinsic optical properties of SWCNTs have been used as a means of signal transduction in bio-optical devices, which has been the focus of much recent research. Single-molecule sensitivity along with indefinite photostability and optical transparency near to infrared fluorescence on biological tissue give way to SWCNTs a wide range of applications; especially in sensing areas.[10] Flexible electrically conductive and optically transparent nanocomposite polymer and multiwalled carbon nanotube (MWCNT) based transparent substrates have the potential to be used in optoelectronic applications.[11]

CDs are a kind of quasi-zero-dimentional (0-D) carbon-based nanomaterial with a size of less than 20 nm, and their intrinsic property is fluorescence. CD development was hampered by low QY and complicated preparation procedures. Yang's group, till 2013 used ethylene diamine and citric acid (CA) as precursors and used a one-step hydrothermal method to create polymer-like CDs with QY of up to 80%. For carbon-based fluorescent materials, this QY was the highest.[12] CDs can be used as functional nanocomposite materials and as inks for printing. CDs sparked as a wide concern and a relevant research topic due to their very minimal toxicity, appreciable QY, and good photobleaching resistance. Taking this into consideration, researchers have obtained various plans and technologies to develop high-performing CDs.

Some of them are near-infrared [(NIR)/multicolor/deep red] emission,[13] two-/multiphoton PL,[15] chirality,[16] narrow full-width half maximum (FWHM),[14] thermally activated delayed fluorescence (TADF)[18] and room temperature phosphorescence (RTP). In the current scenario, based on structural properties, formation mechanisms, degree of carbonization, and varying the graphene layer; CDs can be classified into carbon quantum dots (CQDs) and graphene quantum dots (GQDs) and carbonized polymer dots (CPDs).[19]

6.2 OPTICAL PROPERTIES

1. **Absorption**

   The absorption behavior of CDs varies with changing the synthetic methods and sources of carbon. Strong absorption is observed in the range of 200–400 nm which is in the ultraviolet (UV) region and slightly extends to the visible range. The π–π^* and n–π^* transition is observed in the C=C bonds and C=O or C=N bonds, respectively, and is assigned in the absorption bands.[20] The sp^2 domains have a π–conjugated electron observed in the NIR emission of CDs. This results in absorption in the range 500–800 nm, which has a long wavelength and is usually connected with polymer chains or surface groups. As a result, the absorption characteristics of CDs are primarily influenced by the type and content of surface groups, the difference in the content of nitrogen or oxygen in carbon cores, and the size of the π-conjugated domains.

2. **Photoluminescence (PL)**

   The most crucial feature of CDs, both practically and functionally, is PL. CDs have good light stability, less toxicity, appreciable biocompatibility, good QY, and plenty of low-budget sources. Such properties are compared to other fluorescent materials (such as QDs), which have the presence of lead or cadmium, organic dyes, and other nanomaterials containing rare earth elements. This gives more application in various fields. The brightness of PL values is significantly influenced using synthetic approaches; carbon sources and postpassivation CDs reflect the quantitative QY. Generally, it is said that routes formed from "bottom–up" have high QY when compared to "top–down" routes.[22] Hence, QY of CQD as well as CPD is higher than that of GQD. The low QY of GQD is mainly due to the defects obtained in the oxide cutting carbon resources when viewed from the PL mechanism and formation of carbon dots (CDs).[23]

CPDs have a higher QY than completely carbonized surface modification, which can minimize the nonradiative recombination and improve the unification of the conjugated system of GQDs. CPD has been popularly used in recent years to control the band gaps of GQDs and improve PL behaviors.[24] CDs (GQDs and CQDs) are the crosslinked subfluorophores, molecular state emission, and supramolecular interaction of the CEE effect. For CPDs in solution, a QY of 99% has now been reported.[25] Most CDs emit blue or green fluorescence, limiting their long-term biomedical applications. By rearranging reaction conditions, aromatic compounds showed the presence of NIR and red-emitting CDs in recent years.[26]

Interestingly, while the surface groups, morphology, or the nanostructure of CQD, GQD, or CPD differ, the majority of these groups exhibit similar excitation-dependent energy output, that is, the PL emission often leads to red-shifts with respect to the intensifying excitation wavelength. As a result, the PL of CDs can be controlled easily by changing the excitation wavelength without altering the size or chemical composition. This is very useful in multicolor bioimaging.[27] The multiple PL centers and wide range of energy levels may explain CDs' excitation-dependent PL behaviors and broad PL profile.

3. Phosphorescence

RTP materials are extensively used in areas like sensing, bioimaging and light emission.[28] High toxicity and high cost of metal complexes and inorganic complexes, which are traditional RTP substances, have reduced the applications.[29] As a result, RTP materials that are less expensive and less toxic have a wide range of applications. Although RTP properties have been discovered in rare pure organic compounds, their development has been hampered by their short lifetimes and low intensities.[30] Researchers have developed high-strength, long-life CD RTP materials in recent years.[31]

4 CDs' diverse optical properties lead to a wide range of applications. C=N and C=O relevant moieties have proved to contribute to the making of RTP, due to their low singlet–triplet splitting energy and

strong spin–orbit coupling.[32] Meanwhile, doping heteroatoms such as nitrogen, phosphorous and halogens can help in the n–π* transition of C=N and C=O, thereby supporting the intersystem crossing (ISC) process which takes place. However, to stabilize the produced T1 species, polyvinyl alcohol and filter paper which is used to set the emissive species from quenching to produce RTP.[33] Using appropriate methods, that is, without the usage of additional matrices, the RTP for CPDs can be easily achieved. Additional matrices are avoided because of the covalently cross-linked framework, and supramolecular interaction, and polymer chains are very important in restricting their emissive centers and repressing nonradiative transition.[17] Implanting CDs in different matrices, matrix-free CDs, and enlarging applications in the field of multimodal anticounterfeiting and bioimaging are some of the techniques which are involved in the preparation of CDs with RTP in recent times.[34] However, the processing and application of CDs with RTP are a still developing research area. CDs should be designed in such a way that their QY, stability, and lifetime have to be improved by planning suitable structure, altering reaction parameters, and posttreatment of the prepared CDs. Electrochemiluminescence, chemiluminescence, and two- or multiphoton fluorescence are the other luminescent properties that have been revealed in the recent decade.[35]

4. Fluorescence

The most attractive property of CDs is their ability to emit light. A significant advance was made in the research of PL followed by Stokes shift and up-converted photoluminescence (UCPL) followed by anti-Stokes shift. CDs' diverse luminescence behaviors make them a universally used material in photoelectricity, energy, and biology. The excitation wavelength is shorter than the PL wavelength. The energy that corresponds to the excited wavelength is larger than the emissive wavelength because the energy of the excitation wave is less than that of the emission wave. This is one of the main differences between UCPL and PL. For testing, the ancient research on UCPL of CDs used the same wavelength, that is, monochromatic light which is from a fluorescence spectrophotometer (a type of xenon lamp without filter). In this case, conventional PL's second-order diffraction of wavelength was always misinterpreted as UCPL.[36] The fluorescence properties of CDs suggest that it has the potential of the CD surface.[24]

Some contemporary studies have stated that CD fluorescence is affected by different structures. Ding et al.[13] reported that CDs which possess a QY of 35% replace traditional organic dyes and QDs as a very attractive light-emitting material. Two important parameters that describe the fluorescence properties of CDs are the QY and the emission wavelength.[37] Furthermore, from a structural standpoint, there is no doubt that CD fluorescence characteristics are determined by size, defects, functional groups, and even phase state. Several synthetic techniques can be opted to improve the QY of CDs from 6%. In contrast, most present CD emission wavelengths exhibit strong excitation dependence. For example, CDs are found to be more intense in the blue-to-green spectral region when a UV radiation is made to fall on the sample.[38] Therefore, the emission wavelength can be tuned by adding important groups like complexes that contain nitrogen or fluorine and restricting the oxygen made from hydrothermal methods; by adding oxygen atoms the PL of the above sample can be altered. These results confirm that the oxygen atoms that are bonded and the carbon atoms that are conjugated cause the luminescence of CDs. However, the presence of oxygen was crucial in controlling the band gap of the highest occupied molecular orbital (HOMO) and the lowest occupied molecular orbital (LUMO).[39] Experiments reveal that the band gap and oxygen content at the surface decreased and increased respectively. Also, a red shift of emission peak was observed from 400 to 625 nm. Bao et al.[40] also confirmed the mechanism of surface state emission. They made CDs out of nitric acid-oxidized carbon fibers. The size and surface state of CDs can be altered by controlling the reaction time, temperature, nitric acid concentration, etc. The following conclusions can be attained:

i. Increasing the time of reaction causes more surface oxidation, resulting in a longer emission wavelength.
ii. A long emission wavelength is produced by a high nitric acid concentration and a short reaction time.

Han et al.[41] also investigated the relationship between the surface-state energy gap and fluorescence. Auto-polymerization and self-oxidation were used to create three different types of polymer carbon dots (PCDs). The

CQN contents on the three types of PCDs were 59.29, 20.11, and 13.83%, respectively. According to X-ray photoelectron spectroscopy (XPS) characterization, as the C=N content on the PCD surface increases, the red shift of the PL spectrum also increases. This occurs due to the shortening of the band gap as the presence of C=N on the surface of PCD. The introduction of new energy levels on the PCD's electronic structure is due to the increase in the number of C=N on the PCD surface. This causes a lot of electronic shifts and eventually a red shift in the PL spectrum. Zhu et al.[42] proposed that the fluorescence of CDs was caused by a synergistic effect between C=O and C=N.

Hydrothermal methods can be used to design and fabricate three different kinds of CDs with solvents such as ethanol, dimethylformamide, and water. XPS data and fluorescence spectra were used to validate the results. Particle size is also important to find out the band gap and can affect the change of fluorescence wavelength. Scientist Fan and his group created CDs with narrow bandwidth emission and particularly shaped CDs.[22] Transmission electron microscopy (TEM) image revealed that the CDs were triangular, with a QY of 54–72%. They created four different sizes of triangular CDs. The precursor was chosen to be symmetric phloroglucinol (PG). The adjacent three PG molecules make a six-membered ring with the active –OH and –H groups during the cyclization reaction. For sizes ranging from 3.9 to 3.0, 2.4, and 1.9 nm, bright and multicolored emissions of red (R), yellow (Y), green (G), and blue (B) are observed. When the particle size increases, the band gap narrows, causing fluorescence to redshift.[43]

6.3 OPTICAL APPLICATIONS

CDs are gaining popularity in optical applications such as sensor and information encryption due to their good photostability, low toxicity, aqueous dispersibility, phosphorescence, and strong PL properties, which will be mentioned in the session.

6.3.1 SENSORS

CDs are popularly used for sensing different analytes in biological systems. They are well suited for these applications due to their inherent fluorescence probe properties, such as quick response, simple synthesis procedures, low cost, and high sensitivity to temperature, solvents, and ionic strength. The sensitive and highly reactive nature is mainly due to their large surface area, small size, and abundance of functional groups at the surface. Optical properties such as enhancement or activation referred to as turn-on and quenching referred to as turn-off of fluorescence are some of them. Inner filter effect (IFE) and fluorescence resonance energy transfer (FRET) are the detection mechanisms that are theoretically equivalent.[44] Anions and cations such as ClO^+, Hg^{2+}, Fe3+, Zn^{2+}, Pt^{2+}, etc. can be easily discovered by using CDs. These groups bind to the surface groups such as amino and carboxylate groups which are present on the CDs via electrostatic interaction.[45–47] For the selective sensing of Zn^{2+} ion and EDTA, Scientist Yang and his group created an interchangeable "off–on" fluorescent nanosensor using the repassivation and depasivation of CPD which are passivated using Zn^{2+} made from the compound zinc gluconate.

Captivatingly, orange-emissive CPDs were immobilized in medical cotton cloth with pH-sensitive fluorescence can be used for checking the pH of a wound through visible colorimetric changes and fluorescence.[48] Furthermore, the pH of the wound can be forecasted and estimated for blood impurities and long-period observations by opting for the analytical method and the observed dual nature response to pH change. CDs exhibit a selective nature on a large range of biomolecules, including glutathione, amino acids, formaldehyde, vitamins, DNA, and proteins which are connected with health issues.[49] As a result, CDs can provide valuable insights into disease diagnostics and early prevention.

CDs are also excellent in targeting and locating cancer cells and spotting and sensing bacteria by the modified ligands due to their low toxicity, excellent biocompatibility, and photostability.

CPDs with green emissions can effectively differentiate cancerous cells from healthy cells due to changes in their potentials in the mitochondrial membrane and substance uptake capacities.[50] Scientist Mao and colleagues created Aptamer nanoconjugates CPDs for the selective detection of Yang's group-created acidophilic CPDs from living cancer cells which show blue and red emission to differentiate *Porphyromonas gingivalis* bacteria that exist in the alkaline environment; *Staphylococcus aureus* that exists in *an* acidic environment; *Streptococcus mutans* (Gram-positive bacteria with a teichoic cell wall) and *Escherichia coli* (Gram-negative bacteria with cell walls that lack teichoic acid) which are classified on the basis of different environments and cell types of these bacterial walls.[51] Rather than the ancient indistinguishable bioimaging of these types of bacteria, these researches work as a modern and effective tool to recognize bacteria.

6.3.2 INFORMATION ENCRYPTION

Information encryption helps in preventing important documents or files from replication. In comparison to traditional anticounterfeiting technologies such as plasmonic security labels and fluorescence printing[52] or mediums such as metal-containing complexes like organometallic complexes and metal-free compounds such as pure organic compounds that consist of RTP-containing CDs are always a better option. Applications of CDs lie in the field of anti-counterfeiting of data and encryption due to their environmentally friendly features, simple design, ease of handling, and low cost. In the past years, CD-based RTP materials for anticounterfeiting were obtained by embedding them in various matrices such as PVA, filter paper, and crystalline manganese framework.[53] However, these host matrices are not universal and only allow certain types of CDs to produce RTP. Li et al. described a global method for activating CD's multicolor RTP using a boric acid matrix for information encryption and anticounterfeit.[54] The energy gap between the singlet and triplet states can be reduced by incorporating boron into composites. This improves the ISC between S1 and T1 states. This can result in the guarding of excited triplet states of CDs from nonradioactive deactivation.

The instability of reversible connections in these composites limits their use, and physical immobilization through additional matrices poses difficulties. A novel metal-free RTP CPDs was prepared successfully to overcome this situation by Yang's group from polyacrylic acid and ethylenediamine without using additional matrices, where amide or imide, as well as the CEE effect, were found to help in RTP generation.[17] Furthermore, the images provided demonstrated the potential applications of CPDs in security protection. By controlling the singlet-triplet energy splitting in GQDs through oxidation the transition from RTP to TADF state was achieved.[18] The anticounterfeiting and multilevel information security were facilitated by oxidation-controlled GQDs systems with both RTP and TADF. These analyses on the structure design and photophysical properties of CDs turn in interesting directions for developing high-performance afterglow materials.

6.4 ADVANTAGES AND DISADVANTAGES

CDs are a promising material when compared to toxic metal QDs, because their biocompatibility and excellent optical properties make them indispensable, particularly in the optical fields. Several properties and facts can be observed in multidisciplinary settings even though many issues can be found by using CDs which can provide unique chances that have been proved by recent research.

The research on CDs is still in process when compared with other carbon materials. The production of high-quality CDs with a suitable structure, that is, shape, number of functional groups, size, position of defects, and crystallinity is one of the main critical issues in this research area. Furthermore, due to nonstandard synthetic pathways and impurities, their exact reaction mechanism, nucleation mechanism, and formation process are unknown. As a result, large-scale production of high-performance CDs can be only achieved by opting for suitable reaction conditions and effective precursors. Also, a correct systematic investigation of these processes and purification depending on the polarity and size should be developed for a better application.

6.5 CARBON NANOTUBES (CNTS)

CNTs have remarkable mechanical, electrical, and thermal properties, as well as unique nonlinear optical (NLO) properties. All of the research indicates that CNTs are one among the most promising applicants in the coming generations in the field of photonic components. The two main research hotspots based on CNTs are the use of mode-locking components in ultrashort pulsed lasers by using saturable absorbers and optical limiting in laser protection are of great importance in the 21st century.

Nonlinear optics is a relatively new field of physics that studies the interaction of an intense light field with matter. It has enormous fundamental scientific and technological potential.[55] Under high-intensity light illumination, materials such as solids, gasses, and liquids can show NLO retaliation, the magnitude of which is determined by their intrinsic properties. NLO substances can be classified into bulk and molecular materials. Metals, inorganic crystals, and semiconductors fall in the category of bulk materials. Nonlinearity in the case of molecular materials is based on the structure of the individual molecules of the particular materials, whereas nonlinearity in bulk materials is related to the electronic properties of the same. Nonlinear optics is a relatively new field in physics that studies the interaction of intense light fields with matter. Many materials have been recognized with good NLO properties as a result of the growth in nanoscience and its research. This has been achieved through the usage of photoelectronic and photonic devices by motivating the manufacture and design of these devices.[56]

We can discuss the recent progress in the development of pristine and functionalized nanotubes for NLO applications. Z-scan is a simple but effective technique for characterizing NLO properties such as absorption, scattering, and refraction.[57] An intensity-spatially-varying optical field is created by focusing a laser beam on a convex lens in this method. A change of transmission opposite to incident intensity occurs while moving an optical material along the *z*-axis with respect to its focal point. Using these methods, nonlinear parameters can be easily obtained.

As the intensity of the incident beam increases, the total transmission of nanotubes in liquid dispersions decreases. That is, the nanotube dispersions effectively suppress high-intensity light while allowing high-intensity light to pass through. On the contrary, as the incident power increases, the transmission of the CNT films increases. This implies that low-intensity background noise can be easily subtracted by nanotube films giving an enhanced high-intensity signal. The easily convertible nature of CNT provides multiple nonlinear responses making them favorable for various applications. These diverse nonlinear responses reveal two important applications, regardless of the laser sources used. They are optical limiting for the protection of laser and like a saturable absorber, as a mode-locking element for ultrashort pulsed lasers.

6.6 OPTICAL LIMITING

After the invention of the laser, it became clear that exquisite optical instruments like the human eye could be damaged by the use of intensified laser light. Nowadays, lasers are so common that they have become combined in toys. As a result, laser protection is also a need for public safety rather than a scientific problem.[58] The safeguarding of these instruments has resulted in an extensive study on optical limiting and its applications. While impoverishing intense and life-threatening laser beams, a good optical limiter, which possesses a high transmittance for ambient light with low intensity, should be used. Several organic and inorganic materials, including fullerenes,[59] phthalocyanines,[60] porphyrins,[61] inorganic nanoparticles,[62] metal clusters, and metal complexes,[63] have demonstrated powerful nonlinear absorption or scattering while using a light source of high intensity. Thus, they are applicable as a friendly optical limiter. CNTs have been extensively studied as an optical-limiting material over the last decade.[64] CNTs have a combined effect of fullerene and carbon black which are the allotropes of carbon that make them more demand in the industry. Fullerene functions as a suitable match for functional substances. On the contrary, carbon black is a good optical limiter. Over a huge range of wavelengths, that is, from NIR to visible, CNTs act as a good optical-limiting effect. Importantly, the merit of CNT over other materials is that they can be designed in terms of chemical properties by attaching functional groups such as organic molecules, nanoparticles from metals, and polymers to form convertible optical-limiting composites.[65]

6.7 MECHANISM

The high-intensity beam is dispersed during the scattering process to have larger spatial dimensions, resulting in a reduction in intensity. Nanoscale particles alone cannot scatter the light beam fruitfully, according to the scattering theory proposed by Mie. The formation of scattering centers of sufficient size to scatter the incident laser beam must result in proper scattering. In common, generated dynamic scattering centers are caused by the following processes:

1. The formation of solvent bubbles
2. The ionization of nanoparticles
3. Discontinuity in the refractive index. This is caused by the thermal effect on nanoparticles from the surrounding solvents.

Because the scattered intensity from these types of Mie scattering centers is not reactive toward the incident wavelength, the response wavelength region is broad. Many researchers have looked into the mechanism that causes the optical-limiting effect in pristine nanotubes.[64]

NLS, which is thermally induced, is widely recognized as one of the primary mechanisms for optical limiting. These scattering centers which are induced have two beginnings: the origin and growth of the solvent bubbles. This formation is caused by the transfer of thermal energy from nanotubes to the solvent, and the expansion of carbon microplasmas and its formation caused by nanotube ionization. The former occurs at lower incident energy fluences, while the latter occurs at higher fluences. Vivien et al. used the Mie scattering theory to simulate the extinction scattering profile in the case of suspension in nanotubes.[66] Considering that bubbles are in a uniform spherical shape, the concentration and size of suspension scattering centers can be estimated after ignoring multiple scattering. O'Flaherty et al. observed the scattering at an angular profile with respect to Mie scattering characteristics in CNT or polymer dispersions where the SWCNTs have long cylindrical structures.[67]

However, to the best of our knowledge, there is still no comprehensive conceptual explanation of NLS in nanostructure suspensions. Belousova et al. created a conceptual theory to explain the optical limiting of carbon nanoparticles.[68] The entire limiting process is described conceptually by three main procedures in this model:

i. The dynamic behavior on the formation and expansion of solvent vapor bubbles
ii. Expansion of bubbles on Mie scattering
iii. Nonlinear movement through the medium which has been scattered.

Despite the fact that the objects that the modeling is based on quasi-spherical carbon nanoparticles, the Mie theory-based prediction works only in the case of nanotubes and thus the optical-limiting process in CNT suspensions. The examples revealed that the expansion rate, inside pressure, and radius of gas bubbles as a function of illumination time, as well as the temperature, the onset time for boiling, and radius of gas bubbles as a function of illumination energy density can be determined. Hence, it can be concluded that when vapor bubbles increase, the scattering of the cross-section increases particularly, but the absorption cross-section decreases until it is negligible, beneficially limiting the incident power.

CNTs' optical-limiting performance is typically assessed by comparing them to some classic nonlinear materials, such as C-60. The results after comparison, however, are found to be inconsistent. Vivien et al. for example, demonstrated that SWCNTs' suspensions in either water or surfactant possess good optical-limiting responses when compared with the solutions of C-60.[69] Meanwhile, Mishra et al. discovered that C-60 solutions have better optical-limiting responses than SWCNT suspension in water.[70] Results show that CNT's optical-limiting properties can be affected by a variety of factors, including bundle size, solvent property, wavelength, the incident laser pulse width, and so on. The optical limiting is affected by two intrinsic parameters:

a. The overall structure of CNTs
b. The solvent's thermodynamic property which is used to disperse nanotubes.

However, the length of nanotubes has a comparatively minor impact on optical-limiting properties, the bundle diameter has been shown to have a significant impact.[71] Longer nanotubes with a higher aspect ratio have a better NLO response if the bundle diameter remains constant.[72] The nanotubes with more size on initial scattering center size have the most efficient transfer of heat from nanotubes to solvents. Size on initial scattering is directly dependent on the bundle size of nanotubes. These properties help in quick solvent bubble growth and result in good limiting productivity and a low limiting threshold. However, it is very hard to determine whether SWCNT or MWCNT is excellent for an optical limiter.

Diameter size is still the most important effect for optical limiting in both cases, rather than their specific structures. Furthermore, the thermodynamic properties of the solvents play an important role in limiting the performance of nanotube suspensions. For example, nanotubes dispersed in a solvent with a low boiling point exhibit a low limiting threshold and therefore a better limiting effect.[70] Solvent bubbles which are induced with heat grow quickly in a solvent with minimal surface tension, allowing these bubbles to reach a critical size for effective scattering in a short span of time, which results in a quick limiting response.[71] Furthermore, nanotubes have better optical-limiting performances with a shorter wavelength incident beam, longer pulse duration, and minimal repetition rate, which can be explained easily by an NLS mechanism that is induced thermally.[73] Finally, it can be concluded that all of the effects are strongly related to the dynamic process of scattered centers.

6.8 SATURABLE ABSORBER

Laser and nonlinear optics, as we all know, are always complementary. The laser's invention hastened the transformation of nonlinear optics from theoretical expectations to experiment results. Implementing NLO components into laser cavity controls and modifies laser output performance in an inverse manner. Various organic and inorganic materials with special NLO properties have frequently been used for Q-switching for lasers. Nanomaterials are very weak in the participation of laser modulation until the emergence of CNTs as saturable absorbers.[74]

6.8.1 MODE-LOCKED LASERS

Set et al. demonstrated SWCNT's saturable absorption property for 1 ps ultrashort pulses at 1550 nm for the first time in 2003.[75] A nanotube film that is sprayed on a quartz substrate helps in suppressing background noise with minimal intensity while allowing the high-intensity signal to pass smoothly. This results in fast mode-locking. Following that, Set et al. successfully presented mode-locking from erbium-doped fiber lasers using simple structures such as saturable absorbers.[76] As a result, fiber lasers with ring and linear resonant cavity configurations produced pulse trains of 1.1×10^{-12} s and 318×10^{-15} s, respectively. CNT-based saturable absorbers have since been widely used as mode-locking elements in ultrashort pulse fiber lasers and solid-state lasers. Although various structures of saturable absorbers have been designed in recent years, from the literature, it is clear that saturable absorbers are based on CNTs as Saturable Absorbers Incorporating Nanotubes (SAINTs).[74]

Semiconductor saturable absorber mirrors (SESAM) fabrication is complex, expensive, and environmentally unfriendly, whereas SAINT requires only a cheap and room-temperature fabrication method. SAINTs' nonlinear and linear conditions, such as threshold power, linear transmission, and absorption band can be simply engineered by opting CNTs with suitable radius ranges and varying the thickness and density of the CNT polymer layer. SESAM acts as a reflective device, which works with linear cavity lasers and has a low damage threshold. CNT and nanotube-polymer composites, on the other hand, have more damaging thresholds, and the flexibility of nanotube particles allows for the planning and making of various SAINT structures to meet the various needs. SAINTs have been used effectively in mode-locked fiber lasers and solid-state lasers in linear and ring configurations to date.

6.9 DISCUSSION

Because pristine nanotubes primarily have NLS, most CNT assimilated NLO substances show the NLS effect. Recently, some studies have stated that at 532 nm which is at a single wavelength, the optical-limiting property was observed. Therefore, we can believe that there is a response wavelength for optical limiting in the case of a NLS-dominated CNT system. Optical-limiting effects till 1064 nm have been reported continuously in different nanotubes, while the further has yet to be reported. Nanotubes are best in the effective working of laser pulses due to their NLS nature, which vary from nanoseconds to milliseconds. To achieve the reduction of power in ultrashort pulses from the picosecond to the femtosecond regimes of a continuous wave laser beam with less intensity, functional materials should be integrated with CNTs.

Because of their lower saturation intensity, minimal excitation decay time, and big absorption cross-section at the near-infrared, SWCNTs can be used as mode-locking elements. MWCNTs are typically made of semiconducting and metallic CNTs. This nature gives them a metallic property. In comparison to the less saturation intensity of SWCNTs, MWCNTs need a higher irradiation intensity for saturation.

6.10 GRAPHENE

Graphene, which is a single layer of graphite; was introduced in 2004. Due to its beneficial material characteristics, particularly transparency, high charge mobility, flexibility, and mechanical strength, graphene has drawn a great deal of interest.[77] Graphene is made up of a hybrid nanosheet of a single layer of sp^2 or multiple layers of carbon atoms. These carbon atoms are closely packed together inside hydrogen-atom-free benzene rings.[78] Two comparable sublattices of carbon atoms joined by σ bonds make up the honeycomb structure of graphene. The delocalized network of electrons is influenced by the π orbital of every carbon atom in the lattice.[79]

The fundamental structural ancestor of all carbon allotropes is graphene. Fullerenes, nanorings, graphite, MWCNTs, SWCNTs, nanobuds, double-walled carbon nanotubes (DWCNTs) constitute some of the allotropes of carbon.[80] Graphene could be used to produce CNTs, fullerenes, etc. CNTs are produced when a graphene sheet is rolled up along one axis, and fullerene can be made by forming the CNT into a ball. As a result, graphene is regarded as the starting point for computations of the band structures of CNTs and graphite as well as calculation of the magnetic properties.[81]

Graphene sheets can be generally classified into three types. These include a single layer, a bilayer, and a few layers.[82] They are fabricated using a variety of techniques. The first technique involves mechanically removing individual layers of a bulk graphite sample.[77] Novoselov et al. have used dry cleaving of highly orientated single graphite crystals or pyrolytic graphite to synthesize single, double, and triple-layered flakes that are about 10 μm in size because of its weakly-bound layered structure. This method, often known as the Scotch tape method, involves mechanically rubbing or utilizing adhesive tape to separate graphite flakes from bulk graphite.[83] Raman and optical spectroscopies could be used to count the layers in the resultant flakes.

The monolayer flake of exfoliated graphene can be treated to produce samples of suspended graphene that are not in touch with a substrate. These samples can be created by either etching off the silicon dioxide substrate from below the graphene or by installing metallic electrodes, or by covering the exfoliated graphene with a metal grid.[84]

Another way to fabricate these graphene specimens had been established by de Heer et al.[85] This technique involves heating hexagonal silicon carbide crystals to around 1300°C in a mega-high vacuum, that causes the silicon atoms, to evaporate from the surface and thus leave behind a coating that is entirely graphitic. Depending on the face of the SiC crystal that is used in the process, the residual film's characteristics change. High-quality graphene films are formed using the SiC (0001) (silicon terminated) face and exhibit Dirac cones in the low-energy band structure, while contact between the substrate and the graphene results in *n*-type doping, that raises the Fermi energy above Dirac point.

Lastly, graphene layers and thin graphite films that can be shaped and converted to insulating or semiconducting substrates can be grown via chemical vapor deposition.[86] In this technique, a metal foil surface is in close proximity to a stream of hydrocarbon gas, allowing the hydrocarbon molecules to break down and the carbon atoms to dissolve into the metal. A carbon film may then develop on the foil's surface as it cools (at a specified rate). Chemical etching can be used to transfer this coating to a polymer or semiconducting substrate. By using this technique, graphene monolayers can be totally coherent along step edges in the underlying substrate and have few flaws.[87]

6.11 APPLICATIONS

A potential replacement for conventional electrode materials such as metals and ITO in electrical and optical devices is graphene, a promising next-generation conducting material. Good optical transparency, low sheet resistance and exceptional mechanical properties are some of the positive traits that are offered by it. Because of these characteristics, it is a desirable candidate for use in a variety of components, such as transistors, memory, molecular junctions, touch screens, liquid crystal displays (LCDs), light-emitting diodes (LEDs), and solar cells. Graphene is anticipated to perform a significant role as a transparent electrode in optoelectronic and electrical devices.[88] Transparent electrodes are thus an important component of many widely used electronics, including screens [mobile phones, LCDs, e-papers, etc.], photovoltaic systems, and LEDs.[89]

6.12 OPTICAL DEVICES

Touch screen: There are various technologies of touch screen, including projected capacitance, resistive, surface capacitance, surface acoustic wave, and capacitive.[90] Recently, Bae et.al have created tens of centimeter-sized graphene sheets that meet these criteria for transparency and sheet resistance. As a result, they were able to create transparent electrodes out of large-scale graphene films that were then put into touch-screen panel devices. They created a four-ply graphene stack with 90% transparency, resistance, and endurance using CVD to synthesize 30-inch graphene sheets on top of copper foil. This stack is superior to ITO-based electrodes currently used in displays. The film had a lower electrical resistance than the typical transparent conductor composed of ITO and allowed 90% of light to pass through. When they used graphene in a genuine touch-screen display, they discovered that it worked better than ITO which was used when customers made the credit-card purchase. They demonstrated that a touch panel built with graphene could withstand twice as much stress as a touch screen built with ITO since the IYO was found to be brittle.[91]

Liquid crystal displays: It is a flat panel display that makes use of the liquid crystals' capacity to modulate light. It is made up of optically transparent thin sheets of polymers that have microscopic LC droplets embedded within their pores. LCDs are useful in a variety of applications, such as computer monitors, televisions, instrument panels, aviation cockpit displays, etc. because of their ability to flip between translucent and opaque states.[92] The LCD device constructed has a performance that is comparable to that using ITO electrodes in the case of a thermally reduced sheet of graphene oxide (GO) which is 6-nm thick, according to recent findings.[93] Thus, graphene could be used as a transparent electrode for LCDs. Typically, an LCD's electric field is applied via a transparent electrode made of ITO on glass. However, instability and limited adaptability prevent it from developing further. The transport of ions into active medium and metal oxide's chemical stability are significant issues for the majority of ITO-based LC devices as well as other photonic devices. By injecting oxidation from ITO into the active medium, such processes degrade it and can result in breakdown at lower voltages.[94] The injected ions in LCDs also become trapped at the alignment layer, screening the applied electric field.[95] With graphene, one may generally anticipate that such problems can be avoided because of its chemical stability, which should reduce the amount of ion diffusion.[96]

Light-emitting diodes: LEDs can be utilized as light sources in text and video displays, traffic signs, and flashlights. Even while ITO is still the most widely used transparent conductor today, it has some drawbacks, most notably sharp price swings and being a ceramic material. A novel electrode material that can take the role of ITO in LED devices is thus greatly needed. A large-scale batch production of gallium nitride (GaN) LEDs using transparent electrodes made of patterned graphene is a significant advancement toward more useful applications of graphene films in optoelectronic devices.[97] A graphene layer made through CVD synthesis was used as the top anode for GaN LEDs. Additionally, some groundbreaking research has documented using transparent electrodes made of graphene thin films in organic light-emitting diodes (OLEDs).[98] The graphene-based OLED so far reported, has also created voltage drop in electrodes at high current densities because of the high sheet resistance of graphene. The device performance of graphene-based LED/OLED devices will be improved by engineering the properties of graphene film, for example, to increase the conductivity by doping and adjustment of work function to enhance the interface matching between the LED active layer and graphene. Additionally, the transparent conductor in OLEDs is becoming more and more crucial due to its chemical stability and mechanical flexibility. As a result, there has been rising requests for the creation of different new transparent electrode materials that can replace ITO, which suggests that the study of graphene films as transparent electrodes will continue to advance.

Solar cells: Organic bulk hetero junction (BHJ) photovoltaic cells, dye-sensitized solar cells, Si Schottky junction solar cells, *p-i-n* solar cells, and hybrid ZnO/poly(3-hexylthiophene) (P3HT) solar cells are just a few of the various graphene-based solar cell types that have been reported thus far in various applications.[99] Because graphene has the potential to be used for being inexpensive, flexible, and compatible with organic films with graphene. This makes it potentially profitable for graphene-based photovoltaic device applications when combined with the adaptability of organic molecules.[100] However, one significant hurdle in the development of high-performance solar cells was to efficiently segregate the electron–hole pairs produced by the photons, and then transmit the differentiated charge carriers to the electrodes. It is particularly intriguing that by functionalizing the material during production and/or postprocessing, graphene-based materials have demonstrated programmable surface and interfacial properties as well as customizable work functions.[101] A thin titanium suboxide (TiO_X) layer manufactured via sol–gel and CVD was effectively employed in OSCs as transparent and conductive electrodes.[102] The graphene sheet was then spin coated with a PEDOT:PSS layer. Spin coated on top of the PEDOT:PSS layer was the BHJ composite, which is photoactive and made of [6,6]-phenyl-C61 butyric acid methyl ester (PCBM) and P3HT. Then, a twisted-coated TiO_X coating was applied to select devices. The cathode was eventually created through the thermal evaporation of aluminum. It was found that the power conversion efficiency (PCE) of graphene-electrode OSCs is the main characteristic used to evaluate OSC performance and that the rising temperature of the graphene films is a controlling factor on PCE. The PCE values of OSCs made with graphene electrodes have, up until this point, been reported to be between 0.08 and 2.60%,[103] which is significantly lower than the PCE values of OSCs made with traditional ITO electrodes (8.37%).[104] To make graphene film a strong candidate in OSCs, the PCE of graphene–electrode OSCs must be greatly enhanced.

Sensors: Recent theoretical[110] and experimental[111] research had shown that due to its conductance fluctuations depending on the degree of surface adsorption, large specific surface area, and low Johnson noise, monolayer graphene is a viable candidate to identify a variety of molecules, such as gasses[112] to biomolecules.[113] It is suggested that the charge transfer between graphene and absorbed molecules causes a chemical reaction. As molecules get adsorbed to their surface and experience a charge transmitted with graphene that acts as a donor or acceptor, it changes the electrical resistance of graphene, carrier density, and the Fermi level. According to recent studies, on the graphene surface, CO and NH_3 molecules would act as donors, whereas NO_2 and H_2O molecules would serve as acceptors.[114] Additionally, it has been demonstrated that reduced GO achieves sensitivity levels in parts per billion levels for the identification of chemical warfare agents and explosives and thus, is a good sensor.[115] Both the biocellular and biomolecular scales of CMG have been utilized as sensors in biodevices. It can function as a polarity-specific molecular transistor for protein/DNA adsorption, a label-free, reversible DNA detector, and an interface for identifying individual bacteria.[116]

Transparent Conductive Films: Graphene materials hold the potential for transparent conductive films (TCFs); due to their moderately good optical permeability in visible areas of the spectrum, high carrier mobility, and high electrical conductivity. GO has become a crucial precursor for the creation of TCFs as a result of its easy processing and scalable manufacture. Spin coating,[117] dip coating,[99] vacuum filtration,[118] and Langmuir–Blodgett assembly[119] have all been used to deposit films containing GO platelets. These methods have all been followed by chemical reduction and or thermal annealing. For the creation of TCFs, researchers have also created a number of techniques to generate suspensions of reduced GO or graphene.[120] For instance, by reducing the pH to around 10 before hydrazine reduction, spray-coated TCFs were made from the aqueous dispersions of the reduced GO nanoplatelets.[121]

Clean Energy Devices: Graphene is considered as a promising electrode material because of its large theoretical surface area, which is 2630 m^2g^{-1}, and its capacity to promote hole or electron transfer over the two-dimensional (2D) plane. There have been numerous publications regarding graphene-based electrodes for electrochemical double-layer capacitors and rechargeable lithium–ion batteries (RLBs) [electric double-layer capacitors (EDLCs)]. The most popular material of anode for RLBs, graphite, has the maximum "lithium insertion" capacity of about 372 $mAhg^{-1}$, or the amount of lithium that forms in a molecule called a graphite intercalation compound for every six carbon atoms (LiC_6).[122] To obtain a Li_2C_6 stoichiometry, it has been suggested that lithium ions can attach to monolayer graphene's two surfaces in disordered carbon. Despite the high conductivity and mechanical stability of such paper samples, flexible "graphene paper" electrodes performed electrochemically inferior to the powder made of reduced GO platelets.[123] However, a paper-like substance was created when silicon nanoparticles and reduced GO platelets were combined. For the large storage capacities, it was claimed that the uniform silicon particle dispersion in composites and graphene platelets' reconstitution to form an incredibly conducting and continuous 3D network is crucial. Volume expansion in pure metal/metal oxide electrodes has been solved by enveloping metal oxide or metal particles with graphene-based carbonaceous compounds or graphene.[124] Additionally, it has been claimed that decreased GO's higher conductivity and structural flexibility improve the electrode performance of metal–graphene hybrids. Non-Faradaic ultracapacitors called EDLCs store charge in electric double layers that form where an electrode with a large surface area meets an electrolyte.[125] In EDLCs, high specific surface area activated carbons are frequently combined with a more conductive filler, like carbon black, to form the electrode material.[126]

Graphene–Polymer Nanocomposites: There are many potential uses for graphene-based fillers, which have been used in polymer nanocomposites.[127] GO and some other graphite derivatives, like graphite intercalation compounds (GICs), could be used as precursors to produce a variety of exfoliated CMGs, also known as "graphite nanoplatelets," on a large scale.[120] Poly(vinylidene fluoride),[135] polystyrene,[7] epoxy,[131] PMMA,[128] polyester,[132] polyvinyl alcohol,[129] polycarbonate, polypropylene,[130] polyurethane,[134] and silicone foam[133] are examples of polymer matrix composites with graphene.[136] Layered components like GO must be exfoliated and evenly distributed throughout the polymer matrix to obtain significant property improvements in their nanocomposites.[137] Particularly, mechanically-expanded GO (or "TEGO") is reduced and can be utilized to create composite materials that are electrically conductive,[120] whereas rapidly heating,[138] as well as the ultrasonication of GO has been extensively used to create highly exfoliated platelets for the nanocomposites. Mechanically extended GO still has its insulating chemical structure, therefore reduction must be done separately.[139] Techniques which include mixing of solution, in situ polymerization, or melt blending could be used to distribute these fillers out into polymers; of these, in situ polymerization may provide for better filler dispersion.[130] Graphene or "graphite nanoplatelet" nanocomposites that have significant property enhancements, distinctive architectures, and a few of the lowest percolation thresholds known to date have been created using novel synthesis and processing techniques. Graphite was immediately exfoliated from PP via solid-state shear pulverization, which tripled the modulus when compared with the plain polymer. For instance, layer-by-layer construction produced conductive PVA-CMG nanocomposites.[140] After being reduced with polysulfide ions, GO was decorated with sulfur nanoparticles.[141] After melting this oxide with poly(arylene sulfide), highly exfoliated platelets were created in the matrix.[142] PVA and polystyrene sulfonate-stabilized GO were dissolved in aqueous solutions, which were then frozen and then freeze-dried to create nanocomposite monoliths that include a 3D, porous scaffold-like structures that might be used as catalyst supports and in electronics.[143]

Photodetectors and Phototransistors: An essential component of contemporary optical communications is the photodetector. Graphene is a promising contender for high-frequency, ultrafast photodetectors because of its high charge carrier mobility and short carrier lifetime. A mechanically exfoliated single- and few-layer graphene-based 40 GHz photodetector with a possible inherent bandwidth over 500 GHz has been described.[163] In comparison to traditional semiconductors, photocarriers are produced and transported extremely differently. The internal electric field between the metal electrodes and the graphene drove the ultrafast photodetectors without the requirement for bias. Pure graphene-based photodetectors perform well in high-frequency devices, but their photoresponsivity is rather low since single-layer graphene has a low light absorption coefficient and quick carrier dynamics. As a result, numerous techniques are created to increase photoresponsivity. For instance, to improve graphene's light absorption, semiconductor QDs and nanorods are added.[164] To improve the visible and UV absorption, ZnO nanorods, PbS QDs, and CdS QDs were produced in situ on surfaces of the graphene. Other techniques have also been created by controlling the reduction of reduced GO to tune flaws in graphene.[165] A few-layer, large-scale GO sheet was created via spinning coating, followed by a carefully controlled, low-temperature thermal reduction. It was discovered that the oxygenous flaws had significant effects on the field effect transport of photocarriers.[166] Defects alter the recombination dynamics of excited charge carriers, which causes an increase in photoresponsivity with a loss of high-frequency responses.[167] Another illustration shows that graphene nanoribbon made from CNT splitting exhibits photoresponse and efficiency that are two orders of magnitude higher than those of regular, large-size reduced graphene oxide sheets in infrared photodetectors.[168] To better understand the optoelectronic transport in graphene, phototransistors are being thoroughly researched.[169] The photoimaging of the device's photocurrent distribution has shown that the contact edge-induced internal field significantly affects the optoelectronic performance.[170] The creation and manipulation of ballistic photocurrents in graphene were also investigated.[171] Additionally, it was discovered that the photothermal effect and hot carriers had a significant impact on optoelectronic transport.[172] Recently, high photoresponsive phototransistors with few-layer reduced graphene oxide were also created. Due to their effects on photocarrier transport, it was discovered that the oxygenous groups had a considerable impact on the transfer properties of phototransistors, which is based on graphene oxide which is few-layered and reduced.[166]

Saturable Absorber for Ultrafast Lasers: The ultrafast laser is one of the most important innovations in optoelectronic technology.[173] Saturable absorbers, which convert continuous light into ultrafast brief optical pulses, are widely utilized in mode-locked ultrafast lasers. Due to its ultrafast carrier dynamics, broadband absorption, linear electronic dispersion, and zero bandgap, graphene is an excellent optical material for use as a saturable absorber. Investigations into saturable absorbers in ultrafast lasers have been conducted on both pure graphene and graphene composites.[174] A saturable absorption at a significantly lower threshold than CNTs was demonstrated for atomic layers of graphene, single-layered graphene created by chemical vapor deposition, and random stacks of graphene which is a few and single layered. As a result, such atomic graphene layers could produce ultrashort soliton pulses in mode-locked fiber lasers by acting as saturable absorbers.[173] Similar to this, for saturable absorbers, graphene composites with polymers such as polyvinyl alcohol are a good substitute for pure few-layer graphene.

Since it was extremely solution processable, the graphene employed in the composite was made by exfoliating graphite with inorganic salt. Furthermore, this graphene/polymer hybrid made it possible to produce the material using wet Chemistry and integrate it with a variety of photonic systems with ease.

6.13 LIMITATIONS

In addition, with "intrinsic" corrugations, there are various "defects" that a graphene in a real 3D space can have, such as topological defects (such as pentagons, heptagons, or their combination), adatoms, cracks/edges, adsorbed impurities, and vacancies. Experiments have revealed that defects in a graphene layer, such as adatoms, topological defects, and vacancies, could be created locally by electron irradiation and visible in TEM operating at 120 kV in a material containing single-wall carbon nanotubes (SWCNTs).[105]

6.14 OPTICAL PROPERTIES

The one-atom-thick and 2D honeycomb lattice of carbon is closely related to the optical characteristics. From the infrared to the visible range of the spectrum, it has been asserted that for the Dirac fermions in graphene, high-frequency conductivity is a constant equal to $\pi e^2/2h$.[106] Accordingly, for normal incidence light, the reflectance R and optical transmittance T are $R = 1/4\ \pi^2\alpha^2 T$ and $T = (1 + 1/2\ \pi\alpha)^{-2}$, where $\alpha = 2\pi e^2/hc \approx 1/137$, provided, c is the speed of light, h is the Planck's constant, and e is the electron charge. The opacity could be written as $1 - T \approx \pi\alpha \approx 2.3\%$. The structure and electrical characteristics of graphene are claimed to be the reason for the representation of R and T in relation to fundamental constants which do not involve material factors directly.[107] Infrared spectroscopy was been used to investigate inter-band optical transitions in graphene, and thus optical transitions that are gate-dependent have been discovered.[108] A shift in the Fermi level brought on by the gate creates a considerable fluctuation in charge density, which in turn causes a significant change in transmission since graphene has a low density of states close to the Dirac point. Depending on the carrier concentration of graphene, the recombination and relaxation of electron–hole pairs that are photogenerated, take place over a period of tens of picoseconds.[109] Additionally, it has been discovered that graphene possesses special luminous and PL quenching capabilities, both of which are particularly advantageous for biological applications.[153] It was discovered that graphene oxide sheets are photoluminescent in the visible and infrared spectrums.[154] When it is activated by a laser pulse of 30 fs ultrashort, ultrafast PL appeared in graphene which has a wide range of energy of photons, ranging from 1.7 to 3.5 eV.[155] Due to energy or electron transfer, graphene has been shown to have a strong capability of fluorescence quenching for several QDs and dyes.[156] The transfer of energy from the dyes to graphene gives graphene its special ability to quench. The electrogenerated chemiluminescence (ECL) of graphene oxide in an aqueous suspension of tri-*n*-propylamine (TPrA) and $NaClO_4$ was also seen in addition to PL.[157] While very little ECL was seen without graphene oxide, it is greatly increased when it is present. Colloidal graphene QDs are another intriguing optical feature of graphene. Graphene QDs have inherited the distinctive properties of inorganic colloidal semiconductor QDs, regardless of the method of synthesis used, including top–down cutting (electrochemically or chemically) from graphite, reduced graphene, graphene oxide or carbon fiber,[158] bottom–up synthesis from small molecules,[159] or top-down chemical cutting from graphite, graphene, or graphene oxide.[160] In bulk heterojunction solar cells, graphene QDs, which can glow blue or green, serve as an effective charge carrier separator.[161] Atomic doping can adjust the PL of graphene QDs. In contrast to nondoped ones, nitrogen-doped colloidal graphene QDs from the electrochemical approach had a 4.3% N/C atomic ratio and were intensely blue photoluminescent.[162] A versatile and all-purpose platform for cuttingedge optical and optoelectronic applications could be offered by graphene QDs. Devices such as phototransistors, transparent conductive electrodes, light-emitting devices, photodetectors, saturable absorbers for ultrafast lasers, and photovoltaics as well as other optoelectronic and optical applications have all shown promises for graphene that are significant.

6.15 NANOMATERIALS RELATED TO GRAPHENE

Carbon nanomaterials[144] related to graphene consist of florographene, graphene oxide (GO), doped graphene, graphyne, graphane, generated graphene nanoribbons (GNRs), porous graphene, graphdiyne, and graphone, respectively. The valence and conduction bands of graphene meet at the Dirac point, making it a semimetal with a band gap zero. By cutting and doping the 2D graphene into the 1D GNRs, the band gap can be adjusted.[145] Scientists are very interested in GO, a variant of graphene with oxide functionalization. GO is a hydrophilic substance because of its superior water dispersibility. Due to lowered interfacial energy, it could stick onto surfaces.

GO is also a surfactant which is used to emulsify the organic solvents in water. It can be utilized to disperse CNTs and insoluble graphite in water. This property creates possibilities for the creation of useful hybrid graphene and other P-conjugated systems materials.[146] A theoretically nonmagnetic semiconductor with an energy gap, graphane is a hydrogenated derivative of graphene. Graphane is a carbon-based extended 2D polymer. By demonstrating its enormous potential in nanoelectronics and hydrogen storage applications, graphane enriches the family of carbon-based materials. Along with the graphene sheet, the hydrogen atoms rotate in opposite orientations, changing the carbon lattice from sp^2 to sp^3 hybridization. Another significant structural derivative of graphene with the stoichiometric formula unit of CF is called florigraphene. With each carbon atom coupled to one fluorine atom, florographene has a comparable geometric structure and sp^3 bonding configuration to graphane. Under difficult circumstances, florigraphene can be utilized as a superior solid lubricant for batteries. Another type of graphene is called graphyne, and it has acetylenic links that join the hexagons of graphene. With C_2H Serving as its stoichiometric formula, graphone is projected to be a semihydrogenated (50%) derivative of graphene.[147] On one side of the carbon sheet, there are hydrogen atoms bonded together. Additionally, there is a mix of hybridized sp^2 and sp^3 carbon atoms in it. A brand-new category of lightweight material is porous graphene. Within the graphene's covalent *p* electronic structure, it can be characterized as a dispersed intrinsic structure.

6.15.1 SYNTHESIS OF GRAPHENE-BASED NANOMATERIALS

Mechanical Exfoliation from Graphite: The first high-quality graphene was produced using Scotch tape and mechanical exfoliation from raw graphite.[77] Although it is found to be a crude and ineffective method, it is used to preserve the accuracy of the actual 2D atomic lattices of the graphene. This technique includes, regularly rubbing SiO_2/Si substrates with Scotch tape to exfoliate graphite. Few layered or single-layered graphene which is of high quality, eventhough it is low yielding and laborious, could be detected with reasonable distinction thanks to the distinctive color contrast. The completeness of mechanically exfoliated graphene makes it simple to observe ultra-high mobility and a variety of fascinating electrical characteristics.[148] The first high-quality graphene was successfully mechanically exfoliated from graphite, generating a surge of interest in additional 2D atomic crystals and nanomaterials based on graphene.[83]

Epitaxial Growth from the SiC Thermal Deposition: Because of the exceptional graphene quality, the epitaxial formation from the SiC thermal deposition is another approach that is frequently used to learn about graphene physics.[149] Epitaxial deposition of the graphene from SiC takes place on the terminated Si (0001) face of a single crystal under controlled circumstances at temperatures above 1000°C.[150] On the SiC substrate, few layered and monolayered graphene could be produced. The most important factors of growth are temperature and suction regulation. Because the graphene is on insulating surfaces in this method, it is advantageous to directly fabricate a range of devices on SiC substrates. Depending on the substrate size, epitaxial growth of large-area, homogeneous graphene films is possible. Contrary to CVD-grown graphene on metal, SiC substrates make it extremely difficult to transfer graphene to other substrates.

Chemical Vapor Deposition: One of the most significant developments in large-scale graphene growth technology is chemical vapor deposition. Large-area, homogenous, high-quality graphene can be produced on metal surfaces by catalyzing the decomposition of sources of carbon (such as gaseous methanol or CH_4, or solid poly (methyl methacrylate) using metal (such as Ni, Cu, etc.).[151] After the technique, which commonly takes place at high temperatures of about 1000°C, metal is frequently removed by an etching agent. The metals that are most commonly employed in the CVD development of graphene are nickel and copper. Although nickel-based CVD graphene growth was first developed, copper can have grains that are larger than nickel after the annealing and they allow larger domain expansion in graphene.[152] Even 30-inch graphene transparent electrodes for touch panels were made using CVD growth of graphene on copper.

6.16 CONCLUSIONS

Nanotechnology is a rapidly expanding field. It is one of the fastest-developing industries in the 21st century. Nanomaterials like CDs, CNTs, graphene, etc. are cheaper and can be manufactured without creating environmental pollution. These materials are also highly sensitive to photons, which can give high yields and are highly versatile. They can also be further modified to optical devices and can be modified accordingly. Carbon-based materials have good compatibility in the medical field, analysis of chemicals, and environmental sensors. However, the large-scale production of these materials has to be improvised for these applications. More research and understanding are required to advance this field from its current scenario. The unique features of carbon-based materials will bring a bright future for optical devices and their usage.

KEYWORDS

- **optical properties**
- **carbon dots**
- **carbon nanotubes**
- **graphene**
- **graphite**
- **medium of energy**

REFERENCES

1. Zhang, Y.; Yin, Q. Z. Carbon and Other Light Element Contents in the Earth's Core Based on First-Principles Molecular Dynamics. *Proc. Natl. Acad. Sci. USA* **2012,** *109* (48), 19579–19583. https://doi.org/10.1073/pnas.1203826109
2. Pace, N. R. The Universal Nature of Biochemistry. *Proc. Natl. Acad. Sci. USA* **2001,** *98*, 805–808. https://doi.org/10.1073/pnas.98.3.805
3. Hu, C.; Li, M.; Qiu, J.; Sun, Y. P. Design and Fabrication of Carbon Dots for Energy Conversion and Storage. *Chem. Soc. Rev*. **2019,** *48*, 2315–2337. https://doi.org/10.1039/C8CS00750K
4. Zhang, Z.; Yi, G.; Li, P.; Zhang, X.; Fan, H.; Zhang, Y.; Wang, X.; Zhang, C. A Minireview on Doped Carbon Dots for Photocatalytic and Electrocatalytic Applications. *Nanoscale* **2020,** *12*, 13899–13906. https://doi.org/10.1039/D0NR03163A
5. Wang, X.; Peter, S.; Ullrich, R.; Hofrichter, M; Groves, J. T. *Angew. Chem., Int. Ed.* **2013,** *52*, 9238. https://doi.org/10.1007/978-3-319-03236-85
6. Choi, S. J.; Choi, C.; Kim, S. J.; Cho, H. J.; Hakim, M.; Jeon, S.; Kim, I. D. Highly Efficient Electronic Sensitization of Non-Oxidized Graphene Flakes on Controlled Pore. *Sci. Rep*. **2015,** *5*, 8067. https://doi.org/10.1038/srep08067
7. Stankovich, S.; Dikin, D. A.; Dommett, G. H. B.; Kohlhaas, K. M.; Zimney, E. J.; Stach, E. A.; Piner, R. D.; Nguyen, S. T.; Ruoff, R. S. Graphene Based Composite Materials. *Nature* **2006,** *442*, 282. https://doi.org/10.1038/nature04969
8. Yuan, F.; Wang, Y. K.; Sharma, G.; Dong, Y.; Zheng, X.; Li, P.; Johnston, A.; Bappi, G.; Fan, J. Z.; Kung, H.; Chen, B.; Saidaminov, M. I.; Singh, K.; Voznyy, O.; Bakr O. M.; Lu, Z. H.; Sargent, E. H. Bright High-Colour-Purity Deep-Blue Carbon Dot Light-Emitting Diodes via Efficient Edge Amination. *Nat. Photonics* **2020,** *14*, 171–176. https://doi.org/10.1038/s41566-019-0557-5

9. Samantha, W. Potential Applicability of CNT and CNT/Composites to Implement ASEC Concept: A Review Article. *Solar Energy* **2009,** *83* (8), 1379–1389. https://doi.org/10.1016/j.solener.2009.03.001
10. Liu, Z.; Tabakman, S. M.; Chen, Z.; Dai, H. Preparation of Carbon Nanotube Bioconjugates for Biomedical Applications. *Nat. Protoc.* **2009,** *4* (9), 1372–1381. https://doi.org/10.1038/nprot.2009.146
11. Uday, S. et al. Optically Transparent and Lightweight Nanocomposite Substrate of Poly (Methyl Methacrylate-Co-Acrylonitrile)/MWCNT for Optoelectronic Applications: An Experimental and Theoretical Insight. *J. Mater. Sci.* **2021,** *56* (30), 17040–17061. https://doi.org/10.1007/s10853-021-06390-3
12. Zhu, S.; Meng, Q.; Wang, L.; Zhang, J.; Song, Y.; Jin, H.; Zhang, K.; Sun, H.; Wang, H.; Yang, B. Highly Photoluminescent Carbon Dots for Multicolor Patterning, Sensors, and Bioimaging. *Angew. Chem., Int. Ed.* **2013,** *52*, 3953−3957. https://doi.org/10.1002/ange.201300519
13. Ding, H.; Yu, S. B.; Wei, J. S.; Xiong, H. M. Full-Color Light Emitting Carbon Dots with a Surface-State-Controlled Luminescence Mechanism. *ACS Nano* **2015,** *10*, 484−491. https://doi.org/10.1021/acsnano.5b05406
14. Liu, J.; Li, D.; Zhang, K.; Yang, M.; Sun, H.; Yang, B. One-Step Hydrothermal Synthesis of Nitrogen-Doped Conjugated Carbonized Polymer Dots with 31% Efficient Red Emission for In Vivo Imaging. *Small* **2018,** *14*, 1703919. https://doi.org/10.1002/smll.201703919
15. Li, D.; Jing, P.; Sun, L.; An, Y.; Shan, X.; Lu, X.; Zhou, D.; Han, D.; Shen, D.; Zhai, Y.; Qu, S.; Zboril, R.; Rogach, A. L. Near-Infrared Excitation/Emission and Multiphoton-Induced Fluorescence of Carbon Dots. *Adv. Mater.* **2018,** *30*, 1705913. https://doi.org/10.1002/adma.201705913
16. Zhang, M.; Wang, H.; Wang, B.; Ma, Y.; Huang, H.; Liu, Y.; Shao, M.; Yao, B.; Kang, Z. Maltase Decorated by Chiral Carbon Dots with Inhibited Enzyme Activity for Glucose Level Control. *Small* **2019,** *15*, 1901512. https://doi.org/10.1002/smll.201901512
17. Tao, S.; Lu, S.; Geng, Y.; Zhu, S.; Redfern, S. A. T.; Song, Y.; Feng, T.; Xu, W.; Yang, B. Design of Metal-Free Polymer Carbon Dots: A New Class of Room-Temperature Phosphorescent Materials. *Angew. Chem., Int. Ed.* **2018,** *57*, 2393−2398. https://doi.org/10.1002/anie.201712662
18. Park, M.; Kim, H. S.; Yoon, H.; Kim, J.; Lee, S.; Yoo, S, Jeon, S. Controllable Singlet−Triplet Energy Splitting of Graphene Quantum Dots Through Oxidation: From Phosphorescence to TADF. *Adv. Mater.* **2020,** *32*, 200093. https://doi.org/10.1002/adma.202000936
19. Zhu, S.; Song, Y, Shao, J.; Zhao, X.; Yang, B. Non-Conjugated Polymer Dots with Crosslink-Enhanced Emission in the Absence of Fluorophore Units. *Angew. Chem., Int. Ed.* **2015,** *54*, 14626−14637. https://doi.org/10.1002/anie.201504951
20. Jiang, F.; Chen, D.; Li, R.; Wang, Y.; Zhang, G.; Li, S.; Zheng, J.; Huang, N.; Gu, Y.; Wang, C.; Shu, C. Eco-Friendly Synthesis of Size-Controllable Amine-Functionalized Graphene Quantum Dots with Antimycoplasma Properties. *Nanoscale* **2013,** *5*, 1137−1142. https://doi.org/10.1039/C2NR33191H
21. Qu, S.; Zhou, D.; Li, D.; Ji, W.; Jing, P.; Han, D.; Liu, L.; Zeng, H.; Shen, D.; Toward Efficient Orange Emissive Carbon Nanodots through Conjugated Sp2-Domain Controlling and Surface Charges Engineering. *Adv. Mater.* **2016,** *28*, 3516−3521. https://doi.org/10.1002/adma.201504891

22. Yuan, F.; Yuan, T.; Sui, L.; Wang, Z.; Xi, Z.; Li, Y.; Li, X.; Fan, L.; Tan, Z.; Chen, A.; Jin, M, Yang, S. Engineering Triangular Carbon Quantum Dots with Unprecedented Narrow Bandwidth Emission for Multicolored LEDs. *Nat. Commun.* **2018,** *9*, 2249. https://doi.org/10.1038/s41467-018-04635-5
23. Zhu, S.; Zhang, J.; Qiao, C.; Tang, S.; Li, Y.; Yuan, W.; Li, B.; Tian, L.; Liu, F.; Hu, R.; Gao, H.; Wei, H.; Zhang, H.; Sun, H.; Yang, B. Strongly Green-Photoluminescent Graphene Quantum Dots for Bioimaging Applications. *Chem. Commun.* **2011,** *47*, 6858−6860. https://doi.org/10.1039/C1CC11122A
24. Tetsuka, H.; Asahi, R.; Nagoya, A.; Okamoto, K.; Tajima, I.; Ohta, R.; Okamoto, A.; Optically Tunable Amino-Functionalized Graphene Quantum Dots. *Adv. Mater.* **2012,** *24*, 5333−5338. https://doi.org/10.1002/adma.201201930
25. Zhang, Y.; Liu, X.; Fan, Y.; Guo, X.; Zhou, L.; Lv, Y.; Lin, J. One-Step Microwave Synthesis of N-Doped Hydroxyl-Functionalized Carbon Dots with Ultra-High Fluorescence Quantum Yields. *Nanoscale,* **2016,** *8*, 15281−15287. https://doi.org/10.1039/C6NR03125K
26. Ye, X.; Xiang, Y.; Wang, Q.; Li, Z.; Liu, Z. A Red Emissive Two-Photon Fluorescence Probe Based on Carbon Dots for Intracellular PH Detection. *Small* **2019,** *15*, 1901673. https://doi.org/10.1002/smll.201901673
27. Sun, Y. P.; Zhou, B.; Lin, Y.; Wang, W.; Fernando, K. A. S.; Pathak, P.; Meziani, M. J.; Harruff, B. A, Wang, X.; Wang, H.; Luo, P. G.; Yang, H.; Kose, M. E.; Chen, B.; Veca, L. M.; Xie, S. Y. Quantum Sized Carbon Dots for Bright and Colorful Photoluminescence. *J. Am. Chem. Soc.* **2006,** *128*, 7756−7757. https://doi.org/10.1021/ja062677d
28. Yang, J.; Ren, Z; Xie, Z.; Liu, Y.; Wang, C.; Xie, Y.; Peng, Q.; Xu, B.;. Tian, W.; Zhang, F. AIEgen with Fluorescence–Phosphorescence Dual Mechanoluminescence at Room Temperature. *Angew. Chem., Int. Ed.* **2017,** *56*, 880–884. https://doi.org/10.1002/anie.201610453
29. Zhao, Q.; Huang, C.; Li, F. Phorescent Heavy-Metal Complexes for Bioimaging. *Chem. Soc. Rev.* **2011,** *40*, 2508–2524. https://doi.org/10.1039/C0CS00114G
30. Bolton, O.; Lee, K.; Kim, H. J.; Lin, K. Y.; Kim, J. *Nat. Chem.* **2011,** *3*, 205. https://doi.org/10.1038/nchem.984
31. Schmidt, R.; Krasselt, C.; Go¨hler, C.; von Borczyskowski, C. *ACS Nano* **2014,** *8*, 3506–3521. https://doi.org/10.1021/nn406562a
32. Jiang, K.; Wang, Y.; Li, Z.; Lin, H. Afterglow of Carbon Dots: Mechanism, Strategy and Applications. *Mater. Chem. Front.* **2020,** *4*, 386−399. https://doi.org/10.1039/C9QM00578A
33. Jiang, K.; Zhang, L.; Lu, J.; Xu, C.; Cai, C.; Lin, H. Triple-Mode Emission of Carbon Dots: Applications for Advanced Anti-Counterfeiting. *Angew. Chem., Int. Ed.* **2016,** *55*, 7231−7235. https://doi.org/10.1002/ange.201602445
34. Jiang, K.; Hu, S.; Wang, Y.; Li, Z.; Lin, H. Photo-Stimulated Polychromatic Room Temperature Phosphorescence of Carbon Dots. *Small* **2020,** *16*, 2001909. https://doi.org/10.1002/smll.202001909
35. Shen, C. L.; Lou, Q.; Zang, J. H.; Liu, K. K.; Qu, S. N.; Dong, L.; Shan, C. X. Near-Infrared Chemiluminescent Carbon Nanodots and Their Application in Reactive Oxygen Species Bioimaging. *Adv. Sci.* **2020,** *7*, 1903525. https://doi.org/10.1002/advs.201903525

36. Zhuo, S.; Shao, M.; Lee, S. T. Upconversion and Downconversion Fluorescent Graphene Quantum Dots: Ultrasonic Preparation and Photocatalysis. *ACS Nano* **2012,** *6*, 1059–1064. https://doi.org/10.1021/nn2040395
37. Wang, F.; Pang, S.; Wang, L.; Li, Q.; Kreiter, M.; Liu, C. Y. *Chem. Mater.* **2010,** *22*, 4528–4530. https://doi.org/10.1021/cm101350u
38. Han, M.; Zhu, S.; Lu, S.; Song, Y.; Feng, T.; Tao, S.; Liu, J.; and B Yang, B. *Nano Today* **2018,** *19*, 201–218. https://doi.org/10.1016/j.nantod.2018.02.008
39. Hu, S.; Trinchi, A.; Atkin, P.; Cole, I. *Angew Chem., Int. Ed.* **2015,** *54*, 2970–2974. https://doi.org/10.1002/anie.201411004
40. Bao, L.; Liu, C.; Zhang, Z. L.; Pang, D. W. *Adv. Mater.* **2015,** *27*, 1663–1667. https://doi.org/10.1002/adma.201405070
41. Han, L.; Liu, S. G.; Dong, J. X.; Liang, J. Y.; Li, L. J.; Li, N. B.; Luo, H. Q. *J Mater. Chem. C* **2017,** *5*, 10785–10793. https://doi.org/10.1039/C7TC03314A
42. Zhu, J.; Bai, X.; Bai, J.; Pan, G.; Zhu, Y.; Zhai, Y.; Shao, H.; Chen, X.; Dong, B.; Zhang, H. *Nanotechnology* **2017,** *29*, 085705. DOI :10.1088/1361-6528/aaa321
43. Wang, L.; Wang, Y.; Xu, T.; Liao, H.; Yao, C.; Liu, Y.; Li, Z.; Chen, Z.;. Pan, D.; Sun, L. *Nat. Commun.* **2014,** *5*, 5357. https://doi.org/10.1038/ncomms6357
44. Liu, M. L.; Chen, B. B.; Li, C. M.; Huang, C. Z. Carbon Dots: Synthesis, Formation Mechanism, Fluorescence Origin and Sensing Applications. *Green Chem.* **2019,** *21*, 449−471. https://doi.org/10.1039/C8GC02736F
45. Yang, M.; Tang, Q.; Meng, Y.; Liu, J.; Feng, T.; Zhao, X.; Zhu, S.; Yu, W.; Yang, B. Reversible “off-On” Fluorescence of Zn2+- Passivated Carbon Dots: Mechanism and Potential for the Detection of EDTA and Zn2+. *Langmuir* **2018,** *34*, 7767−7775. https://doi.org/10.1021/acs.langmuir.8b00947
46. Gao, W.; Song, H.; Wang, X.; Liu, X.; Pang, X.; Zhou, Y.; Gao, B.; Peng, X. Carbon Dots with Red Emission for Sensing of Pt2+, Au3+, and Pd2+ and Their Bioapplications in Vitro and in Vivo. *ACS Appl. Mater. Interfaces* **2018,** *10*, 1147−1154. https://doi.org/10.1021/acsami.7b16991
47. Wei, Z.; Li, H.; Liu, S.; Wang, W.; Chen, H.; Xiao, L.; Ren, C.; Chen, X. Carbon Dots as Fluorescent/Colorimetric Probes for Real-Time Detection of Hypochlorite and Ascorbic Acid in Cells and Body Fluid. *Anal. Chem.* **2019,** *91*, 15477−15483. https://doi.org/10.1021/acs.analchem.9b03272
48. Yang, P.; Zhu, Z.; Zhang, T.; Zhang, W.; Chen, W.; Cao, Y.; Chen, M.; Zhou, X. Orange-Emissive Carbon Quantum Dots: Toward Application in Wound PH Monitoring Based on Colorimetric and Fluorescent Changing. *Small* **2019,** *15*, 1902823. https://doi.org/10.1002/smll.201902823
49. Yang, D.; Guo, Z.; Wang, J.; Jin, Y.; Mei, Q.; Miao, P. Carbon Nanodot−Based Fluorescent Method for Virus DNA Analysis with Isothermal Strand Displacement Amplification. Part. Syst. Charact. **2019,** *36*, 1900273. https://doi.org/10.1002/ppsc.201900273
50. Cheng, W.; Xu, J.; Guo, Z.; Yang, D.; Chen, X.; Yan, W.; Miao, P. Hydrothermal Synthesis of, N.;S Co-Doped Carbon Nanodots for Highly Selective Detection of Living Cancer Cells. *J. Mater. Chem. B* **2018,** *6*, 5775−5780. https://doi.org/10.1039/C8TB01271G
51. Zhao, X.; Tang, Q.; Zhu, S.; Bu, W.; Yang, M.; Liu, X.; Meng, Y.; Yu, W.; Sun, H.; Yang, B. Controllable Acidophilic Dual-Emission Fluorescent Carbonized Polymer Dots for Selective Imaging of Bacteria. *Nanoscale* **2019,** *11*, 9526−9532. https://doi.org/10.1039/C9NR01118H

52. Cui, Y.; Hegde, R. S.; Phang, I. Y.; Lee, H. K.; Ling, X. Y. Encoding Molecular Information in Plasmonic Nanostructures for Anti-Counterfeiting Applications. *Nanoscale* **2014,** *6*, 282–288. https://doi.org/10.1039/C3NR04375D
53. Wang, B.; Yu, Y.; Zhang, H.; Xuan, Y.; Chen, G.; Ma, W.; Li, J.; Yu, J.; Carbon Dots in a Matrix: Energy-Transfer-Enhanced Room Temperature Red Phosphorescence. *Angew. Chem., Int. Ed.* **2019,** *58*, 18443–18448. https://doi.org/10.1002/ange.201911035
54. Li, W.; Zhou, W.; Zhou, Z.; Zhang, H.; Zhang, X.; Zhuang, J.; Liu, Y.; Lei, B.; Hu, C.; A Universal Strategy for Activating the Multicolor Room-Temperature Afterglow of Carbon Dots in a Boric Acid Matrix. *Angew. Chem., Int. Ed.* **2019,** *58*, 7278–7283. https://doi.org/10.1002/ange.201814629
55. Prasad, P. N.; Williams, D. J. *Introduction to Nonlinear Optical Effects in Molecules and Polymers*; Wiley-Interscience, 1991.
56. Prasad, P. N. *Nanophotonics*; Wiley Blackwell, 2004.
57. Sheik Bahae, M.; Said, A. A.; Wei, T. H.; Hagan, D. J.; Vanstryland, E. W. *IEEE J. Quantum Electron*. **1990,** *26*, 760–769. https://doi.org/10.1109/3.53394
58. Chen, Y.; Lin, Y.; Liu, Y.; Doyle, J.; He, N.; Zhuang, X. D.; Bai, J. R.; Blau, W. J. *J Nanosci. Nanotechnol.* **2007,** *7*, 1268–1283. https://doi.org/10.1166/jnn.2007.308
59. Blau, W. J.; Byrne, H. J.; Cardin, D. J.; Dennis, T. J.; Hare, J. P.; Kroto, H. W.; Taylor, R.; Walton, D. R. M. *Phys. Rev. Lett.* **1991,** *67*, 1423–1425. https://doi.org/10.1103/PhysRevLett.67.1423
60. Perry, J. W.; Mansour, K.; Lee, I. Y. S.; Wu, X. L.; Bedworth, P. V.; Chen, C. T.; Ng, D.; Marder, S. R.; Miles, P.; Wada, T.; Tian, M.; Sasabe, H. *Science* **1996,** *273*, 1533–1536. https://doi.org/10.1126/science.273.5281.1533
61. O'Flaherty, S. M.; Hold, S. V.; Cook, M. J.; Torres, T.; Chen, Y.; Hanack, M.; Blau, W. J. *Adv. Mater.* **2003,** *15*, 19–32. https://doi.org/10.1002/adma.200390002
62. Sun, Y. P.; Riggs, J. E. *Int. Rev. Phys. Chem.* **1999,** *18*, 43–90. https://doi.org/10.1080/014423599230008
63. Sun, Y. P.; Riggs, J. E.; Henbest, K. B.; Martin, R. B. *J. Nonlinear Opt. Phys. Mater.* **2000,** *9*, 481–503. https://doi.org/10.1142/S0218863500000315
64. Chen, P.; Wu, X.; Sun, X.; Lin, J.; Ji, W.; Tan, K. L. *Phys. Rev. Lett.* **1999,** *82*, 2548–2551. https://doi.org/10.1103/PhysRevLett.82.2548
65. Wang, J.; Blau, W. J. *J. Opt. A: Pure Appl. Opt.* **2009,** *11*, 024001. DOI: 10.1088/1464-4258/11/2/024001
66. Vivien, L.; Riehl, D.; Delouis, J. F.; Delaire, J. A.; Hache, F.; Anglaret, E. *J. Opt. Soc. Am. B* **2002,** *19*, 208–214. https://doi.org/10.1364/JOSAB.19.000208
67. O'Flaherty, S. M.; Hold, S. V.; Brennan, M. E.; Cadek, M.; Drury, A.; Coleman, J. N.; Blau, W. J. *J. Opt. Soc. Am. B* **2003,** *20*, 49–58. https://doi.org/10.1364/JOSAB.20.000049
68. Belousova, I. M.; Mironova, N. G.; Yur'ev, M. S. *Opt. Spectroscopy* **2003,** *94*, 86–91. https://doi.org/10.1134/1.1540206
69. Vivien, L.; Anglaret, E.; Riehl, D.; Hache, F.; Bacou, F.; Andrieux, M.; Lafonta, F.; Journet, C.; Goze, C.; Brunet, M.; Bernier, P. *Opt. Commun.* **2000,** *174*, 271–275. https://doi.org/10.1016/S0030-4018(99)00656-2
70. Mishra, S. R.; Rawat, H. S.; Mehendale, S. C.; Rustagi, K. C.; Sood, A. K.; Bandyopadhyay, R.; Govindaraj, A.; Rao, C. N. R. *Chem. Phys. Lett.* **2000,** *317*, 510–514. https://doi.org/10.1016/S0009-2614(99)01304-4
71. Wang, J.; Blau, W. J. *J. Phys. Chem. C* **2008,** *112*, 2298–2303. https://doi.org/10.1021/jp709926r

72. Jin, Z. X.; Huang, L.; Goh, S. H.; Xu, G. Q.; Ji, W. *Chem. Phys. Lett.* **2002,** *352*, 328–333. https://doi.org/10.1016/S0009-2614(01)01468-3
73. Vivien, L.; Riehl, D.; Lancon, P.; Hache, F.; Anglaret, E. *Opt. Lett.* **2001,** *26*, 223–225. https://doi.org/10.1364/OL.26.000223
74. Set, S. Y.; Yaguchi, H.; Tanaka, Y.; Jablonski, M. *IEEE J. Sel. Top. Quantum Electron.* **2004,** *10*, 137–146. https://doi.org/10.1109/JSTQE.2003.822912
75. Set, S. Y.; Yaguchi, H.; Jablonski, M.; Tanaka, Y.; Sakakibara, Y.; Rozhin, A.; Tokumoto, M.; Kataura, H.; Achiba, Y.; Kikuchi, K. in: *Proc. Optical Fiber Communication Conf.'03*, Atlanta, GA, p. FL2. https://doi.org/10.1109/OFC.2003.316026
76. Set Sze, Y. et al. "Laser Mode Locking Using a Saturable Absorber Incorporating Carbon Nanotubes." *J. Light Wave Technol.* **2004,** *22* (1), 51. https://opg.optica.org/jlt/abstract.cfm?URI=jlt-22-1-51
77. Novoselov, K. S.; Geim, A. K.; Morozov, S. V.; Jiang, D.; Zhang, Y.; Dubonos, S. V.; Grigorieva I V and Firsov, A. A. *Science* **2004,** *306*, 666. DOI: 10.1126/science.1102896
78. Geim, A. K.; Novoselov, K. S. *Nat. Mater.* **2007,** *6*, 183. DOI: 10.1038/nmat1849
79. Fasolino, A.; Los, J. H.; Katsnelson, M. I. *Nat. Mater.* **2007,** *6*, 858. https://doi.org/10.1038/nmat2011
80. Raza, H. *Graphene Nanoelectronics*; Springer: Berlin, 2012; pp 15–586. https://doi.org/10.1007/978-3-642-22984-8
81. Wallace, P. R. *Phys. Rev.* **1947,** *71*, 622. https://doi.org/10.1103/PhysRev.71.622
82. Kuilla, T.; Bhadra, S.; Yaoa, D.; Kim, N. H.; Bose, S.; Lee, J. H. *Prog. Polym. Sci.* **2010,** *35*, 1350. 10.1016/j.progpolymsci.2010.07.005
83. Novoselov, K. S.; Jiang, D.; Schedin, F.; Booth, T. J.; Khotkevich, V. V.; Morozov, S. V.; Geim, A. K. *PNAS* **2005,** *102*, 10451. 10.1073/pnas.0502848102
84. Meyer, J. C.; Geim, A. K.; Katsnelson, M. I.; Novoselov, K. S.; Booth, T. J.; Roth, S. *Nature* **2007,** *446*, 60. DOI: 10.1038/nature05545
85. Berger, C.; Song, Z.; Li, T.; Li, X.; Ogbazghi, A. Y.; Feng, R.; Dai, Z.; Marchenkov, A. N.; Conrad, E. H.; First, P. N.; de Heer, W. A. *J. Phys. Chem. B* **2004,** *108*, 19912. DOI: 10.1021/jp040650f
86. Obraztsov, A. N.; Obraztsova, E. A.; Tyurnina, A. V.; Zolotukhin, A. A. *Carbon* **2007,** *45*, 2017. https://doi.org/10.1016/j.carbon.2007.05.028
87. Yu, Q.; Lian, J.; Siriponglert, S.; Li, H.; Chen, Y. P.; Pei, S. S. *Appl. Phys. Lett.* **2008,** *93*, 113103. https://doi.org/10.1063/1.2982585
88. Wassei, J. K.; Kaner, R. B. *Mater. Today* **2010,** *13*, 52–59. https://doi.org/10.1016/S1369-7021(10)70034-1
89. Bullen, P.; Isele, R. *German Flat Panel Display Forum (DFF) European Technology: Flat Panel Displays*, 6th edn.; VDMA: Germany: Germany, 2008.
90. Bonaccorso, F.; Sun, Z.; Hasan T and Ferrari, A. C. *Nat. Photon.* **2010,** *4*, 611. https://doi.org/10.1038/nphoton.2010.186
91. Bae, S. et al. *Nat. Nanotechnol.* **2010,** *5*, 574. https://doi.org/10.1038/nnano.2010.132
92. Hecht, D. S.; Hu, L.; Irvin, G. *Adv. Mater.* **2011,** *23*, 1482. https://doi.org/10.1002/adma.201003188
93. Nordendorf, G.; Kasdorf, O.; Kitzerow, H. S.; Liang, Y.; Feng, X.; Muellen, K. *Japan. J. Appl.Phys.* **2010,** *49*, 100206. DOI: 10.1143/JJAP.49.100206
94. Scott, J. C.; Kaufman, J. H.; Brock, P. J.; DiPietro, R.; Salem, J.; Goitia, J. A. *J. Appl. Phys.* **1996,** *79*, 2745. https://doi.org/10.1063/1.361096

95. Bremer, M.; Naemura, S.; Tarumi, K. *Japan. J. Appl. Phys.* **1998,** *37*, L88. DOI: 10.1143/JJAP.37.L88
96. Blake, P. et al. *Nano Lett.* **2008,** *8*, 1704. https://doi.org/10.1021/nl080649i
97. Jo, G. et al. *Nanotechnology* **2010,** *21*, 175201. DOI: 10.1088/0957-4484/21/17/175201
98. Wu, J.; Agrawal, M.; Becerril H c, A.; Bao, Z.; Liu, Z.; Chen, Y.; Peumans, P. *ACS Nano* **2010,** *4*, 43. https://doi.org/10.1021/nn900728d
99. Wang, X.; Zhi, L.; Mullen, K. *Nano Lett.* **2008,** ,*8*, 323. https://doi.org/10.1021/nl072838r
100. Arco, L. G. D.; Zhang, Y.; Schlenker, C. W.; Ryu, K.; Thompson, M. E.; Zhou, C. *ACS Nano* **2010,** *4*, 2865. https://doi.org/10.1021/nn901587x
101. Guo, C. X.; Guai, G. H.; Li, C. M. *Adv. Energy Mater.* **2011,** *1*, 448. https://doi.org/10.1002/aenm.201100119
102. Choe, M. et al. *Org. Electron.* **2010,** *11*, 1864. https://doi.org/10.1016/j.orgel.2010.08.018
103. Wan, X.; Long, G.; Huang, L.; Chen, Y. *Adv. Mater.* **2011,** *23*, 5342. https://doi.org/10.1002/adma.201102735
104. He, Z.; Zhong, C.; Huang, X.; Wong, W.; Wu, H.; Chen, L.; Su, S.; Cao, Y. *Adv. Mater.* **2011,** *23*, 4636. https://doi.org/10.1002/adma.201103006
105. Hashimoto, A.; Suenaga, K.; Gloter, A.; Urita, K.; Iijima, S. *Nature* **2004,** *430*, 870. https://doi.org/10.1038/nature02817
106. Peres, N. M. R.; Guinea, F.; Neto, A. H. C. *Phys. Rev.*, B. **2006,** *73*, 125411. https://doi.org/10.1103/PhysRevB.73.125411
107. Peres, N. M. R. *J. Phy. Condens. Matter*. **2009,** *21*. DOI: 10.1088/0953-8984/21/32/323201
108. Wang, F.; Zhang, Y. B.; Tian, C. S.; Girit, C.; Zettl, A.; Crommie, M.; Shen, Y. R. *Science* **2008,** *320*, 206. DOI: 10.1126/science.1152793
109. George, P. A.; Strait, J.; Dawlaty, J.; Shivaraman, S.; Chandrashekhar, M.; Rana, F.; Spencer, M. G. *Nano Lett.* **2008,** *8*, 4248. https://doi.org/10.1021/nl8019399
110. Schedin, F.; Geim, A. K.; Morozov, S. V.; Hill, E. W.; Blake, P.; Katsnelson, M. I.; Novoselov, K. S. *Nat. Mater.*, **2007,** *6*, 652. https://doi.org/10.1038/nmat1967
111. Huang, B.; Li, Z. Y.; Liu, Z. R.; Zhou, G.; Hao, S. G.; Wu, J.; Gu, B. L.; Duan, W. H. *J. Phys. Chem. C* **2008,** *112*, 13442. https://doi.org/10.1021/jp8021024
112. Dan, Y. P.; Lu, Y.; Kybert, N. J.; Luo, Z. T.; Johnson, A. T. C. *Nano Lett.* **2009,** *9*, 1472. https://doi.org/10.1021/nl8033637
113. Ohno, Y.; Maehashi, K.; Yamashiro, Y.; Matsumoto, K. *Nano Lett.* **2009,** *9*, 3318. https://doi.org/10.1021/nl901596m
114. Lee Naerts, O.; Partoens, B.; Peeters, F. M. *Phys. Rev. B* **2008,** *77*, 125416. https://doi.org/10.1103/PhysRevB.77.125416
115. Robinson, J. T.; Perkins, F. K.; Snow, E. S.; Wei, Z. Q.; Sheehan, P. E. *Nano Lett.* **2008,** *8*, 3137. https://doi.org/10.1021/nl8013007
116. Mohanty, N.; Berry, V. *Nano Lett.* **2008,** *8*, 4469. https://doi.org/10.1021/nl802412n
117. Watcharotone, S.; Dikin, D. A.; Stankovich, S.; Piner, R.; Jung, I.; Dommett, G. H. B.; Evmenenko, G.; Wu, S. E.; Chen, S. F.; Liu, C. P.; Nguyen, S. T.; Ruoff, R. S. *Nano Lett.* **2007,** *7*, 1888. https://doi.org/10.1021/nl070477+
118. Eda, G.; Fanchini, G.; Chhowalla, M. *Nat. Nanotechnol.* **2008,** *3*, 270. https://doi.org/10.1038/nnano.2008.83
119. Cote, L. J.; Kim, F.; Huang, J. X. *J. Am. Chem. Soc.* **2009,** *131*, 1043. https://doi.org/10.1021/ja806262m

120. Park, S.; Ruoff, R. S. *Nat. Nanotechnol.* **2009,** *4*, 217. https://doi.org/10.1038/nnano.2009.58
121. Li, D.; Muller, M. B.; Gilje, S.; Kaner, R. B.; Wallace, G. G. *Nat. Nanotechnol.* **2008,** *3*, 101. https://doi.org/10.1038/nnano.2007.451
122. Winter, M.; Besenhard, J. O.; Spahr, M. E.; Novak, P. *Adv. Mater.* **1998,** *10*, 725. https://doi.org/10.1002/(SICI)1521-4095(199807)10:10%3C725::AID-ADMA725%3E3.0.CO;2-Z
123. Wang, C. Y.; Li, D.; Too, C. O.; Wallace, G. G. *Chem. Mat.* **2009,** *21*, 2604. https://doi.org/10.1021/cm900764n
124. Liang, M. H.; Zhi, L. J. *J. Mater. Chem.* **2009,** *19*, 5871. https://doi.org/10.1039/B901551E
125. Conway, B. E. *Electrochemical Supercapacitors: Scientific Fundamentals and Technological Applications*; Plenum Publishers: New York, 1999. https://doi.org/10.1007/978%2D1%2D4757%2D3058%2D6
126. Simon, P.; Gogotsi, Y. *Nat. Mater*. **2008,** *7*, 845. https://doi.org/10.1038/nmat2297
127. Jang, B. Z.; Zhamu, A. *J. Mater. Sci.* **2008,** *43*, 5092. https://doi.org/10.1007/s10853-008-2755-2
128. Ramanathan, T.; Abdala, A. A.; Stankovich, S.; Dikin, D. A.; Herrera Alonso, M.; Piner, R. D.; Adamson, D. H.; Schniepp, H. C.; Chen, X.; Ruoff, R. S.; Nguyen, S. T.; Aksay, I. A.; Prud'homme, R. K.; Brinson, L. C. *Nat. Nanotechnol.* **2008,** *3*, 327. https://doi.org/10.1038/nnano.2008.96
129. Liang, J. J.; Huang, Y.; Zhang, L.; Wang, Y.; Ma, Y. F.; Guo, T. Y.; Chen, Y. S. *Adv. Funct. Mater.* **2009,** *19*, 2297. https://doi.org/10.1002/adfm.200801776
130. Kalaitzidou, K.; Fukushima, H.; Drzal, L. T. *Compos. Sci. Technol.* **2007,** *67*, 2045. https://doi.org/10.1016/j.compscitech.2006.11.014
131. Yu, A. P.; Ramesh, P.; Itkis, M. E.; Bekyarova, E.; Haddon, R. C. *J. Phys. Chem. C* **2007,** *111*, 7565. https://doi.org/10.1021/jp071761s
132. Kim, H.; Macosko, C. W. *Macromolecules* **2008,** *41*, 3317. https://doi.org/10.1021/ma702385h
133. Verdejo, R.; Barroso Bujans, F.; Rodriguez Perez, M. A.; de Saja, J. A.; Lopez Manchado, M. A. *J. Mater. Chem.* **2008,** *18*, 2221. https://doi.org/10.1039/B718289A
134. Cai, D.; Yusoh, K.; Song, M. *Nanotechnology* **2009,** *20*, 5. DOI: 10.1088/0957-4484/20/8/085712
135. Ansari, S.; Giannelis, E. P. *J. Pol. Sci., Part B Polym. Phys.* **2009,** *47*, 888. https://doi.org/10.1002/polb.21695
136. Kim, H.; Macosko, C. W. *Polymer* **2009,** *50*, 3797. https://doi.org/10.1016/j.polymer.2009.05.038
137. Thostenson, E. T.; Li, C. Y.; Chou, T. W. *Compos. Sci. Technol.* **2005,** *65*, 491. https://doi.org/10.1016/j.compscitech.2004.11.003
138. McAllister, M. J.; Li, J. L.; Adamson, D. H.; Schniepp, H. C.; Abdala, A. A.; Liu, J.; Herrera Alonso, M.; Milius, D. L.; Car, R.; Prud'homme, R. K.; Aksay, I. A. *Chem. Mater.* **2007,** *19*, 4396. https://doi.org/10.1021/cm0630800
139. Dreyer, D. R.; Park, S.; Bielawski, C. W.; Ruoff, R. S. *Chem. Soc. Rev.* **2010,** *39*, 228. https://doi.org/10.1002/marc.200900641
140. Wu, J. H.; Tang, Q. W.; Sun, H.; Lin, J. M.; Ao, H. Y.; Huang, M. L.; Huang, Y. F. *Langmuir* **2008,** *24*, 4800. https://doi.org/10.1021/la800095z

141. Wakabayashi, K.; Pierre, C.; Dikin, D. A.; Ruoff, R. S.; Ramanathan, T.; Brinson, L. C.; Torkelson, J. M. *Macromolecules* **2008,** *41*, 1905. DOI: 10.1021/ma071687b
142. Du, X. S.; Yu, Z. Z.; Dasari, A.; Ma, J.; Mo, M. S.; Meng, Y. Z.; Mai, Y. W. *Chem. Mater.* **2008,** *20*, 2066. https://doi.org/10.1021/cm703285s
143. Vickery, J. L.; Patil, A. J.; Mann, S. *Adv. Mater.* **2009,** *21*, 2180. https://doi.org/10.1002/adma.200803606
144. Li, N.; Wang, Z.; Shi, Z. *Synthesis of Graphene with Arc Discharge Method* (Ed. S. Mikhailov); Intech: Rijeka, 2011; pp 23–36. http://dx.doi.org/10.5772/14961
145. Huang, X.; Yin, Z.; Wu, S.; Qi, X.; He, Q.; Zhang, Q.; Yan, Q.; Boey; Zhang, H. *Small* **2011,** *7*, 1876. https://doi.org/10.1002/smll.201002009
146. Cote, L. J.; Kim, J.; Tung, V. C.; Luo, J.; Kim, F.; Huang, J. *Pure Appl. Chem.* **2011,** *83*, 95. https://doi.org/10.1351/PAC-CON-10-10-25
147. Peng, Q.; Dearden, A. K.; Crean, J.; Han, L.; Liu, S.; Wen, X.; De, S. *Nanotechnol. Sci. Appl.* 2014. https://doi.org/10.2147/NSA.S40324
148. Zhang, Y. B.; Tan, Y. W.; Stormer, H. L.; Kim, P. *Nature* **2005,** *438*, 201. https://doi.org/10.1038/nature04235
149. Berger, C.; Song, Z. M.; Li, X. B.; Wu, X. S.; Brown, N.; Naud, C.; Mayou, D.; Li, T. B.; Hass, J.; Marchenkov, A. N.; Conrad, E. H.; First, P. N.; de Heer, W. A. *Science* **2006,** *312*, 1191. https://doi.org/10.1126/science.1125925
150. Emtsev, K. V.; Bostwick, A.; Horn, K.; Jobst, J.; Kellogg, G. L.; Ley, L.; McChesney, J.; Ohta, T.; Reshanov, S. A.; Röhrl, J.; Rotenberg, E.; Schmid, A. K.; Waldmann, D.; Weber, H. B.; Seyller, T. *Nat. Mater.* **2009,** *8*, 203. https://doi.org/10.1038/nmat2382
151. Reina, A.; Jia, X. T.; Ho, J.; Nezich, D.; Son, H. B.; Bulovic, V.; Dresselhaus, M. S.; Kong, J. *Nano Lett.* **2009,** *9*, 30. https://doi.org/10.1021/nl801827v
152. Li, X. S.; Cai, W. W.; An, J. H.; Kim, S.; Nah, J.; Yang, D. X.; Piner, R.; Velamakanni, A.; Jung, I.; Tutuc, E.; Banerjee, S. K.; Colombo, L.; Ruoff, R.S. *Science* **2009,** *324*, 1312. https://doi.org/10.1126/science.1171245
153. Shan, C. S.; Yang, H. F.; Song, J. F.; Han, D. X.; Ivaska, A.; Niu, L. *Anal. Chem.* **2009,** *81*, 2378. https://doi.org/10.1021/ac802193c
154. Sun, X. M.; Liu, Z.; Welsher, K.; Robinson, J. T.; Goodwin, A.; Zaric, S.; Dai, H. J. *Nano Res.* **2008,** *1*, 203. https://doi.org/10.1007/s12274-008-8021-8
155. Lui, C. H.; Mak, K. F.; Shan, J.; Heinz, J. T. *Phys. Rev. Lett.* **2010,** *105*, 127404. https://doi.org/10.1103/PhysRevLett.105.127404
156. Chang, H. X.; Tang, L. H.; Wang, Y.; Jiang, J H.; Li, J. H. *Anal. Chem.* **2010,** *82*, 2341. https://doi.org/10.1021/ac9025384
157. Fan, F. R. F.; Park, S.; Zhu, Y. W.; Ruoff, R. S.; Bard, A. J. *J Am. Chem. Soc.* **2009,** *131*, 937. https://doi.org/10.1021/ja8086246
158. Li, Y.; Hu, Y.; Zhao, Y.; Shi, G. Q.; Deng, L. E.; Hou, Y. B.; Qu, L. T. *Adv. Mater.* **2011,** *23*, 776, https://doi.org/10.1002/adma.201003819
159. Yan, X.; Cui, X.; Li, B. S.; Li, L. S. *Nano Lett.* **2010,** *10*, 1869. https://doi.org/10.1021/nl101060h
160. Li, L. S.; Yan, X. *J. Phys. Chem. Lett.* **2010,** *1*, 2572. https://doi.org/10.1021/jz100862f
161. Pan, D. Y.; Zhang, J. C.; Li, Z.; Wu, M. H. *Adv. Mater.* **2010,** *22*, 734. https://doi.org/10.1002/adma.200902825
162. Li, Y.; Zhao, Y.; Cheng, H. H.; Hu, Y.; Shi, G. Q.; Dai, L. M.; Qu, L. T. *J. Am. Chem. Soc.* **2012,** *134*, 15. https://doi.org/10.1021/ja206030c

163. Xia, F. N.; Mueller, T.; Lin, Y. M.; Valdes-Garcia, A.; Avouris, P. *Nat. Nanotechnol.* **2009,** *4*, 839. https://doi.org/10.1038/nnano.2009.292
164. Chang, H. X.; Lv, X. J.; Zhang, H.; Li, J. H. *Electrochem. Commun.* **2010,** *12*, 483. https://doi.org/10.1016/j.elecom.2010.01.025
165. Ghosh, S.; Sarker, B. K.; Chunder, A.; Zhai, L.; Khondaker, S. I. *Appl. Phys. Lett.* **2010,** *96*, 163109. https://doi.org/10.1063/1.3415499
166. Chang, H. X.; Sun, Z. H.; Yuan, Q. H.; Ding, F.; Tao, X. M.; Yan, F.; Zheng, Z. J. *Adv. Mater.* **2010,** *22*, 4872. https://doi.org/10.1002/adma.201002229
167. Chitara, B.; Panchakarla, L. S.; Krupanidhi, S. B.; Rao, C. N. R. *Adv. Mater.* **2011,** *23*, 5419. https://doi.org/10.1002/adma.201101414
168. Lv, X.; Huang, Y.; Liu, Z. B.; Tian, J. G.; Wang, Y.; Ma, Y. F.; Liang, J. J.; Fu, S. P.; Wan, X. J.; Chen, Y. S. *Small* **2009,** *5*, 1682. https://doi.org/10.1002/smll.200900044
169. Gabor, N. M.; Song, J. C. W.; Ma, Q.; Nair, N. L.; Taychatanapat, T.; Watanabe, K.; Taniguchi, T.; Levitov, L. S.; Jarillo-Herrero, P. *Science* **2011,** *334*, 648. https://doi.org/10.1126/science.1211384
170. Lee, E. J. H.; Balasubramanian, K.; Weitz, R. T.; Burghard, M.; Kern, K. *Nat. Nanotechnol.* **2008,** *3*, 486. https://doi.org/10.1038/nnano.2008.172
171. Sun, D.; Divin, C.; Rioux, J.; Sipe, J. E.; Berger, C.; de Heer, W. A.; First, P. N.; Norris, T. B. *Nano Lett.* **2010,** *10*, 1293. https://doi.org/10.1021/nl9040737
172. Xu, X. D.; Gabor, N. M.; Alden, J. S.; van der Zande, A. M.; McEuen, P. L. *Nano Lett.* **2010,** *10*, 562. https://doi.org/10.1021/nl903451y
173. Bao, Q. L.; Zhang, H.; Wang, Y.; Ni, Z. H.; Yan, Y. L.; Shen, Z. X.; Loh, K. P.; Tang, D. Y. *Adv. Funct. Mater.* **2009,** *19*, 3077. https://doi.org/10.1063/1.3244206
174. Sun, Z. P.; Hasan, T.; Torrisi, F.; Popa, D.; Privitera, G.; Wang, F. Q.; Bonaccorso, F.; Basko, D. M.; Ferrari, A. C. *ACS Nano* **2010,** *4*, 803. https://doi.org/10.1021/nn901703e

CHAPTER 7

An Overview of Nanocellulose-Based Nanocomposites for Water Purification

EAPEN PHILIP[1], P. H. VAISAKH[1], SNEHA SARA VARGHESE[1], TRESSIA ALIAS PRINCY PAULOSE[1], LINDA E. JACOB[1], and H. AKHINA[2,3]

[1]*Post Graduate and Research Department of Chemistry, Bishop Moore College, Mavelikara, Kerala, India*

[2]*International and Interuniversity Centre for Nanoscience and Nanotechnology, Mahatma Gandhi University, Kottayam, Kerala, India*

[3]*Govt.Technical High School, Krishnapuram, Kerala, India*

ABSTRACT

Every day, there is an increased need for filtered water. More effective technologies, such as membrane separation and adsorbent materials, have been shown to be better productive and less expensive than traditional industrial processes. A promising option for use in water filtration is nanocellulose (NC), such as nontoxic, biodegradable, and renewable nanofiller with good mechanical characteristics, a large surface area, and altered surface morphology. Recently, there has been a significant amount of attention on biopolymer-based nanocomposite membranes and films for purifying water. Comparatively to its untreated condition, the synthesized NC improves material compatibility and forms excellent network structures more quickly, both of which are necessary for the creation of pathways for improved pollutant adsorption and greater water flux. This chapter has discussed some current research that focuses on developing and exploiting nanofiltration membranes made from NC composites and their various water purification applications.

Resilience and Sustainability for Energy and Water: New Advancements, Opportunities, and Framework.
Cristóbal Noé Aguilar, A. K. Haghi, Amritanshu Shukla, Ann Rose Abraham, Swati Gokul Talele (Eds.)

7.1 INTRODUCTION

Nowadays, sanitation and industrial water treatment and purification are essential for human and environmental life. The majority of the water on Earth—roughly 97.5% of it—is salty and unusable. Just 1% of the freshwater that is still present, or 2.5%, is suitable for human consumption. Ten to 20% of the total wastewater produced by human activity is treated and repurposed. As a result, access to clean water and sanitary facilities is a problem for billions of people worldwide.[1] Human activity has contaminated freshwater, which has led to an influx of alien and potentially hazardous species. The four types of these contaminants are microbes, inorganic substances, organic compounds, and pollutants. All of them cause changes in natural aquatic habitats and organisms, endanger human health and the environment, and have an impact on biodiversity and water quality. Certain water treatment procedures have been established, but due to the complexity of wastewater types and the enormous variety of pollutants, it is necessary to consult professionals when choosing the best filtering methods as well as membrane substances.[2] Nanocelluloses (NCs) have become a viable substitute for traditional treatment of wastewater components. Phloem fibers, grasses, soft and hardwood species, as well as bacteria, fungi, algae, and marine crustaceans are some of the sources from which they can be produced.[3,4]. NC-based wastewater treatment materials are advised as best choices for industrial water treatment systems due to their inherent qualities.[5] Because of their high aspect ratio, surface area, surface charge, and mechanical strength, they can achieve this goal by offering extremely effective materials for wastewater treatment. The effectiveness of NC-based wastewater treatment components must serve a variety of purposes. Nevertheless, the best possible balance among porosity, high permeability, selective binding, durability, and particular filtration processes must be offered to NC functions. Moreover, it is necessary to manage the size, shape, and surface chemistry of NCs in order to concurrently tune their affinity for particular pollutants and prevent microbial development and fouling.[6]

This chapter discusses the special characteristics of NCs that are based on water purification applications, as well as contemporary methods for controlling their sizes, aspect ratios of geometric shapes, and pathways to rationalize their surface morphology for filtration such as through size exclusion for the heavy metals, oils, organic pollutants, toxic dyes, and desalination. The filtration of nanocomposites and their function in the eradication of pollutants from water as well as the intrinsic adsorption capabilities of NC is discussed.

7.2 CHARACTERISTICS OF NANOCOMPOSITES

By choosing particular raw materials, production methods, and processing settings, one can modify the properties of NC. The unique physicochemical properties of NCs, such as their high specific surface area, significant aspect ratio, high mechanical stiffness, high crystalline degree, susceptibility to surface functionalization, stability in water, and high surface tension, are what best explain their favorable characteristics for use in wastewater treatment. In addition to these qualities, improved rheological properties, alignment, orientation, biodegradability, biocompatibility, renewability and abundance, chemical inertness, low coefficient of thermal expansion, low density, and dimensional stability are particularly intriguing for water treatment applications. The NCs exhibit extremely small pores, a large surface area, and a significant quantity of hydroxyl groups.[7] The hydroxyl groups could be chemically altered to increase the NCs' affinity for anionic and nonionic contaminants. The hydroxyl groups are negatively charged and exhibit electrostatic attraction toward cationic wastewater pollutants, including heavy metals and cationic dyes. By surface grafting and decorating with inorganic nanomaterials, carbon nanomaterials, and other biopolymers, the interactions of water contaminants with the NC surface may be further tailored. As NCs are materials that are safe for the environment, nontoxic, sustainable, affordable, and highly effective for a range of purposes, especially water treatment, their industrial-scale manufacturing and market is expanding quickly on a global scale. The specific characteristics of nanocomposites for water purification are depicted in Figure 7.1.

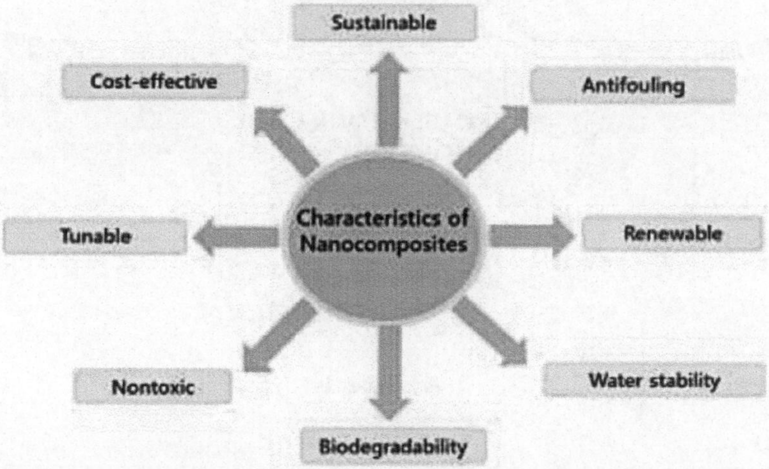

FIGURE 7.1 Specific features of nanocomposites for water purification.

Several types of NCs are already in the market, and Markets & Markets predicts that the market revenue will reach USD 783 million by 2025. From 2020 to 2026, the NC market is projected to develop at a compound yearly rate of growth of 21.4%.[8]

7.3 NANOCELLULOSE-BASED WATER PURIFICATION

To remove various pollutants and toxins from water, such as organic/inorganic pollutants, ionic metals, and other pollutants from bad water, NC-based materials can be used as innovative techniques.[9] The manufacture of green nanoparticles (NPs) actively incorporates organic polymers and natural components. Green NPs offer an alternate method for cleaning up pollution in aquatic bodies.[10] A detailed pictorial representation of the types of nanocomposites utilized in the design of water purification applications is shown in Figure 7.2. The use of green NPs offers cost-effective, practical, and ecofriendly methods for treating wastewater.

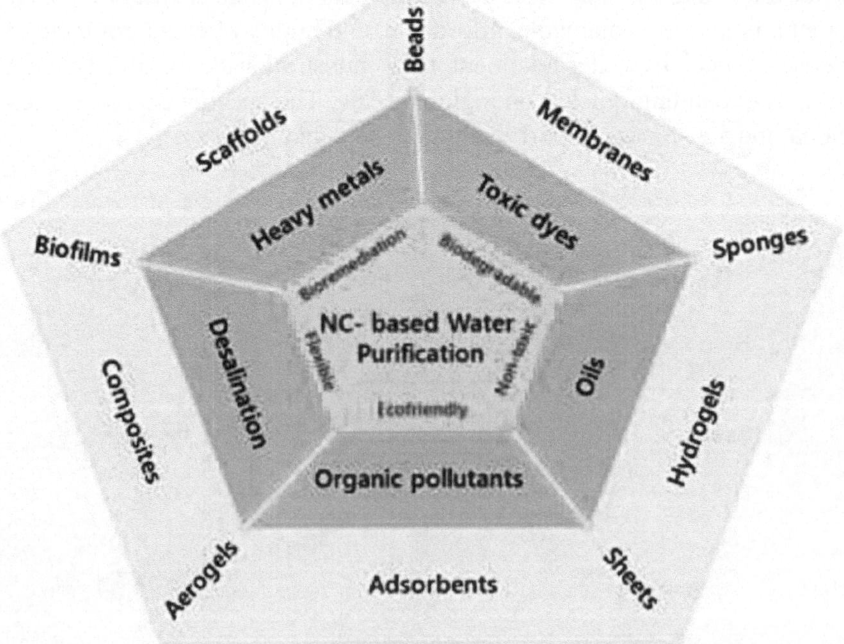

FIGURE 7.2 Pictorial representation of major water purification applications.

7.3.1 *REMOVAL OF HEAVY METAL IONS*

Heavy metal contamination can be harmful for human health. The usage of copper in a number of sectors is linked to the contamination of water sources, particularly groundwater sources, potentially as a result of its leaking. Consumption of copper ions in excess would have negative effects on human health.[11] The same holds true for other widely used metals in industries, like silver, iron, and chromium. Many methods, including chemical precipitation, photocatalysis, ion exchange, physical adsorption, and biological treatment, have been used in the past to remove heavy metal water pollution.[12] However, physical adsorption is currently the method of choice due to its simplicity and environmental friendliness. For the adsorption of cadmium and lead metal ions, as well as mercury metal ions, Jalali et al.[13] and Johari et al.[14] reported that cellulose is produced from sunflower stalks and coconut husk wastes, respectively. In regard to metal ion adsorption, both groups achieved satisfactory results. In contrast, Sun et al.[15] reported that they created a cellulosic adsorbent by halogenating microcrystalline cellulose and then functionalizing it with pyridone diacid to remove Pb^{+2} and Co^{+2}. In addition to this, a number of studies focus on the modification of CNCs using various chemical substances so that they can be used as adsorbents for water purification. Rafieian et al.[16] reported the fabrication and characterization of a membrane made of (3 aminopropyl)triethoxysilane, also known as APTES-modified CNC (MCNC), and a nonwoven substrate at the bottom for the purpose of removing heavy metals from water, such as copper ions. According to observations, MCNC was significant in improving. This demonstrated how the nanocomposite's ability to remove copper (II) was enhanced by the inclusion of MCNCs in the membrane structure. Further, filtration–regeneration testing on the PES-1 wt. % membrane revealed a minimal loss in the membrane's efficiency. CNFs, which may also be derived from industrial crops, have recently been developed as an adsorbent for various ions in still media.[17,18] This lignocellulosic biomass was discovered to be abundantly present in agave bagasse. Agave has lately been investigated for its potential to produce nanofibers and cellulosic crystals.[19] Voniatis et al.[20] combined polycaprolactone (PCL) with agave bagasse CNFs at various v/v ratios in order to create electrospun organic membranes. Due to its negative surface charge, NC has also been claimed to be helpful in the adsorption of heavy metals. Therefore, it was anticipated that these membranes would be utilized to filter out specific heavy metal pollutants as well as to make drinking water clearer and less conductive. The electrospun PCL50:CNF50 (vol/vol ratio) films successfully retained a sizable portion of the toxic substances that were present in the tap water in addition to completely eliminating conductivity and turbidity. The study's findings showed that CNF is helpful in the creation of PCL/CNF electrospun membranes, which might be used in environmentally friendly filtration for water purification. CNFs can be used to create cellulose nanopapers, which have also been used in membrane technology. The aforementioned nanopapers, though, appeared to be too dense, which prevented high water permeance and decreased efficiency. Yet, it appeared that decorating a typical natural fiber substrate with charged cellulose could be able to overcome the problem and enable high permeance while also providing enough area of contact for the adsorption of dissolved ions as water passed through it. Mautner et al.[21] reported using natural adsorbent material to penetrate a substrate made of agave and flax fibers. They combined agave and flax fibers to create a nonwoven substrate and then added anionic TEMPO [(2,2,6,6-tetramethylpiperidin-1-yl)oxyl]-oxidized CNFs to it. As a result, there was a significant net negative surface charge, which aided in the adsorption of cationic metal ions. These characteristics made it possible to use these membranes very effectively for the removal of heavy metals in aqueous systems.

7.3.2 REMOVAL OF OILS

Due to the rising need for petroleum resources, the petrochemical industry has experienced significant growth. Oil spills have frequently been caused by the disposal of oily waste materials into water bodies or, more specifically, by the transportation of goods made from crude oil. The marine ecosystem and human health could both be at risk from oil spills. Oil/water emulsions are thought to be best separated via membrane separation.[22] Commercial filter papers, however, lack the wettability and large pore size necessary to effectively isolate oil/water emulsions. Tunicate CNCs (TCNCs) are segregated from the tunicate's mantle, which gives them a high aspect ratio and strong reinforcing capacity. TCNCs have been applied to polymeric materials as a 1D nanofiller to strengthen them.[23,24] Due to the inherent design and chemical makeup of TCNCs, membranes made of them exhibit superhydrophilic and superoleophobic surfaces. Membranes made of TCNCs can also be created by coassembling TCNCs with other nanomaterials. Huang et al.[25] reported that they modified filter sheets with both the physical and chemical methods. Their stability under various situations, as well as their wettability and separation capabilities against various oil/water mixes, was measured. Even after 25 abrasion cycles, neither the morphology nor the oil contact angles (OCAs) under the water of physically or chemically modified filter sheets changed significantly, falling to 148 from 157. After 20 cycles, the filter sheets' underwater OCA remained high, proving their superoleophobicity. These enhanced filter sheets demonstrated superhydrophilicity, nanoporous structure, and superoleophobicity, according to the results, aiding in the effective separation of emulsions and oil/water combinations. Moreover, it demonstrated remarkable antiabrasion, UV, and environmental tolerance. The membrane prepared via chemical modification outperformed the other two preparation techniques in terms of antiabrasion stability and oil/water combination separation effectiveness.

7.3.3 REMOVAL OF TOXIC DYES

Anionic, cationic, or nonionic characteristics can be seen in organic dye pollutants. Using NC functionalized with anionic moieties, cationic dyes are adsorbed. At pH = 9, the carboxylated NC generated by TEMPO-mediated oxidation produced an expressively increased dye absorption of cationic methylene blue.[26] Regarding anionic dyes, Jin et al.[27] found that the largest amount of acid red GR could be adsorbed by cationic NC created by sequential oxidation using sodium periodate, followed by a reaction with ethylenediamine. Heavy metal, oil, and dye adsorption systems for water have been created using BNC and its composites.[28] In order to create scal able, reusable films that could adsorb harmful dyes including rhodamine 6G, methyl orange, methylene blue, and heavy metal ions, Gholami et al.[29] used BNC and polydopamine (PDA). In the fabrication of three-dimensional (3D)-networked BNC with cross-linked poly (2-methacryloyloxyethyl phosphorylcholine) (PMPC) nanocomposite membranes for the elimination of organic dyes, the relationship between a zwitterionic polymer and BNC had not been investigated.[30] Because of its methacrylic functional group as well as the zwitterionic phosphorylcholine moiety, which has been made up of a trimethylammonium cation and even a phosphate anion, 2-methacryloyloxyethyl phosphorylcholine (MPC) was chosen as the nontoxic, polymerizable polymer.[31] Moreover, this MPC polymer exhibited bioinert, antibacterial, and antifouling qualities as well as a distinct hydration state, all of which would be significant assets in limiting the growth of microorganisms in polluted water. They assessed the ability of different bacteria to absorb water and remove organic dyes, as well as their thermal stability, antibacterial traits, and mechanical characteristics.[32] With the greater zwitterionic polymer comprising the membrane, it was discovered that they showed dye adsorption amounts of 4.44 ±0.32 mg g^{-1} of methylene blue and 4.56 ± 0.43 mg g^{-1} of methyl orange. The findings demonstrated the effectiveness of the proposed nanocomposite membranes in adsorbing both anionic and cationic natural dyes. The membranes additionally exhibited outstanding mechanical qualities, good bacterial growth inhibition, and good heat stability up to 250^0C.

Several carbon-based compounds, including graphene oxide, multidoped carbon fibers, and others, have been employed in studies pertaining to the purification of wastewater. One of them is graphene oxide (GO) sheets, which serve as a selective membrane to let ions and molecules permeate into aqueous solutions.[33] Recent research has shown that GO, which is made from graphite by chemical exfoliation, is a nanostructure with great potential for adsorption due to its large surface area, extremely reactive surface caused by the profusion of carboxyl as well as hydroxyl functional groups, as well as its remarkably hydrophobic interactions. GO demonstrates a variety of advantageous traits, such as hydrophilicity, antifouling capabilities, and high surface activity. GO sheets do have two significant disadvantages, though, namely limited structural strength in the wet condition with poor permeation flux.[34] It has been discovered that adding a layer of GO sheets to CNF membranes by successive water filtering enhances the membrane's functionality. More than 90% of dyes were likewise rejected by these membranes, regardless of charge. Also, it has been found that adding a thin layer of GO to CNF as a functional layer may produce synergistic membranes. This was primarily because the two could interact strongly due to the construction of inclusive networks of inter- and intramolecular hydrogen connections. Liu et al.[35] created double-layered CNF–GO membranes that incorporate an incredibly thin GO layer of CNF membrane without the use of chemical crosslinkers. They created the membrane using Exilva, a CNF of industrial grade, and investigated the effects of the GO layer on its surface features, mechanical capabilities, and dye rejection behavior. Methyl Violet 2B (MV2), Rhodamine 6 G (R6G), and VictoriaBlue 2B (VBB) were used as contaminants in the research. Thus, they were able to demonstrate how to create a stable GO barrier layer on a CNF membrane without the need for a chemical crosslinker. Compared to monolayered GO membranes, this membrane's GO layer had a thickness of about 100 nm and showed greater water and dye permeability. The GO–CNF membranes are a strong candidate for membranes currently used in the food industry, biomedicine, and water remediation because of their excellent rejection rate, structural stability, high flux, and simple preparation process. Valencia et al.[36] revealed that the improved wet mechanical properties resulted from the GO layer's inclusion, which further emerged to preserve the cellulose chain from different variants in the framework during redrying, which prevented dramatic implosion as it takes place in an immaculate CNF membrane, which prevented the CNF network from expanding during purification of water.

Research findings have shown that NC exhibits significant adsorption efficiency for dyes, nitrates, humic acid, metal ions, and other pollutants from industrial effluents, whether it has increased surface complexation like enzymatic phosphorylation, TEMPO oxidation, cationization, and so forth or even in its native form.[37] The manufacture of a membrane with controlled pore size and structure in addition to the layering of many membranes atop one another presents difficulties without sacrificing the NC's ecofriendliness, flux, or rejection capacity. The process of producing membranes with a large surface area and an open, interlinked pore structure using electrospinning is getting more popular. At the micron scale, electrospun membranes have indeed been utilized to absorb contaminants from water. According to Goetz et al.,[38] cellulose acetate-based bio-based and freestanding electrospun membranes with appropriate mechanical characteristics, water flux, and antifouling performance can be produced. Goetz et al.'s[39] goal was to show that it was possible to create high flux, renewable membranes for water filtering. On such a cellulose acetate electrospun mesh template, CNCs were inserted. The membranes worked well at rejecting contaminating particles with a size range of 0.5 to 2.0 m, and they had remarkable antifouling qualities. The creation of this material made from renewable membranes, which incorporate adsorption, size exclusion, and hydrophilicity was made possible by the functionalization of cellulose acetate membranes utilizing CNCs. The water flux has been found to be extremely high, but it gradually decreased even more as concentration increased and the nanocrystal network became more continuous. Moreover, they showed a rise in hydrophilicity and potential antifouling against hydrophobic substances. Moreover, its pore structure revealed a moderate refusal of fragments between 0.5 and 2.0 m. The study illustrated a straightforward and effective technique for functionalizing electrospun fibers with CNCs. The developed bio-based membranes had better mechanical characteristics, were more hydrophilic, could move more water, and could reject pollutants by size rejection as well as adsorption. A composite membrane made of polyethersulfone (PES) and CNCs that have been given amine functionality was created. They also investigated its potential for color removal from wastewater, a byproduct of industrial licorice processing, and for reducing chemical oxygen demand. APTES was applied to modify CNCs. The composite's active sites, which are in charge of purging water's pollutants, are multiplied by the APTES. With a water contact angle (WCA) of 72.6, the purest PES membrane displayed a higher hydrophobic characteristic. The WCAs of PES/0.1 wt. % MCNC, PES/0.5 wt. % MCNC, and PES/1 wt. % MCNC were decreased, indicating that the membrane surfaces of nanocomposite materials are hydrophilic. This may be related to the amine's hydrophilic hydroxyl groups and the CNC groups from APTES-modified CNC. It led them to draw the conclusion that being able to prepare mixed matrix nanocomposite might offer a straightforward procedure, devoid of any solvent requirement, for the surface modification of CNC with the goal of enhancing the color adsorption process for industrial effluents from the handling of licorice.[40]

7.3.4 *REMOVAL OF ORGANIC POLLUTANTS*

The water bodies could be contaminated by harmful organic pollutants. The use of materials depends on NCs for the treatment of wastewater polluted with harmful organic contaminants. Because there are not as many active sites for interacting with organic contaminants, natural cellulose microfibers often have an affinity for them that is 100–500 times lower than that of common nanomaterials like zeolite and activated carbon.[41] As a substitute, surface-modified NCs were evaluated as adsorption supports for a range of organic contaminants. Its strong mechanical characteristics, a large specific surface area, which enables the creation of an effective interaction and a proper implementation synthesis process, and its filters and membrane micropore sizes all contribute to this. To increase the adsorption and filtering capacity of NC because its inherent hydrophilicity is unsuitable for the absorption of organic compounds, surface functionalization and the creation of nanocomposites are essential.[42,43] Even in low quantities, organic pollutants like pesticides, medicines, and dyes are harmful to the environment. Humans could suffer a number of health issues from pollutants in contaminated water, like diarrhea, difficulties in inhaling, skin rashes, muscle and joint discomfort, headaches, and an irregular heartbeat.[44] Acute toxicity refers to the conditions that arise from the oral consumption and inhalation of organic pollutants, which can cause allergic reactions, skin rashes, and various malignancies.[45] Lately, NCs have been studied as viable adsorbents to remove pollutants like pesticides, fertilizers, and pharmaceuticals. Bouli et al.[46] reported that carboxymethyl NC hydrogels containing nanoclays are useful adsorbents, soil conditioners, and nutrient transporters, mainly to boost cucumber development. Mautner et al.[47] reported that cationic cellulose nanopaper membranes were effectively beneficial for the complete removal of nitrates produced from the chemical fertilizer in the agricultural soil and groundwater. Similar to this, various studies have demonstrated that NC can potentially act as an adsorbent in the elimination of pesticides including diuron, nitenpyram, tetraconazole, organophosphorus, and chlorpyrifos. They are also used to adsorb and remove harmful pharmaceuticals from industrial effluents, including ciprofloxacin, diclofenac, metronidazole, and ceftriaxone.[48] It can be used to adsorb cationic organic contaminants from wastewater both on their own and after being anionically functionalized. CNFs made from kenaf fibers were examined for cationic dye adsorption. The unique CNF shape, which relies on the chosen source, can be used to describe the very large capacity for adsorption. The rapid expansion of the textile industry, which uses nearly 7×105 tonnes of dyes annually, is a major factor in the dangers of dyes.[49] Because dyes are often not biodegradable, they are a major threat to plants and food. Anionic direct dyes have a great affinity for NCs and are water-soluble. Acid dyes are anionic dyes that are water-soluble and have a very low affinity for binding NCs. Direct and acid dyes with NCs' chromophores are attracted to one another by dipolar forces, hydrogen bonds, and nonspecific hydrophobic interactions. The absorption of direct and acid dyes is typically increased by the cationic alteration of NCs.

Surface modification of NCs improves the organic adsorptive characteristics and leads to the development of useful NC-based materials. Chemical alterations, like the addition of positively charged materials, are used to create NC-adsorbent materials for anionic dyes. Using cationic CNFs, which are created by surface quaternizing with glycidyl trimethyl ammonium chloride, anionic dyes can be segregated.[50] Several investigations have focused on the preferential adsorption of negatively charged aqueous organic contaminants onto cationic CNFs. Extremely porous NC gels were created from NCs, and a variety of organic contaminants' adsorption characteristics were investigated.[51] Reduced graphene oxide (rGO), vitamin C, and CNCs were used to create a brand-new, incredibly durable porous 3D sponge. The utilization of vitamin C as a lowering agent and a soft template enabled the construction of a special hierarchical system with a vast array of interconnected and evenly dispersed pores, even in the sponge's core. This particular arrangement produced a relatively large surface area for methylene blue absorption in water.[52] Compared to what was obtained employing cellulose/rGO fibers made using the wet-spinning method, this is significantly better. To remove cationic and anionic dyes, hybrid cellulose aerogels made of CNFs with carbon nanotubes and graphene nanosheets were used.[53] The pseudo second-order adsorption kinetics and the monolayer Langmuir adsorption isotherm are used to explain the adsorption kinetics for methylene blue and Congo red in the single dye system. When compared to virgin CNFs and graphene nanosheets, the CNF/graphene nanosheet hybrid exhibits greater dye adsorption capability. Furthermore, employing ethanol as a desorption agent, nearly 80% of methylene blue and Congo red may be desorbed from CNF/GnP. In the end, recyclable adsorption media for dye removal, such as aerogels constructed from nanohybrid materials containing carbon nanotubes and graphene, are viable.

7.3.5 DESALINATION

Thin-film composites (TFCs) were fabricated of polyamide and showed potential for water purification. The efficiency of polyamide-based TFC NF membranes (NFMs) has been improved using a variety of techniques over the past few decades, including changing the reaction monopolymers, altering the physical and chemical properties of porous substrates, integrating nanomaterials into polyamide film and porous substrates, and adding a layer between both the polyamide film and substrate.[54,55] To change the permeability characteristics of the polyamideselective layers, it is more common to incorporate NPs into membranes. In order to increase the antifouling, chlorine, and antibacterial resistance, TFC NFMs were employed. Many organic and inorganic nanomaterials were fabricated; unfortunately, their processing methods have a tendency to be challenging. Hence, it must have been feasible to identify an organic nanocomposite with a facile manufacturing process, excellent durability for long-term uses, and high compatibility with the polyamide layer. One potential contender is CNC, which is easily generated and available in large quantities when cellulose is acid-hydrolyzed.[56,57] When added to reverse osmosis membranes, it also exhibited impressive antifouling capabilities. Huang et al.[58] reported the impacts of CNCs on the characteristics of TFN NFMs. A rougher and much more hydrophilic membrane would have had better efficiency than a conventional TFN NFM.

The water flux increased sharply by about 35% with an increase in CNCs. Na_2SO_4 and $MgSO_4$ rejection were above 98% and above 96%, indicating strong interaction between CNCs and the polyamide matrix. It also exhibited antichlorine characteristics. The film continued to have a consistent water penetration rate even after being exposed to a NaClO solution. Their innovation offers outstanding ecofriendly nanomaterials that can be used to fabricate TFN NFMs with increased water flux without affecting their ability to remove salt.

7.4 CONCLUSION AND FUTURE PERSPECTIVES

The development of ecofriendly treatment technologies for wastewater with enhanced efficiency has received significant academic and commercial attention. To date, a wide range of strategies have been created to remedy clean water, which is undoubtedly one of the biggest concerns addressing the society in the 21st century. In terms of effectiveness and efficiency, among the various NPs, NCs are promising renewable sources for wastewater treatment. They are also lighter in weight, chemically inert, and have a variety of surface chemistry possibilities. In addition, they are produced at low costs and can be handled safely when compared to inorganic nanomaterials. Several researchers have created NC-based biomaterials for effluent treatment as an outcome of advancements in NC development. Especially in membranes and filters as well as in adsorbent materials, various NC types have been successfully investigated for water purification. Such NC based materials possess excellent scalability and regeneration capacity, as shown by their continued great performance. NC surface modification through surface grafting and mixing with various inorganic, polymeric, and carbon components enables the more affordable mass manufacture of novel, durable, and regenerative cellulose composites. NCs are suitable for water treatment through adsorption, nanofiltration, removal, antifouling, ultrafiltration, and reverse osmosis due to their special characteristics and great compatibility with several modifiers. However, more work needs to be done before NCs may be used and commercialized as a practical nanocomposite for remediation methods. Another challenge is their polydispersity and dispersion durability, particularly for industrial and commercial production. To enhance the creation and effective application of NCs in environmental remediation, the key difficulties and restrictions associated with interacting with NC-based components should be recognized. Because NPs can effortlessly infiltrate the human body, the most effective technique is to use NCs as support materials for NP immobilization. However, effects related to the removal of NPs from the liquid during the interaction and the usage of functional NPs in cooperation with NCs for treating wastewater must be established. This field of research is expanding quickly in preparation for the potential commercial application of NC/inorganic hybrids for a variety of industrial uses, particularly wastewater treatment.

KEYWORDS

- **membrane**
- **nanocellulose**
- **water purification**
- **dyes**
- **heavy metal**

REFERENCES

1. Ranade, V. V.; Bhandari, V. M. *Industrial Wastewater Treatment, Recycling, and Reuse: An Overview*; Elsevier Ltd., 2014.
2. Ahankari, S.; George, T.; Subhedar, A.; Kar, K. K. Nanocellulose as a Sustainable Material for Water Purification. *SPE Polym.* **2020,** *1* (2), 69–80. DOI: 10.1002/pls2.10019.
3. Barhoum, A. et al. Plant Celluloses, Hemicelluloses, Lignins, and Volatile Oils for the Synthesis of Nanoparticles and Nanostructured Materials. *Nanoscale* **2010,** *12* (45), 22845–22890. DOI: 10.1039/d0nr04795c.
4. Trache, D. et al. *Nanocellulose: From Fundamentals to Advanced Applications*, vol. 8, no. May, 2020.
5. Harish, V. et al. Nanoparticle and Nanostructure Synthesis and Controlled Growth Methods. *Nanomaterials* **2022,** *12* (18), 1–30. DOI: 10.3390/nano12183226.
6. Hassan, M. L. et al. Water Purification Ultrafiltration Membranes Using Nanofibers from Unbleached and Bleached Rice Straw. *Sci. Rep.* **2020,** *10* (1), 1–9. DOI: 10.1038/s41598-020-67909-3.
7. Barhoum, A.; Pal, K.; Rahier, H.; Uludag, H.; Kim, I. S.; M. Bechelany. Nanofibers as New-Generation Materials: From Spinning and Nano-Spinning Fabrication Techniques to Emerging Applications. *Appl. Mater. Today* **2019,** *17*, 1–35. DOI: 10.1016/j.apmt.2019.06.015.
8. Barhoum, A. *Handbook of Nanofibers*; Springer International Publishing, 2019.

9. Reshmy, R. et al. Nanocellulose as Green Material for Remediation of Hazardous Heavy Metal Contaminants. *J. Hazard. Mater.* **2022,** *424*, 127516. DOI: 10.1016/j.jhazmat.2021.127516.
10. Ahankari, S. S.; Mohanty, A. K.; Misra, M. *Nanomaterials from Renewable Resources for Emerging Applications*, 2023.
11. Bilal, M et al. Waste Biomass Adsorbents for Copper Removal from Industrial Wastewater—A Review. *J. Hazard. Mater.* **2013,** *263*, 322–333. DOI: 10.1016/j.jhazmat.2013.07.071.
12. Jain, R. et al. Preferential Adsorption of Cu in a Multi-Metal Mixture onto Biogenic Elemental Selenium Nanoparticles. *Chem. Eng. J.* **2016,** *284*, 917–925. DOI: 10.1016/j.cej.2015.08.144.
13. Jalali, M.; Aboulghazi, F. Sunflower Stalk, an Agricultural Waste, as an Adsorbent for the Removal of Lead and Cadmium from Aqueous Solutions. *J. Mater. Cycles Waste Manag.* **2013,** *15* (4), 548–555. DOI: 10.1007/s10163-012-0096-3.
14. Johari, K.; Saman, N.; Song, S. T.; Chin, C. S.; Kong, H.; Mat, H. Adsorption Enhancement of Elemental Mercury by Various Surface Modified Coconut Husk as Eco-Friendly Low-Cost Adsorbents. *Int. Biodeterior. Biodegrad.* **2016,** *109*, 45–52. DOI: 10.1016/j.ibiod.2016.01.004.
15. Sun, C. et al. Preparation of a Cellulosic Adsorbent by Functionalization with Pyridone Diacid for Removal of Pb(II) and Co(II) from Aqueous Solutions. *Cellulose* **2017,** *24* (12), 5615–5624. DOI: 10.1007/s10570-017-1519-z.
16. Rafieian, F.; Jonoobi, M.; Yu, Q. A Novel Nanocomposite Membrane Containing Modified Cellulose Nanocrystals for Copper Ion Removal and Dye Adsorption from Water. *Cellulose* **2019,** *26* (5), 3359–3373. DOI: 10.1007/s10570-019-02320-4.
17. Tibolla, H.; Pelissari, F. M.; Rodrigues, M. I.; Menegalli, F. C. Cellulose Nanofibers Produced from Banana Peel by Enzymatic Treatment: Study of Process Conditions. *Ind. Crops Prod.* **2017,** *95*, 664–674. DOI: 10.1016/j.indcrop.2016.11.035.
18. Robles, E.; Fernández-Rodríguez, J.; Barbosa, A. M.; Gordobil, O.; Carreño, N. L. V.; Labidi, J. Production of Cellulose Nanoparticles from Blue Agave Waste Treated with Environmentally Friendly Processes. *Carbohydr. Polym.* **2018,** *183*, 294–302. DOI. 10.1016/j.carbpol.2018.01.015.
19. Alemán-Nava, G. S.; Gatti, I. A.; Parra-Saldivar, R.; Dallemand, J. F.; Rittmann, B. E.; Iqbal, H. M. N. Biotechnological Revalorization of Tequila Waste and By-Product Streams for Cleaner Production—A Review from Bio-Refinery Perspective. *J. Clean. Prod.* **2018,** *172*, 3713–3720. DOI: 10.1016/j.jclepro.2017.07.134.
20. Voniatis, C.; Barczikai, D.; Gyulai, G.; Jedlovszky-Hajdu, A. Fabrication and Characterisation of Electrospun Polycaprolactone/Polysuccinimide Composite Meshes. *J. Mol. Liq.* **2021,** *323*. DOI: 10.1016/j.molliq.2020.115094.
21. Mautner, A. et al. Natural Fibre-Nanocellulose Composite Filters for the Removal of Heavy Metal Ions from Water. *Ind. Crops Prod.* **2018,** *133*, 325–332. DOI: 10.1016/j.indcrop.2019.03.032.
22. Sun, F.; Liu, W.; Dong, Z.; Y. Deng. Underwater Superoleophobicity Cellulose Nanofibril Aerogel Through Regioselective Sulfonation for Oil/Water Separation. *Chem. Eng. J.* **2017,** *330*, 774–782. DOI: 10.1016/j.cej.2017.07.142.
23. Sacui, I. A. et al. Comparison of the Properties of Cellulose Nanocrystals and Cellulose Nano Fibrils Isolated from Bacteria, Tunicate, and Wood Processed Using Acid,

Enzymatic, Mechanical, and Oxidative Methods. *ACS Appl. Mater. Interfaces* **2014,** 1–7.

24. Zhang, T.; Zuo, T.; Hu, D.; Chang, C. Dual Physically Cross-Linked Nanocomposite Hydrogels Reinforced by Tunicate Cellulose Nanocrystals with High Toughness and Good Self-Recoverability. *ACS Appl. Mater. Interfaces* **2017,** *9* (28), 24230–24237. DOI: 10.1021/acsami.7b06219.
25. Huang, Y.; Zhan, H.; Li, D.; Tian, H.; C. Chang. Tunicate Cellulose Nanocrystals Modified Commercial Filter Paper for Efficient Oil/Water Separation. *J. Memb. Sci.* **2019,** *591*. DOI: 10.1016/j.memsci.2019.117362.
26. Batmaz, R.; Mohammed, N.; Zaman, M.; Minhas, G.; Berry, R. M.; Tam, K. C.; Cellulose Nanocrystals as Promising Adsorbents for the Removal of Cationic Dyes. *Cellulose* **2014,** *21* (3), 1655–1665. DOI: 10.1007/s10570-014-0168-8.
27. Jin, L.; Li, W.; Xu, Q.; Sun, Q. Amino-Functionalized Nanocrystalline Cellulose as an Adsorbent for Anionic Dyes. *Cellulose* **2015,** *22* (4), 2443–2456. DOI: 10.1007/s10570-015-0649-4.
28. Torres, F. G.; Arroyo, J. J.; Troncoso, O. P. Bacterial Cellulose Nanocomposites: An All-Nano Type of Material. *Mater. Sci. Eng. C* **2019,** *98*, 1277–1293. DOI: 10.1016/j.msec.2019.01.064.
29. Gholami Derami, H. et al. A Robust and Scalable Polydopamine/Bacterial Nanocellulose Hybrid Membrane for Efficient Wastewater Treatment. *ACS Appl. Nano Mater.* **2019,** *2* (2), 1092–1101. DOI: 10.1021/acsanm.9b00022.
30. Vilela, C.; Moreirinha, C.; Almeida, A.; A. J. Silvestre, D.; Freire, C. S. R. Zwitterionic Nanocellulose-Based Membranes for Organic Dye Removal. *Materials (Basel).* **2019,** *12* (9). DOI: 10.3390/ma12091404.
31. Ishihara, K.; Mu, M.; Konno, T.; Inoue, Y.; Fukazawa, K. The Unique Hydration State of Poly(2-methacryloyloxyethyl Phosphorylcholine). *J. Biomater. Sci. Polym. Ed.* **2017,** *28* (10–12), 884–899. DOI: 10.1080/09205063.2017.1298278.
32. Kwon, J. S. et al. Novel Anti-Biofouling Bioactive Calcium Silicate-Based Cement Containing 2-Methacryloyloxyethyl Phosphorylcholine. *PLoS One* **2019,** *14* (1), 1–14. DOI: 10.1371/journal.pone.0211007.
33. Banerjee, S.; Benjwal, P.; Singh, M.; Kar, K. K. Graphene Oxide (rGO)-Metal Oxide (TiO2 /Fe3O4) Based Nanocomposites for the Removal of Methylene Blue. *Appl. Surf. Sci.* **2018,** *439*, 560–568. DOI: 10.1016/j.apsusc.2018.01.085.
34. Yeh, C. N.; Raidongia, K.; Shao, J.; Yang, Q. H.; Huang, J. On the Origin of the Stability of Graphene Oxide Membranes in Water. *Nat. Chem.* **2015,** *7* (2), 166–170. DOI: 10.1038/nchem.2145.
35. Liu, P.; Zhu, C.; Mathew, A. P. Mechanically Robust High Flux Graphene Oxide—Nanocellulose Membranes for Dye Removal from Water. *J. Hazard. Mater.* **2019,** *371*, pp. 484–493. DOI: 10.1016/j.jhazmat.2019.03.009.
36. Valencia, L. et al. Nanocellulose/Graphene Oxide Layered Membranes: Elucidating Their Behaviour During Filtration of Water and Metal Ions in Real Time. *Nanoscale* **2019,** *11* (46), 22413–22422. DOI: 10.1039/c9nr07116d.
37. Gopakumar, D. A.; Pasquini, D.; Henrique, M. A.; L. C. De Morais, Grohens, Y.; Thomas, S. Meldrum's Acid Modified Cellulose Nanofiber-Based Polyvinylidene Fluoride Microfiltration Membrane for Dye Water Treatment and Nanoparticle Removal. *ACS Sustain. Chem. Eng.* **2017,** *5* (2), 2026–2033. DOI: 10.1021/acssuschemeng.6b02952.

38. Goetz, L. A.; Jalvo, B.; Rosal, R.; Mathew, A. P. Superhydrophilic Anti-Fouling Electrospun Cellulose Acetate Membranes Coated with Chitin Nanocrystals for Water Filtration. *J. Memb. Sci.* **2016,** *510*, 238–248. DOI: 10.1016/j.memsci.2016.02.069.
39. Goetz, L. A.; Naseri, N.; Nair, S. S.; Karim, Z.; Mathew, A. P. All Cellulose Electrospun Water Purification Membranes Nanotextured Using Cellulose Nanocrystals. *Cellulose* **2018,** *25* (5), 3011–3023. DOI: 10.1007/s10570-018-1751-1.
40. Jonoobi, M.; Ashori, A.; Siracusa, V. Characterization and Properties of Polyethersulfone/Modified Cellulose Nanocrystals Nanocomposite Membranes. *Polym. Test.* **2019,** *76*, 333–339. DOI: 10.1016/j.polymertesting.2019.03.039.
41. Barhoum, A. et al. 3d Self-Supported Nitrogen-Doped Carbon Nanofiber Electrodes Incorporated Co/Coox Nanoparticles: Application to Dyes Degradation by Electro-Fenton-Based Process. *Nanomaterials* **2021,** *11* (10). DOI: 10.3390/nano11102686.
42. Salama, A.; Shoueir, K. R.; Aljohani, H. A. Preparation of Sustainable Nanocomposite as New Adsorbent for Dyes Removal. *Fibers Polym.* **2017,** *18* (9), 1825–1830. DOI: 10.1007/s12221-017-7396-0.
43. Shahnaz, T.; Vishnu Priyan, V.; Pandian, S.; Narayanasamy, S. Use of Nanocellulose Extracted from Grass for Adsorption Abatement of Ciprofloxacin and Diclofenac Removal with Phyto, and Fish Toxicity Studies. *Environ. Pollut.* **2021,** *268*, 115494. DOI: 10.1016/j.envpol.2020.115494.
44. Salama, A. *Cellulose/Calcium Phosphate Hybrids: New Materials for Biomedical and Environmental Applications*, Vol. 127; Elsevier B.V., 2019.
45. Lu, M.; Li, G.; Yang, Y.; Yu, Y. A Review on In Vitro Oral Bioaccessibility of Organic Pollutants and Its Application in Human Exposure Assessment. *Sci. Total Environ.* **2021,** *752*, 142001. DOI: 10.1016/j.scitotenv.2020.142001.
46. Bauli, C. R.; Lima, G. F.; A. G. de Souza, Ferreira, R. R.; Rosa, D. S. Eco-Friendly Carboxymethyl Cellulose Hydrogels Filled with Nanocellulose or Nanoclays for Agriculture Applications as Soil Conditioning and Nutrient Carrier and Their Impact on Cucumber Growing. *Colloids Surfaces A Physicochem. Eng. Asp.* **2021,** *623*, 126771. DOI: 10.1016/j.colsurfa.2021.126771.
47. Mautner, A.; Kobkeatthawin, T.; Bismarck, A. Efficient Continuous Removal of Nitrates from Water with Cationic Cellulose Nanopaper Membranes. *Resour. Technol.* **2017,** *3* 1, pp. 22–28. DOI: 10.1016/j.reffit.2017.01.005.
48. Salama, A. et al. Nanocellulose-Based Materials for Water Treatment: Adsorption, Photocatalytic Degradation, Disinfection, Antifouling, and Nanofiltration. *Nanomaterials* **2021,** *11* (11). DOI: 10.3390/nano11113008.
49. He, X.; Male, K. B.; Nesterenko, P. N.; Brabazon, D.; Paull, B.; Luong, J. H. T. Adsorption and Desorption of Methylene Blue on Porous Carbon Monoliths and Nanocrystalline Cellulose. *ACS Appl. Mater. Interfaces* **2013,** *5* (17), 8796–8804. DOI: 10.1021/am403222u.
50. Pei, A.; Butchosa, N.; Berglund, L. A.; Zhou, Q. Surface Quaternized Cellulose Nanofibrils with High Water Absorbency and Adsorption Capacity for Anionic Dyes. *Soft Matter* **2013,** *9* (6), 2047–2055. DOI: 10.1039/c2sm27344f.
51. Maatar, W.; Alila, S.; Boufi, S. Cellulose Based Organogel as an Adsorbent for Dissolved Organic Compounds. *Ind. Crops Prod.* **2013,** *49*, 33–42. DOI: 10.1016/j.indcrop.2013.04.022.

52. Yousefi, N. et al. Hierarchically Porous, Ultra-Strong Reduced Graphene Oxide-Cellulose Nanocrystal Sponges for Exceptional Adsorption of Water Contaminants. *Nanoscale* **2018,** *10* (15), 7171–7184. DOI: 10.1039/c7nr09037d.
53. Yu, Z.; Hu, C.; Dichiara, A. B.; Jiang, W.; Gu, J. Cellulose Nanofibril/Carbon Nanomaterial Hybrid Aerogels for Adsorption Removal of Cationic and Anionic Organic Dyes. *Nanomaterials* **2020,** *10* (1), 1–20. DOI: 10.3390/nano10010169.
54. Veerababu, P.; Vyas, B. B.; Singh, P. S.; Ray, P. Limiting Thickness of Polyamide – Polysulfone Thin- Film-Composite Nano Filtration Membrane A **2014,** *346*, 19–29.
55. Bi, R.; Zhang, Q.; Zhang, R.; Su, Y.; Jiang, Z. Thin Film Nanocomposite Membranes Incorporated with Graphene Quantum Dots for High Flux and Antifouling Property. *J. Memb. Sci.* **2018,** *553*, 17–24. DOI: 10.1016/j.memsci.2018.02.010.
56. Chen, L.; Zhu, J. Y.; Baez, C.; Kitin, P.; Elder, T. Highly Thermal-Stable and Functional Cellulose Nanocrystals and Nanofibrils Produced Using Fully Recyclable Organic Acids. *Green Chem.* **2016,** *18* (13), 3835–3843. DOI: 10.1039/c6gc00687f.
57. Yu, H. Y.; Zhang, H.; Song, M. L.; Zhou, Y.; Yao, J.; Ni, Q. Q. From Cellulose Nanospheres, Nanorods to Nanofibers: Various Aspect Ratio Induced Nucleation/ Reinforcing Effects on Polylactic Acid for Robust-Barrier Food Packaging. *ACS Appl. Mater. Interfaces* **2017,** *9* (50), 43920–43938. DOI: 10.1021/acsami.7b09102.
58. Huang, S. et al. Polyamide Nanofiltration Membranes Incorporated with Cellulose Nanocrystals for Enhanced Water Flux and Chlorine Resistance. *ACS Sustain. Chem. Eng.* **2019,** *7* (14), 12315–12322. DOI: 10.1021/acssuschemeng.9b01651.

CHAPTER 8

Nanocellulose: A Versatile Green Sustainable Material for Water Remediation

REKHA MIRLE

Assistant Professor, Department of Chemistry and Biochemistry, School of Sciences, Jain (Deemed-To-Be) University, Bangalore-27, India

ABSTRACT

The urge for cleansed water has been escalating gradually. Many technologies have emerged out to be more efficient and cheaper than conventional methods. One of the potential candidates is nanocellulose materials such as cellulose nanocrystals and cellulose nanofibrils that serve as renewable-based materials that gained importance in treating waste water due to their unique properties of biocompatibility and renewability, which comprise higher material steadiness and a larger surface area. It possesses flexible hydrophilic surface properties but is lethargic to chemicals. This chapter provides a broad overview of the structural organization of cellulose, its nomenclature, various methods to produce nanocellulose, and its distinct properties. In addition, recent advancements and the latest applications of nanocellulose in nanocomposite as a prospective tool for treating contaminated water are focused upon. Lastly, the vital challenges and opportunities of cellulose-based nanomaterials in the treatment of waste water are discussed, which offers proper perceptions for future researchers in this field.

8.1 INTRODUCTION

Water is essential for the entire living system in spite of feeding other essential nutrients. Water is a clear, tasteless, odorless, and colorless chemical

Resilience and Sustainability for Energy and Water: New Advancements, Opportunities, and Framework.
Cristóbal Noé Aguilar, A. K. Haghi, Amritanshu Shukla, Ann Rose Abraham, Swati Gokul Talele (Eds.)

component making a chief constituent of Earth's hydrosphere. It is termed a universal solvent and an important fluid for all living organisms.[1] But in recent years, the quality of water sources has diminished due to population explosion, urbanization, and industrialization, which discharge various kinds of pollutants including heavy metal ions, dyes, oils, and organic and inorganic substances into water. These pollutants have a detrimental impact on the ecological system and human life.[2] Challenges lie in addressing the universal issue of water pollution that demands different novel technologies to degrade water pollutants from water sources and ensure the supply of pure water globally. In this direction, the updation of conventional methods is most desirable, and hence, advanced nanotechnological processes provide great opportunities for the development of various innovative techniques for treating waste water. This chapter intends to give a complete review of the latest advancements in nanocellulose as an effective tool to attain water sustainability globally. It gives insight on the recent advancement of nanotechniques toward water sustainability.[3] The advantages of various nanomaterials and their practical glitches over conventional methods are also discussed.

Nanocellulose as a nanoscale material has drawn enormous attention due to its ample accessibility, biodegradability, larger surface area, exceptional mechanistic behavior, good selective adsorption of pollutants from waste water and other various industrial applications, etc. It also briefs on the recent efforts of nanocellulose as a potential material to protect our nature;

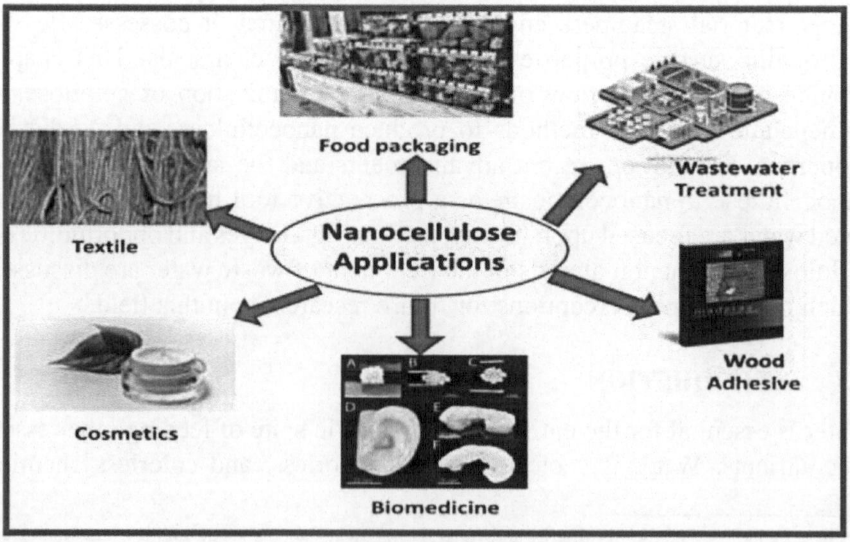

FIGURE 8.1 Applications of nanocellulose in multiple disciplines

in particular, it adsorbs heavy metals, purifies water, effectively acts as a catalyst, has the ability to combat microbes, and offers additional strength for decomposable materials. One of the promising applications of nanocellulose-based nanocomposites as catalysts is that they facilitate the breakdown of organic pollutants, thus getting rid of contaminants from polluted water.[4]

The exceptional cellulosic nature of the larger surface area, inactivity toward chemicals, hydrophilic nature of the surface, and extraordinary stability, including its nanoscale dimension and biosorptivity features, substantiate nanocellulose as a proficient tool for the selective removal of pollutants from various fouled water resources in the form of separable membranes and filters.

Nanocellulose, as a new potential candidate, resolves the issues related to water pollution in an ecofriendly manner. Nanocellulose and its membranes, which are porous and semipermeable, are used in the purification of water. It basically depends on the ability of nanocellulose to adsorb, store, and desorb the contaminants from waste water selectively.[5] Among various renewable sources, cellulose is the most abundant compound that originated in the ecosystem in diverse forms.[6,7] The latest advancement in the field of nanomaterials offers new prospects in the purification of water. This nanocellulose can be widely incorporated in several water treatment processes due to its low cost and sustainability in purifying water.

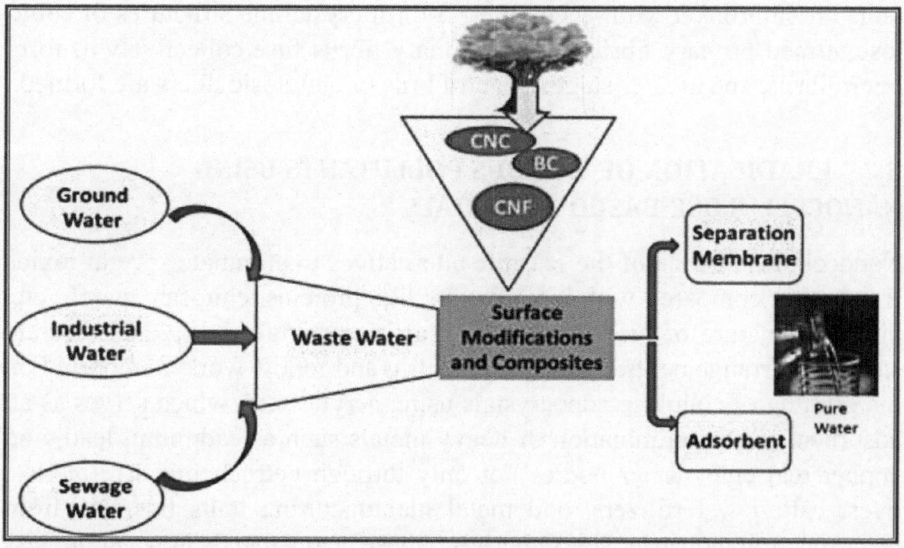

FIGURE 8.2 Separation of pollutants from wastewater using nanocellulose-based membranes

These cellulose nanocrystals serve as excellent and ideal adsorbents that remove heavy metals, and their ecofriendly nature provides wide applications in environmental protection.[8,9] In order to enhance the capacity of adsorption, magnetic cellulose nanocrystals (MCNs) and their derived composites comprising metal-organic frameworks are developed to remove lead content from polluted water.[10] Literature surveys have reported ecofriendly methods of oxidizing cellulose into cellulose nanocrystals in the presence of ferrous/H_2O_2, which helps in the ejection of copper ions and dyes (methylene blue) from contaminated water by high carboxylation of cellulose nanocrystals.[11] Several studies reported many approaches to increase the adsorption ability of cellulose nanocrystals and their derivatives, comprising biopolymers, for the effective ejection of methylene blue dye from industrial effluent.[12,13] Also, carboxycellulose nanofibers derived from nanocellulose have a higher surface charge and greater attraction to water, making them more capable of eliminating lead (II) and cadmium (II) ions from effluents.[14,15]

Cellulose is the most abundant material obtained as a prospective candidate in different resources such as plants, algae, microorganisms, etc.[16,17] This cellulose, being an organic material, forms a prime root for the production of nanocomposites.[18] The main commercial sources of cellulosic supplies are wood and cotton. It is extracted from feedstock using a suitable reagent to degrade the lignin under pressure.[19] Cellulose is an extended form of carbon chain comprising β 1,4-D glucopyranosyl units. These generated cellobiose units are interlinked with each other to form crystalline structures of cellulose termed primary fibrils. These primary fibrils fuse collectively to form microfibrils, and at later stages, macrofibrils or cellulosic fibers are formed.

8.2 ERADICATION OF VARIOUS POLLUTANTS USING NANOCELLULOSE-BASED MATERIALS

Nanocellulose is one of the feasible alternatives to eliminate several toxins from water compared with biopolymers like proteins, chitosan, starch, etc. due to its nature of biodegradability, sufficient obtainability, and discrete structural arrangements.[20] For instance, Liu and fellow workers reported on the grafting of cellulose nanocrystals using acrylic acid, which serves as an adsorbent for the elimination of heavy metals such as cadmium, lead, and copper that enter water bodies not only through petrochemical refineries, dyes, batteries, fertilizers, and metal manufacturing units but also from electroplating industries.[21,22] Otherwise, these heavy metals in water are less mobile, nonbiodegradable, and accumulate in biological systems leading to environmental hazards.[23] Recent studies have proposed modified cellulose with imidazole as an excellent adsorbent for the discharge of chromium from the effluent.[24]

Luo and fellow workers developed an effective method for the exclusion of methylene blue dye present in waste water using solutions of lignin, hemicellulose, and hybrid films of nanocellulose, which was evidenced by enhanced mechanical characteristics.[25] Cellulose nanocrystals have also been used for the removal of other dyes like blue B-RN, reactive light-yellow K-4G, acid red GR, and Congo red 4BS, B-4RFN. Further modification of nanocellulose by polyhedral oligomeric silsesquioxane with multi-N-methylol (R-POSS) exhibited better adsorption of yellow B-4RFN and blue B-RN reactive dyes; hence, nanocellulose has proven to be more effective in the removal of reactive dyes from waste water in comparison with pure cellulose.[26,27] A literature survey reveals that the nanocomposite materials such as magnetic (Fe_3O_4) recyclable cellulose act as good adsorbents for Congo red dye removal. Many reports confirmed the removal of anionic dye by nanocellulose, acid red GR, reactive light-yellow K-4G, and Congo red 4BS, which were effectively removed by cellulose nanocrystals with amphoteric polyvinyl amine.[28] Recent studies have shown that hairy nanocellulose was more effective in the adsorption of methylene blue.[29] Also, a greener one-step process was possible by zinc oxide/cellulose nanocrystal hybrids, which is not only efficient in the removal of malachite green and methylene blue but is also a cheap adsorbent of cationic dyes from waste water.[30]

Recent studies reveal that cellulose nanocrystals are used as nanofillers in the development of membranes due to their hydrophilic nature, which is quite economical and also offers good mechanistic properties. Cellulose nanocrystals are less toxic, inexhaustible, and able to be tailored merely to enrich their surface properties, thus serving as a desirable candidate to substitute traditional inorganic fillers.[31] Cellulose nanocrystals were incorporated into a polyamide layer to increase hydrophilicity and penetrability due to the formation of thin nanofiltration membrane composites.[32] It has been reported that carbon nanofibers or cellulosic membranes are utilized in the reverse osmosis process during water desalination.[33] These membranes also execute antifouling and antibacterial properties on under water surfaces.[34]

8.3 CONCLUSION AND FUTURE SCOPE

Nanocellulose and its derivatives have proven to be efficient in the elimination of several impurities present in effluents, in particular, organic and inorganic impurities as they contain carboxyl and hydroxyl functional groups on the surface. These are ecofriendly, inexhaustible, and quite economical; hence, they substitute for other existing approaches used for waste water treatment. Nanocellulose exhibits better adsorption capabilities toward the ejection of heavy metals, dyes, etc. than the normal adsorption materials. Nanocellulose exhibits a prominent role in the purification of water like membrane filtration, flocculation, catalytic degradation, etc., following various strategies such as electrostatic attraction, Vander Waal's forces, hydrogen bonding, and so on. The major challenge lies not only in exploiting nanocellulose for the discharge of toxins from effluents industrially at a larger scale in a profitable manner but also in the intricacy found during rejuvenation, recycle of nanocellulose components, and gathering of toxins after each phase of treatment, which is a major drawback. Nonetheless, these nanocellulose derivatives have an abundant potential for treating industrial waste water in the forthcoming years as the level of pollutants in water will be enhanced due to population explosion, urbanization, and industrialization. It is predictable that recent progress in nanocellulose-based materials will extricate the issues incompatible with water remediation on an immense scale.

KEYWORDS

- **renewable materials**
- **nanocellulose**
- **nanocrystals**
- **nanofibrils**
- **nanocomposite**
- **waste water treatment**

REFERENCES

1. Water Q&A: Why Is Water the "universal solvent"? Water Science School. United States Geological Survey, U.S. Department of the Interior. 20 June 2019. Archived from the original on 6 Feb 2021. Accessed on 15 Jan 2021.
2. "10.2: Hybrid Orbitals in Water" Chemistry Libre Texts, 18 March 2020. Archived from the Original on 30 July 2022. Accessed on 11 April 2021.

3. Gehrke, I.; Geiser, A.; Somborn-Schulz, A. Innovations in Nanotechnology for Water Treatment. *Nanotechnol. Sci. Appl.* **2015,** *8*, 1–17. https://doi.org/10.2147/NSA.S43773
4. Wilson, R.; Joy, J.; George, G.; Anuraj, V. Nanocellulose: A Novel Support for Water Purification. In: *Advanced Environmental Analysis*, 2016; pp 456–476.
5. Lee, K. Y., Ed. *Nanocellulose and Sustainability: Production, Properties, Applications, and Case Studies*; CRC Press, 2018.
6. Vázquez, M.; Vélez, D.; Devesa, V.; Puig, S. Participation of Divalent Cation Transporter DMT1 in the Uptake of Inorganic Mercury. *Toxicology* **2015,** *331*, 119–124.
7. Trache, D.; Hussin, M. H.; Chuin, C. T. H.; Sabar, S.; Fazita, M. N.; Taiwo, O. F.; Haafiz, M. M. Microcrystalline Cellulose: Isolation, Characterization and Bio-Composites Application—A Review. *Int. J. Biol. Macromol.* **2016,** *93*, 789–804.
8. Liu, Q.; Li, Y.; Chen, H.; Lu, J.; Yu, G.; Möslang, M.; Zhou, Y. Superior Adsorption Capacity of Functionalised Straw Adsorbent for Dyes and Heavy-Metal Ions. *J. Hazard. Mater.* **2020,** *382*, 121040.
9. Liu, Q.; Zhou, Y.; Lu, J.; Zhou, Y. Novel Cyclodextrin-Based Adsorbents for Removing Pollutants from Wastewater: A Critical Review. *Chemosphere* **2020,** *241*, 125043.
10. Miao, P.; Tang, Y.; Wang, L. (2017). DNA Modified Fe3O4@Au Magnetic Nanoparticles as Selective Probes for Simultaneous Detection of Heavy Metal Ions. *ACS Appl. Mater. Interfaces* **2017,** *9* (4), 3940–3947.
11. Fan, X. M.; Yu, H. Y.; Wang, D. C.; Mao, Z. H.; Yao, J.; Tam, K. C. Facile and Green Synthesis of Carboxylated Cellulose Nanocrystals as Efficient Adsorbents in Wastewater Treatments. *ACS Sustain. Chem. Eng.* **2019,** *7* (21), 18067–18075.
12. Qiao, H.; Zhou, Y.; Yu, F.; Wang, E.; Min, Y.; Huang, Q. et al. Effective Removal of Cationic Dyes Using Carboxylate-Functionalized Cellulose Nanocrystals. *Chemosphere* **2015,** *141*, 297–303.
13. Mohammed, N.; Grishkewich, N.; Waeijen, H. A.; Berry, R. M.; Tam, K. C. Continuous Flow Adsorption of Methylene Blue by Cellulose Nanocrystal-Alginate Hydrogel Beads in Fixed Bed Columns. *Carbohydr. Polym.* **2016,** *136*, 1194–1202.
14. Sharma, P. R.; Chattopadhyay, A.; Sharma, S. K.; Geng, L.; Amiralian, N.; Martin, D.; Hsiao, B. S. Nanocellulose from Spinifex as an Effective Adsorbent to Remove Cadmium (II) from Water. *ACS Sustain. Chem. Eng.* **2018,** *6* (3), 3279–3290.
15. Sharma, P. R.; Chattopadhyay, A.; Zhan, C.; Sharma, S. K.; Geng, L.; Hsiao, B. S. Lead Removal from Water Using Carboxycellulose Nanofibers Prepared by Nitro-Oxidation Method. *Cellulose* **2018,** *25* (3), 1961–1973.
16. Trache, D.; Tarchoun, A. F.; Derradji, M.; Hamidon, T. S.; Masruchin, N.; Brosse, N.; Hussin, M. H. Nanocellulose: From Fundamentals to Advanced Applications. *Front. Chem.* **2020,** *8*, 392.
17. Trache, D.; Tarchoun, A. F.; Derradji, M.; Mehelli, O.; Hussin, M. H.; Bessa, W. Cellulose Fibers and Nanocrystals: Preparation, Characterization, and Surface Modification. In: *Functionalized Nanomaterials I*; CRC Press, 2020; pp 171–190.
18. Klemm, D.; Heublein, B.; Fink, H. P.; Bohn, A. Cellulose: Fascinating Biopolymer and Sustainable Raw Material. *Angew. Chem. Int. Ed.* **2005,** *44* (22), 3358–3393.
19. Sjostrom, E. *Wood Chemistry: Fundamentals and Applications*; Elsevier, 2013.
20. Yang, L.; Hsiao, W. W.; Chen, P. Chitosan–Cellulose Composite Membrane for Affinity Purification of Biopolymers and Immunoadsorption. *J. Membr. Sci.* **2002,** *197* (1–2), 185–197.

21. Meena, A. K.; Mishra, G. K.; Rai, P. K.; Rajagopal, C.; Nagar, P. N. Removal of Heavy Metal Ions from Aqueous Solutions Using Carbon Aerogel as an Adsorbent. *J. Hazard. Mater.* **2005,** *122* (1–2), 161–170.
22. O'Connell, D. W.; Birkinshaw, C.; O'Dwyer, T. F. Heavy Metal Adsorbents Prepared from the Modification of Cellulose: A Review. *Bioresour. Technol.* **2008,** *99* (15), 6709–6724.
23. Pan, Y.; Liu, Z.; Wang, W.; Peng, C.; Shi, K.; Ji, X. Highly Efficient Macroporous Adsorbents for Toxic Metal Ions in Water Systems Based on Polyvinyl Alcohol–Formaldehyde Sponges. *J. Mater. Chem. A* **2016,** *4* (7), 2537–2549.
24. Peng, X.; Yan, Z.; Hu, L.; Zhang, R.; Liu, S.; Wang, A.; Chen, L. Adsorption Behavior of Hexavalent Chromium in Aqueous Solution by Polyvinylimidazole Modified Cellulose. *Int. J. Biol. Macromol.* **2020,** *155*, 1184–1193.
25. Luo, J.; Huang, K.; Zhou, X.; Xu, Y. Hybrid Films Based on Holistic Celery Nanocellulose and Lignin/Hemicellulose with Enhanced Mechanical Properties and Dye Removal. *Int. J. Biol. Macromol.* **2020,** *147*, 699–705.
26. Xie, K.; Jing, L.; Zhao, W.; & Zhang, Y. (2011). Adsorption Removal of Cu2+ and Ni2+ from Waste Water Using Nano-Cellulose Hybrids Containing Reactive Polyhedral Oligomeric Silsesquioxanes. *J. Appl. Polym. Sci.* **2011,** *122* (5), 2864–2868.
27. Xie, K.; Zhao, W.; He, X. Adsorption Properties of Nano-Cellulose Hybrid Containing Polyhedral Oligomeric Silsesquioxane and Removal of Reactive Dyes from Aqueous Solution. *Carbohydr. Polym.* **2011,** *83* (4), 1516–1520.
28. Jin, L.; Sun, Q.; Xu, Q.; Xu, Y. Adsorptive Removal of Anionic Dyes from Aqueous Solutions Using Microgel Based on Nanocellulose and Polyvinylamine. *Bioresour. Technol.* **2015,** *197*, 348–355.
29. Tavakolian, M.; Wiebe, H.; Sadeghi, M. A.; van de Ven, T. G. Dye Removal Using Hairy Nanocellulose: Experimental and Theoretical Investigations. *ACS Appl. Mater. Interfaces* **2019,** *12* (4), 5040–5049.
30. Guan, Y.; Yu, H. Y.; Abdalkarim, S. Y. H.; Wang, C.; Tang, F.; Marek, J.; Yao, J. M. Green One-Step Synthesis of ZnO/Cellulose Nanocrystal Hybrids with Modulated Morphologies and Superfast Absorption of Cationic Dyes. *Int. J. Biol. Macromol.* **2019,** *132*, 51–62.
31. Tayeb, A. H.; Amini, E.; Ghasemi, S.; Tajvidi, M. Cellulose Nanomaterials—Binding Properties and Applications: A Review. *Molecules* **2018,** *23* (10), 2684.
32. Bai, L.; Liu, Y.; Bossa, N.; Ding, A.; Ren, N.; Li, G.; ... Wiesner, M. R. (2018). Incorporation of Cellulose Nanocrystals (CNCs) into the Polyamide Layer of Thin-Film Composite (TFC) Nanofiltration Membranes for Enhanced Separation Performance and Antifouling Properties. *Environm. Sci. Technol.* **2018,** *52* (19), 11178–11187.
33. Dabaghian, Z.; Rahimpour, A.; Jahanshahi, M. Highly Porous Cellulosic Nanocomposite Membranes with Enhanced Performance for Forward Osmosis Desalination. *Desalination* **2016,** *381*, 117–125.
34. Xu, C.; Chen, W.; Gao, H.; Xie, X.; Chen, Y. Cellulose Nanocrystal/Silver (CNC/Ag) Thin-Film Nanocomposite Nanofiltration Membranes with Multifunctional Properties. *Environ. Sci.: Nano* **2020,** *7* (3), 803–816.

CHAPTER 9

Reliable, Resilient, and Sustainable Water Management of Pharmaceutical Industry

MANSI P. BANDAWANE, DEEPTI N. CHAUDHARI, PRIYANKA R. GUPTA, SWATI G. TALELE, and LAXMIKANT B. BORSE

Department of Pharmaceutics, Sandip Institute of Pharmaceutical Sciences, Nashik, Maharashtra, India

ABSTRACT

The most common raw material utilized in the production of finished dosage forms, intermediates, and active pharmaceutical ingredients (API) is water. High-purity water is distinctive as a raw material in that it is the only element that must be manufactured by the maker since it cannot be purchased in a ready-to-use state from a vendor. Water that has been purified is used to prepare pharmaceutical goods other than those that need sterile or apyrogenic water. The pharmaceutical business invests significant resources in the creation and upkeep of water purification systems primarily to control water quality, particularly the microbiological purity, ionic contaminants, and total organic carbon (TOC). For all varieties of water used in the manufacturing of medications, the USP provides descriptions and instructions. PW, WFI, sterile water for injection, and bacteriostatic water for injection are among the specific monographs in the USP. In a lot of cases, potable water from wells or other surface sources is used in the manufacture of API. This chapter includes various ion removal and water purification methods. This chapter specifically addresses the chemical, bacterial, bacterial endotoxin, and other factors that are related to each of the pharmaceutical grades of water.

Resilience and Sustainability for Energy and Water: New Advancements, Opportunities, and Framework. Cristóbal Noé Aguilar, A. K. Haghi, Amritanshu Shukla, Ann Rose Abraham, Swati Gokul Talele (Eds.)

9.1 INTRODUCTION

9.1.1 PHARMACEUTICAL WATERS DEFINITION

Several pharmacopoeias categorize the water and steam utilized in the pharmaceutical industry and associated professions. Compendial waters are categorized by the U.S. Pharmacopeia (USP) as follows:

- Water for injection
- Bacteriostatic water for injection
- Sterile water for inhalation
- Sterile water for injection
- Sterile water for irrigation
- Purified water
- Sterile purified water
- Water for hemodialysis
- Pure steam
- Drinking water (indirectly)

The classifications above refer to "packaged water"[1] with the exception of Drinking Water, USP Purified Water, USP Water for Injection, and USP Pure Steam. The three main fluids utilized for the majority of pharmaceutical applications are drinking water, USP purified water, and USP water for injection. Except for drinking water, all compendial water systems that produce USP Purified Water or USP Water for Injection must undergo validation. Drinking water used in a particular application typically needs to be "qualified" to a "internal" specification, which verifies that the water's quality, from a chemical and microbiological standpoint, does not change over time from set internal specifications. This certification procedure is frequently used to create and execute documentation that describes the nature of the system in a manner similar to that used for compendial water systems, as well as to maintain control over the product water. A qualified system's internal standards may, of course, correspond to a specific USP official monograph specification, as the one for purified water. Finally, some applications might increase the USP specifications for a specific grade of water. For instance, many biotechnology water criteria call for purified water with "minimal bacterial endotoxin." The biotechnology business would incorporate an internal bacterial endotoxin definition and verify the system as a USP Purified Water system for such an application. This chapter specifically addresses the chemical, bacterial, bacterial endotoxin, and other factors that are related to each of the pharmaceutical grades of water mentioned above.

The United States Pharmacopeial Convention is a private organization that creates and publishes USP. "Expert Committees" create the content for USP, which is then released for public review and comment before being accepted. The U.S. Food and Drug Administration provides feedback, reviews, and approval for the Expert Committees and review procedures. Since the USP is updated on a regular basis with new volumes, it is advised that references to the USP include the volume number of the most recent addition and/or edition, which includes all "Supplements."[2]

9.1.1.1 PHARMACEUTICAL WATERS DEFINITION: EP, JP, BP, ETC.

As said, various pharmacopeias, such as the European Pharmacopeia (EP), Japanese Pharmacopeia (JP), and the British Pharmacopeia, classify water and steam utilized in the pharmaceutical industry and associated disciplines (BP). There have been numerous attempts over the past few years to "harmonize" the descriptions, requirements, and production methods for compendial waters. Even though there has been a lot of improvement, this chapter will focus on several important discrepancies.

9.1.2 SPECIFICATION OF WATER

TABLE 9.1 Specification of Distilled Water/Water for Injection

Sr. No.	Parameters	Specifications
1.	Appearance	Clear, colorless, no visible particles
2.	Odor	Odorless
3.	pH	5.0–7.0
4.	Acidity or alkalinity	NMT 0.1 mL of 0.01M NaOH/HCl
5.	Chloride	0 ppm
6.	Oxidizable substances	0 ppm
7.	Sulfates	0 ppm
8.	Total hardness	0 ppm
9.	Total dissolved solid (TDS)	NMT 10.0 ppm
10.	Conductivity	NMT 0.1 ms
11.	Microbial count	fu/100 mL

TABLE 9.2 Specification of Purified Water.

Sr. No.	Parameters	Specifications
1.	Appearance	Clear, colorless, no visible particles
2.	Odor	Odorless
3.	pH	5.0–7.0
4.	Acidity or alkalinity	NMT 0.1 mL of 0.01 M NaOH/HCl
5.	Chloride	0 ppm
6.	Oxidizable substances	0 ppm
7.	Sulfates	0 ppm
8.	Total hardness	0 ppm
9.	Ammonia	0.2 ppm
10.	Heavy metal	0.1 ppm
11.	Nitrate	2.0 ppm
12.	Total dissolved solid (TDS)	NMT 1.0 ppm
13.	Conductivity	NMT 1.0 ms
14.	Microbial count	100 cfu/mL and absence of pathogenic bacteria

9.1.3 CHEMICAL SPECIFICATIONS

9.1.3.1 DRINKING WATER

Chemically speaking, water designated as "Drinking Water," for uses like some initial rinsing operations and active pharmaceutical ingredient manufacturing operations, must adhere to the National Primary Drinking Water Regulations (NPDWR) of the United States Environmental Protection Agency (EPA), or equivalent regulations of the European Union, Japan, and/or World Health Organization, as applicable. The parameters would be one example, but not the only one. It is crucial to remember that the NPDWR will evolve over time, adding new criteria or altering the concentrations of regulated items. It should be highlighted that communication identifying changes to these standards should be available to all verified USP systems and systems using Drinking Water.

Consideration should be given to supplemental drinking water analysis, including feedwater to a USP purified water or USP water for injection system. The intended use of the drinking water determines the kind and quality of analyses. It might be appropriate to cleanse the water using a particulate removal filter and/or a water softening system if groundwater is used for the initial rinse stage during applications like "clean in place" (CIP) or the manufacturing of an active pharmaceutical ingredient, for instance. Whenever water softening is employed, it is important to determine whether the feedwater contains high molecular weight multivalent cations such as barium, strontium, and aluminum. These substances will have an impact on standard operating procedures (SOPs), particularly the concentration and dosage of regeneration salts during water softening system regeneration. Although multivalent cations like calcium and magnesium are not included in the NPDWR, they have an impact on how well the system works.

The operation of various water-purifying units depends on additional particular components. The quantity of naturally occurring organic material (NOM) in a surface water supply to a USP Purified Water system serves as another illustration. Organic substances will clog reverse osmosis (RO) membranes and anion resin. In addition to dictating the type of pretreatment equipment, the NOM level in feedwater will also set up an analytical monitoring program that will amply show that the chosen pretreatment activities "guard" the anion resin within the ion exchange system or RO membranes from fouling.

9.1.3.2 PURIFIED WATER

Physical Testing chapters for conductivity and total organic carbon are used to describe the chemical requirements for USP Purified Water in the Official Monograph (TOC). The TOC specification, capabilities of the TOC analyzer, "system appropriateness" requirements, and calibration requirements are provided in Physical Tests Section 643. The provision makes no distinction between "grab" sampling and laboratory analysis and online measurement requirements. The frequency of analysis is not stated in the section either. The USP Purified Water TOC limit is 0.50 mg/L. The requirements are in line with the existing EP requirements.

The specification for conductivity, method of measurement, instrument (meter and probe), calibration criteria, etc. is outlined in USP Physical Tests Section 645. This section describes a three-stage test procedure that accounts for carbon dioxide and pH. According to the EP specification, the most stringent specification, referred to as "Stage 1," is 1.3 mS/cm at 258°C or 1.1 mS/cm at 208°C. The section makes no distinction between online measurement as opposed to grab sampling and laboratory analysis. The frequency of analysis is not stated in the section either.[3]

There may be extra testing needs in some pharmacopeias. For instance, the EP specifies a maximum nitrate concentration of 0.2 mg/L. Reviewing the various pharmacopeial specifications for the nations where products will be distributed is crucial. As previously said, some systems could need further sampling and monitoring to look for significant pollutants that are not covered in the pharmacopeia. Controlling lingering disinfection byproducts like trihalomethanes, for instance, may be crucial for a purified water system that uses ozone for microbial control (storage and distribution system). Trihalomethane molecules will be converted to carbon dioxide by the action of ozone. This carbon dioxide then reacts with water to produce the hydronium and bicarbonate ions, which increases the conductivity of purified water.

9.1.3.3 USP STERILE PURIFIED WATER

Sterilized and properly packaged USP Purified Water is simply USP Purified Water. Antimicrobial agents are absent. The Official Monograph defines the chemical requirements for USP Sterile Purified Water, which at this time consist of the following:

- Oxidizable materials
- conducting ability

Packaged waters must pass chemical tests similar to those used in a lab, unlike drinking water, USP purified water, or USP water for injection.

9.1.3.4 USP WATER FOR INJECTION

The chemical requirements for USP Water for Injection and USP Purified Water are the exact same. Nitrate specifications (0.2 mg/L, maximum) are part of the EP chemical specification for Water for Injection.[4]

9.1.3.5 USP STERILE WATER FOR INJECTION

Sterilized and properly packaged USP Water for Injection is USP Water for Injection. There are no additional ingredients or antimicrobials in it. It must pass the chemical and physical evaluations outlined in the most recent USP version with all supplements.

9.1.3.6 USP BACTERIOSTATIC WATER FOR INJECTION

The USP Water for Injection is sterile and properly packed and bacteriostatic. One or more appropriate antibacterial agents are present. It must pass the chemical and physical tests outlined in the most recent USP version with all supplements.

9.1.3.7 USP STERILE WATER FOR IRRIGATION

USP Sterile Water for Irrigation is packaged properly and sterilized to USP Water for Injection standards. It does not include any additional ingredients or antimicrobials. It must pass the chemical and physical evaluations outlined in the most recent USP version with all supplements.

9.1.3.8 USP STERILE WATER FOR INHALATION

USP Sterile Water for Inhalation is packaged properly and sterilized to USP Water for Injection standards. Except when used in humidifiers or other similar devices that may be susceptible to contamination over time, it does not contain any antibacterial agents or other additives. It must pass the chemical and physical evaluations outlined in the most recent USP version with all supplements.

9.1.4 ALERT LIMITS AND BACTERIA ACTION

The USP Official Monographs for pharmaceutical waters do not contain any bacterium standards, with the exception of the mention of sterility for packaged waters. Water for drinking has a definition. The NPDWR of the EPA makes special reference to regulations concerning the absence of coliform bacteria. The USP Purified Water and Water for Injection Official Monographs, on the other hand, solely specify that the feedwater to the water purification system producing Purified Water or Water for Injection must fulfill the NPDWR (or other appropriate European, Japanese, or World Health Organization standards). Therefore, it is crucial to note that the USP General Information Section 1231, "Water for Pharmaceutical Purposes," contains information about bacterial levels. Although this section contains useful information to complement the Official Monographs, it is implied that the information in this area is only for "general information." The EP Purified Water and Water for Injection Monographs[5] directly incorporate suggested maximum total viable bacteria levels and the enumeration procedure.

The following definitions are given when referring to the "Alert" and "Action" levels for bacteria in the USP General Information section[2]:

Events or levels known as alert levels, when they take place or are exceeded, show that a process may have deviated from its intended state. Excursions at the Alert Level are a warning and do not always call for a course correction. However, alarm level excursions typically result in the alerting of QA and water system operation employees. Excursions to the alert level may also trigger extra monitoring, which entails a closer examination of the associated and nearby data as well as other process indicators.

Events or higher levels known as "action levels" show that a process is likely straying from its regular operating range when they happen or are exceeded. Examples of Action Level "events" include continually exceeding Alert Levels, exceeding a higher microbial level in numerous sites at once, or repeatedly recovering a particular group of unfavorable bacteria. In order to take quick remedial action to return the process to its normal operating range, exceeding an action level should result in notice of both QA and staff involved in water system operations.

While it should be noted that the definitions given above are taken straight from the USP General Information section, they serve as a great "guideline" for the rest of the USP's content and give a clear explanation of what an alert and action level is, as well as the appropriate course of action.

According to the USP General Information section, the recommended action levels at their maximum are as follows:

- Water for drinking: 500 CFU/mL
- Purified water: 100 cfu/mL
- Water for injection: 10 cfu/100 mL

Although the suggested EP upper limits for purified water and water for injection are comparable, there are differences in the microbiological enumeration methodology. Actual microorganism Warning and Action Limits for Purified Water and Water for Injection Systems are typically lower than the specified levels, as mentioned in this text.[6]

9.1.5 SPECIFICATION OF BACTERIAL ENDOTOXINS

The following is a list of the bacterial endotoxin requirements for USP Waters (bulk and packaged):

- Drinking Water: none

- USP Purified Water: none
- USP Sterile Purified Water: none
- USP Water for Injection: <0.25 EU/mL
- USP Sterile Water for Injection: <0.25 EU/mL
- USP Bacteriostatic Water for Injection: <0.5 EU/mL
- USP Sterile Water for Irrigation: <0.25 EU/mL
- USP Sterile Water for Inhalation: <0.5 EU/Ml

9.1.6 ADDED SUBSTANCES

An "added substance" is what the General Notices section of USP refers to. An added substance is described as:

Only those particular Added Substances that are approved by the Official Monograph may be present in Official Substances.

It includes no additional component, according to the official monographs for USP Purified Water and Water for Injection.

There is an exception to the added ingredient regulation that should be thoroughly explained for USP Purified Water systems. Ozone is frequently used in USP Purified Water systems to control microbes. These systems include a storage system, distribution pump, inline ultraviolet units that destroy dissolved ozone, and a distribution loop. Ozone breaks down quickly into oxygen. Ozone is not regarded as an additional ingredient when it is used as an oxidizing agent to kill bacteria or when it is broken down to produce oxygen. Nonetheless, it is crucial to take into account the interaction between ozone and organic material, specifically NOM, that may be present in USP Purified Water after ozone treatment. Many references to unfavorable, partially oxidized NOM by-products of the ozonation process as well as "stable" chemicals, such as formaldehyde and chloroform, connected to the ozonation of NOM can be found in the literature.[7–12] The makeup water's TOC concentration, contact time, and dissolved ozone concentration are undoubtedly significant factors that need to be taken into consideration. If the Purified Water generating system does not contain a procedure for eliminating NOM, it is highly probable that the unwanted fragmented compounds of NOM may be created. It is impossible to determine the rate and extent of the reaction of the NOM with ozone, with the possible exception of a recorded pH decrease and an increase in carbon dioxide linked with the reaction of the NOM and ozone.

Additional analysis may be necessary to identify additional compounds connected to the above-described condition in the USP Purified Water system using ozone. These analyses may make use of mass spectrometry and gas chromatography (GC) (MS). It is challenging to gather, move, handle, and analyze samples. The chemicals created by the reaction frequently do not exist in the "library" of compounds kept by the analytical laboratory that was used. These substances are frequently recognized as a "peak" with an estimated molecular weight and/or chemical makeup. With seasonal and climatic changes in the water supply to the USP Purified Water production system, there will be variations in the concentration and chemical profile of fragmented organic material associated with these phenomena. The General Notices portion of USP is an "enforceable" addition to the Official Monographs for USP Purified Water and USP Water for Injection, in contrast to USP's General Information section.

9.2 ION REMOVAL TECHNIQUES—REVERSE OSMOSIS

Reverse osmosis has, and will continue to, supersede all other ion removal technologies as a unit operation in water filtration systems. It is suggested that most pharmaceutical water systems use reverse osmosis as the primary method of ionic removal due to the development of RO membranes that can reject more than 99% of the ions in the feedwater, the development of composite type membranes, and the increasing use of hot water sanitizable membranes. As mentioned in Chapter 4B, ion exchange (rechargeable canisters for smaller applications and regenerative units for high-volume applications) may still be used for systems requiring infrequent or small daily volumes of water as well as for some active pharmaceutical ingredient applications requiring significant amounts of water in a very short period of time. Regrettably, many of the technical terms, fundamental operating principles, and RO "jargon" demand that those working with pharmaceutical water purification systems have a working knowledge of the RO process. Although the majority of these people have a fundamental understanding of the ion exchange procedure, they view reverse osmosis as a "black box." This is plainly shown by the contrast between the availability of technical courses on ion exchange and the number of courses accessible each year that discuss reverse osmosis.

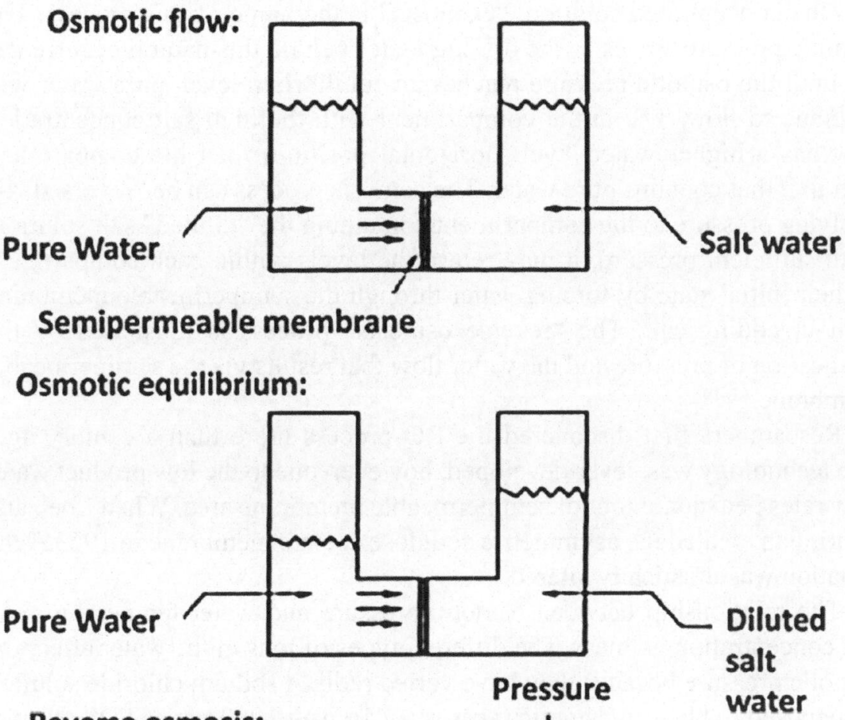

FIGURE 9.1 Osmosis and reverse osmosis processes. (The pure water level can be increased further by increasing the pressure, which further concentrates the salt solution).

It is frequently used to illustrate the reverse osmosis theory with the following example. On one side of a semipermeable membrane in an apparatus, a water solution with a high salt content is added. On the opposite side of the semipermeable membrane, in the adjacent compartment of the same apparatus, pure water is placed. The starting water levels are set up so that there is an equal amount of water in each of the two compartments. The semipermeable membrane allows water to pass from the diluted solution into the concentrated solution. "Osmosis" is the name of this process. The osmotic pressure serves as the driving factor behind this natural occurrence. Up until the osmotic pressure reaches an equilibrium level, pure water will continue to flow. The initial compartment with the high salt concentration now has a higher water level (horizontal position) than the compartment next to it that contains pure water. The osmosis process can be "reversed" by applying pressure to the compartment containing the "diluted" salt solution. With sufficient pressure, it may return the levels within each compartment to their initial state by forcing water through the semipermeable membrane without adding salt. The "reverse osmosis" process is illustrated by the application of pressure and the water flow that results via the semipermeable membrane.[13–19]

Researchers first discovered the RO process more than a century ago. The technology was never developed, however, due to the low product water flow rates per square foot of semipermeable membrane area. When Loeb and Sourirajan created the asymmetric cellulose acetate membrane in 1963,[20] the situation was drastically altered.

The relationship between osmotic pressure and water temperature and salt concentration is linear. The different types of ions in the water affect the osmotic pressure because they have varied radii. A sodium chloride solution is supposed to have an osmotic pressure of 10 psig/1000 mg of TDS at room temperature.

The process of reverse osmosis allows for the nearly complete removal of multivalent and monovalent ions from water. As a result of ongoing research and development, RO elements can now have membranes that can reject more than 99% of the dissolved ions in a feedwater stream. Organic material having a molecular weight larger than 150 Da and practically all reactive silica will be removed by conventional RO membranes. According to the literature, a study of 200 RO units carried out in 1994 found that the average water recovery rate (the proportion of feedwater recovered as product water) was 71%.[21] The majority of RO membrane systems used in pharmaceutical water systems are built to function at a feedwater recovery rate of roughly 75%. According to published research, "brackish water RO membranes," which use membranes resembling those used in pharmaceutical water systems, have a water recovery rate of 15% to 20% per membrane element.[22]

In a double-pass reverse osmosis process, two separate RO systems are run in parallel. The second unit receives its feedwater from the first unit's product water. The number of pharmaceutical water systems using double-pass reverse osmosis as their primary ion removal method increased as a result of the change in USP Purified Water and Water for Injection chemical test methods in 1996 (USP 23 Fifth Supplement conductivity specification of 1.3 mS/cm at 258°C for Stage 1). However, within a few years, the utilization of double-pass reverse osmosis for pharmaceutical water primary removal applications fell sharply with the development of "secondgeneration" continuous electrodeionization (CEDI) as a polishing method for RO product water.

9.2.1 REVERSE OSMOSIS AND POLISHING SYSTEM DESIGN

The continual operation of RO units is strongly advised. Moreover, the continuous recirculation system ought to include downstream polishing components like CEDI, which is covered later in the chapter. Recirculation of pretreatment components or the pretreatment system as a whole is recommended to reduce bacterial growth. An RO break tank should be taken into consideration to make the suggested "RO/polishing component loop" easier to construct, run, and maintain. This tank serves the following purposes:

- The tank gives the RO unit a depressurized area where pretreated makeup water can flow.
- The tank creates an "air break" between the makeup water that has been previously treated and the water that is being held there, preventing the entrance of germs from biofilm from the piping or tubing carrying the pretreated water to the tank.
- When product water is not being transported to a downstream tank (Purified Water) or giving makeup to a distillation unit, the tank serves as a location for the recirculation of RO/polishing component water (Water for Injection).
- The tank offers a site for the entrance of the chemical sanitizing agent for chemically sanitized systems. The RO/polishing component may be completely cleaned by a sanitizing chemical that then flows back to the tank.
- A 316 L stainless steel tank may be used with a heating jacket or a downstream heat exchanger for routine hot water sanitization in RO/polishing component systems.

9.2.2 THE ELIMINATION OF REMAINING CHLORINE OR REMAINING CHLORAMINES FOR USES WITH THIN-FILM COMPOSITE MEMBRANES

As previously mentioned, thin-film composite polyamide types make up a very large portion of RO membranes utilized in pharmaceutical water systems. Although these membranes have a number of advantages over cellulostic membranes, they are not resistant to residual disinfection agent. Conventional thin-film composite membranes, which have a three-year membrane life, can only withstand a maximum chlorine residual concentration of 0.038 mg/L, or 200 to 1000 ppm-hours of chlorine. When opposed to chlorine, RO is typically more tolerant to chloramines. This is partially mitigated by the fact that feedwater contains two to three times as much chloramine as residual chlorine. A crucial step in the pretreatment process for RO units using thin-film composite membranes is the removal of chloramine. With an activated carbon unit running at a face velocity of 3 gpm/ft^2 of cross-sectional bed area and a volumetric flow rate of roughly 1.0 gpm/ft^3 of activated carbon medium, 100% residual chlorine removal is possible. Nevertheless, contact time is crucial to guarantee the elimination of chloramines, necessitating a volumetric flow of activated carbon media of approximately 0.5 gpm/ft^3. Moreover, activated carbon type, feedwater pH, and TOC should be taken into account. At the absolute least, media should be changed every six months. Thin-film composite membranes will fail if they are exposed to chloramines and residual chlorine.

9.2.2.1 SAMPLING PROVISIONS FOR REVERSE OSMOSIS UNITS

An RO unit's sample provisions should make it possible for the operational staff to assess the unit's performance, prevent frequent unscheduled maintenance procedures, such as chemical sanitization for removing microbiological fouling, and assess system parameters while on excursions.

Sample-taking rules could include the following:

- Feedwater, product water, and wastewater monitoring for conductivity, TOC, pH, bacteria, and bacterial endotoxins.
- A representative sample of the feedwater for the RO unit is provided by the SDI measuring provisions, which are located downstream of the prefilters in the RO system. The sample location for SDI assessment should be placed downstream of the recycle lines if permeate and/or trash recycling are utilized.

- There should be sampling provisions for each pressure vessel in systems with numerous pressure vessels. Provision for "probing," a technique that allows insertion of a sample hose or tube "down" the permeate collector in an effort to identify one or more membranes that have lost integrity or interconnector O-ring failure, may be incorporated for pressure vessels holding several RO membranes (Figure 9.2).
- A potential "breach" of system integrity, specifically the entry of bacteria to the clean side of the RO membranes, should be taken into account while designing the sampling arrangements in the product water line from a RO unit. Since germs will reproduce on the clean side of the membranes, this is very concerning. While probing provisions are helpful, the risk of bacterial introduction could outweigh their advantages. If available, post-probing chemical or hot water sanitization is strongly advised.
- Chemical sanitization should be possible with the RO system, as previously stated. Instead of carrying out a three-step cleaning/sanitization program on-site, it is advised that periodic membrane "rotation" be combined with an off-site RO membrane-cleaning program that is hired. Off-site cleaning of RO membranes offers "individual" cleaning as opposed to cleaning using waste from an upstream membrane, offers customization for an application, minimizes downtime, and offers "individual" cleaning. The low and high labor-intensive cleaning phases with rinse are not necessary, but chemical or hot water sanitization should be carried out after installing "rotated" cleansed RO membranes.
- If chemical cleaning is done on-site, a tank, heater, pump, and filter are recommended cleaning accessories, which will be covered later. The cleaning agent's potency is crucial. Thus, the sampling tank should have provisions for sampling. This is particularly significant when using acid to remove calcium carbonate scalants since maintaining the proper pH balance is essential to eliminating the scale from the membranes.

9.2.3 REVERSE OSMOSIS MEMBRANES

In the US, there are numerous manufacturers of RO membrane components. Nonetheless, it is thought that only five membrane manufacturers supply 90% of the RO membranes used in pharmaceutical applications. The vast majority of these membrane producers run extremely aggressive research initiatives and, more significantly, offer computer tools for choosing the RO system design (number of membranes, configuration, and array) for a particular application. As previously mentioned, recent developments in the design and production of RO membranes have led to the creation of membranes with exceptional rejection qualities. Significant commercialization and research have been conducted in the past ten years to improve cleaning, notably for microbiological fouling. The capacity to provide superior microbiological control has significantly increased because of the development of hot water sanitizable RO membranes and the availability of pharmaceutical grade full fit RO membranes. In the "Theory and Application" section, the parameters pertaining to the creation of RO membranes, including restrictions pertaining to the materials utilized to make the membranes, were presented. An RO membrane obtained from one of the businesses that supplies the bulk of membranes for pharmaceutical applications would therefore be suitable for a particular use. Nonetheless, a person building a system would want to evaluate the most recent developments in hot water sanitization and membrane cleaning, which are covered in manufacturer's literature. Even though ion rejection rates are typically quoted to be greater than 99%, reactive gases like carbon dioxide or ammonia will lower actual ionic rejection for the majority of operational systems. With RO/polishing component recirculating systems, the observed ionic rejection may actually decrease even though the product RO water conductivity continues to decline following cycling for an extended period of time (>1–2 h). A standard single-pass RO unit will ultimately be unable to generate theoretically "ion-free" product water due to the existence of the sodium ion, which has a short ionic radius.

9.2.4 REVERSE OSMOSIS PRESSURE VESSELS

A fiberglass material (filament wrapped epoxy/glass composite with high gloss polyurethane paint on the exterior) or stainless steel may be used to make the RO pressure vessel. The performance of the system will generally not be significantly impacted by the material choice. Since they improve the unit's look, several pharmaceutical firms prefer stainless steel pressure vessels. The pressure vessels used for an application should have a safety margin factor of at least two to three and be able to withstand the highest projected pressure demands, as determined by the three-year RO system estimate. Design and construction of pressure vessels should adhere to the ASME Code for "Unfired Pressure Vessels," which may be code-stamped. The following pressure ratings are available for vessels: 300, 450, 600, 1000, and 1200 psig. For pharmaceutical applications, pressure-rated vessels with ratings of 450 or 600 psig are advised. The installation qualification (IQ) for the RO system should be prepared using information from the pressure vessel manufacturer's literature.

9.2.5 REVERSE OSMOSIS UNIT INTEGRAL CLEANING SYSTEM

A responsive regulated off-site membrane chemical cleaning is advocated, as was previously mentioned. However, various design elements need to be taken into account if three-step on-site cleaning is needed. The following details connected to the cleaning system's design could be mentioned:

- The cleaning pump capacity (pressure and flow rate) should be specified for cleaning the first array of the system, which contains the greatest number of membranes.
- The cleaning system piping/tubing should be designed to allow operating personnel to clean each array separately, increasing the effectiveness of the cleaning operation.
- The cleaning piping/tubing should be arranged, as indicated, to allow cleaning of separate arrays.
- The cleaning tank should have heating provisions and be the appropriate size for the cleaning application, as was previously discussed. Several RO unit cleaning systems feature this form of heating provision, despite the fact that it has been indicated that electrical heaters in plastic tanks are dubious in terms of safety. Composite-type tanks made of fiberglass-reinforced polyester or vinylester are strongly advised as a possible replacement for polyethylene tanks for the cleaning process.
- The piping and tubing of the cleaning system should be set up to permit cleaning of the lead elements in the opposite flow direction. This is crucial because the flow channel will enable the clearance of debris from the lead elements' feedwater spacers. Historically, the feedwater spacers' "shadow side" has been described as this region.

9.3 ION REMOVAL TECHNIQUES—ION EXCHANGE

In the past, the manufacturing of medicinal grades of water has relied heavily on the ion exchange method of ion removal. Although reverse osmosis has become increasingly popular as a principal method for removing ions during the past 20 years, ion exchange is still used in some applications. Ion exchange technology is still used for water softening, a pretreatment method for reverse osmosis (see chap. 3), or applications that need softened water for direct usage in products, such as numerous applications for active medicinal ingredients. Ion exchange may also be used by USP Purified Water systems with high water instantaneous flow rates (200–300 gpm and above) for batching applications to reduce the amount of storage space needed. On the other hand, USP Purified Water systems that use small amounts of water every day (like 500 gal or less) can rely on rechargeable ion exchange canisters and require little maintenance, negating the need for extensive pretreatment, recurring flushing, and membrane cleaning for a small RO-based system.

9.3.1 ION EXCHANGE RESINS: DESCRIPTION AND CLASSIFICATION

Ion exchange resin typically includes spherical beads with a diameter ranging from 0.5 to 1.2 mm. The resin can be any color, although it usually has an opaque yellow hue. Resin, especially anionic resin, will have an amine-like smell. Anion and cation are the two types of ion exchange resin. Although the anion resin has functional sites that can exchange negative ions, the cation resin has functional sites that can only exchange positive ions. Strong acid cation exchange resin, weak acid cation exchange resin, strong base anion exchange resin, and weak base anion exchange resin are the four main subtypes of resin. Each kind of resin can be created in a macroreticular (porous) or gelular form.[23,24]

- Strong Acid Cation Exchange Resin
- Weak Acid Cation Exchange Resin
- Strong Base Anion Exchange Resin
- Weak Base Anion Exchange Resin
- Gelular and Macroreticular (Macroporous) Resins
- Styrenic and Acrylic Resins

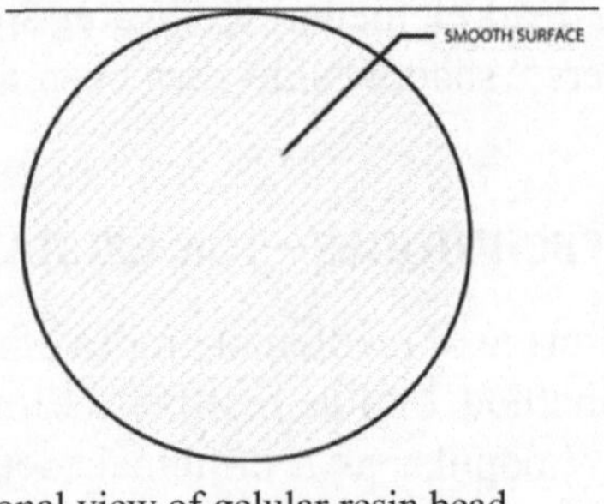

FIGURE 9.2a Cross-sectional view of gelular resin bead.

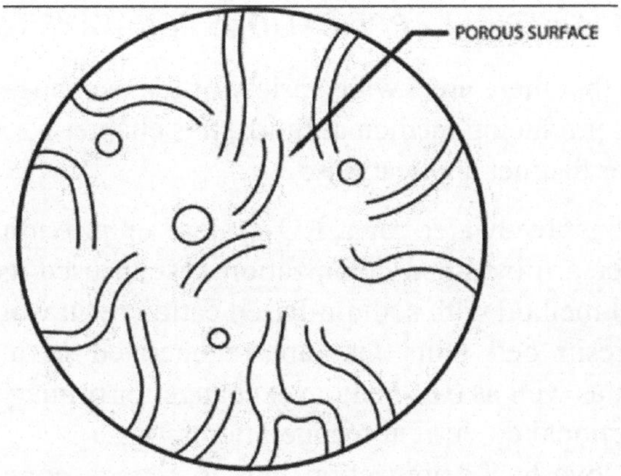

FIGURE 9.2b Cross-sectional view of macroreticular resin bead

9.3.2 PHYSICAL ELEMENTS THAT INFLUENCE ION EXCHANGE RESIN

- Anion resin, not cation resin, controls the maximum continuous operating temperature for ion exchange resin. Strong base anion resin should not typically be run continuously at temperatures higher than 1208 F.
- Ion exchange resin should not be subjected to temperatures below 328°F because the freezing of water trapped in the resin matrix will cause the resin, especially macroreticular resin, to physically fracture.
- Resin ought to be kept wet at all times. It ought to be kept in airtight plastic bags. Frequent wetting and drying of the resin causes expansion and contraction, which ultimately causes the resin beads to break and resin fines to be produced.
- Resin should not be subjected to chemicals in large quantities, especially acids and caustics. Chemical shock is the effect of this.
- If the resins are exposed to an oxidizing substance, like a residual disinfectant, there will be a slow oxidation of the resin and concomitant de-cross-linking.
- Iron will cause cation resins to foul.
- Organic material has a tendency to contaminate anion resins.
- Shock from a mechanical or physical source will cause resin to break apart. This could be brought on by excessive backwash, for instance, where resin particles hit the top or side walls of an ion exchange column.

9.3.3 PHARMACEUTICAL APPLICATIONS FOR WATER SYSTEMS

Despite the fact that there are a wide variety of ion exchange topologies and uses for systems producing medicinal fluids, this chapter's section will focus primarily on four distinct applications:

1. Pretreating feedwater for a RO system or providing regeneration water for a mixed-bed deionization system used as a primary ion removal method with sodium-based cation resin water softening.
2. Fixed resin bed units that are regenerated in a countercurrent manner, as well as two-bed ion exchange for primary ion removal in the traditional co-current regeneration mode.
3. After a two-bed deionization unit to reduce conductivity, cation polishers operated in the hydrogen form to remove trace sodium concentrations, providing water that complied with the USP Purified Water conductivity criteria.
4. Applications of mixed bed ion exchange for initial ion removal following two-bed deionization or RO systems.

These ion exchange applications share a lot of the design, operating, and maintenance requirements as well as potential issues. For instance, in any ion exchange vessel, dispersion is a concern. Comments mentioned in the original extensive description of ion exchange therefore generally apply to other applications and will not be repeated in an effort to prevent repetition. There will be appropriate references to the opening conversation.

9.4 ADDITIONAL ION REMOVAL TECHNIQUES

Several ion removal methods than traditional ion exchange and reverse osmosis are covered in the chapter's concluding section. Continuous electro-deionization is the main subject of this section (CEDI). We will briefly describe electrodialysis and electrodialysis reversal, two ion removal processes that make use of the concepts that helped shape CEDI. Even though electrodialysis and electrodialysis reversal are rarely used, it is appropriate to give a brief history of each method.

This chapter's part has now been separated into two main sections. A brief historical overview of electrodialysis and electrodialysis reversal will be given in the first section. The design, operation, and maintenance aspects of CEDI systems are covered in the second section. As said, electrodialysis and, to a lesser extent, electrodialysis reversal were the two processes from which CEDI "developed" technically. It is theoretically possible to talk about the minor changes connected with the development of CEDI once the theory of electrodialysis and electrodialysis reversal is known.

9.4.1 ELECTRODIALYSIS AND ELECTRODIALYSIS REVERSAL

By using an electronic force to physically pull the ions out of the solution and through a semipermeable membrane, electrodialysis removes ions from water. An "ion-concentrating" portion and an "ion-depleting" section are produced by this electrochemical separation process, in which ions are transported via anion- and cation-selective membranes. The electrochemical technique uses semipermeable membranes and a direct current to achieve the separation necessary to extract ions from water. Electrodialysis performs deionization by "pulling" the ions away from the water, as opposed to pushing the water away from the ions, making it distinct from other membrane technologies. Furthermore, in contrast to other membrane processes like reverse osmosis, semipermeable ion exchange membranes work at low pressure. The procedure does not fit the definition of a dewatering process and does not call for pushing water up against and through a membrane surface.

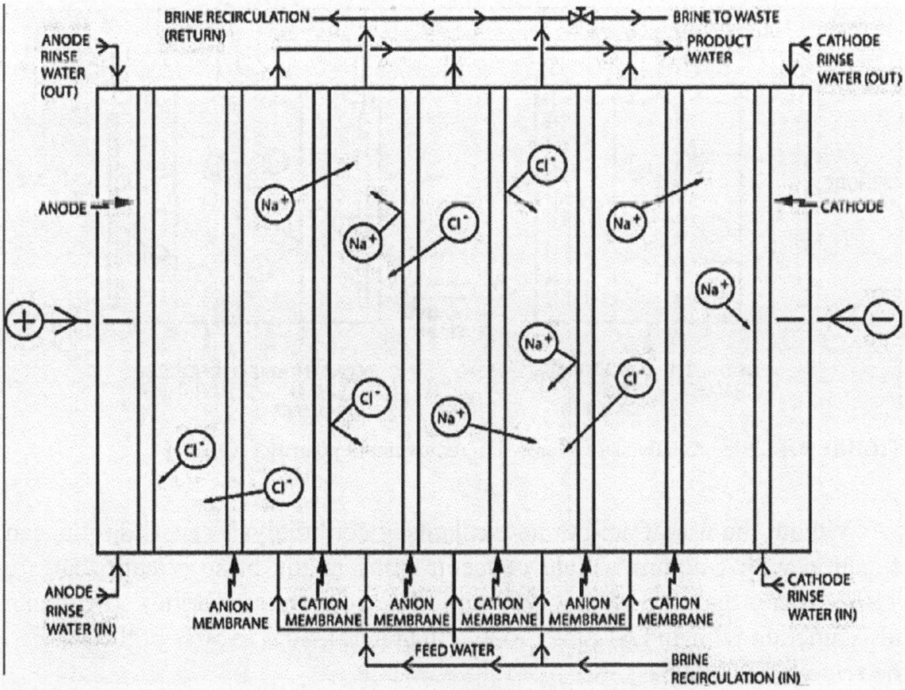

FIGURE 9.3 Typical electrodialysis cell arrangement with flow paths.

9.4.2 ELECTRODIALYSIS REVERSAL

With the exception that the polarity of the anode and cathode supplying the electric current via the ion exchange membranes is periodically reversed to remove fouling and scaling, electrodialysis reversal is very similar to electrodialysis. The early 1970s saw the development of electrodialysis reversal technology, which significantly enhances the electrodialysis procedure. With a few exceptions, electrodialysis has been supplanted by this method.

The foulants are "lifted" off the anion exchange membranes' surface upon polarity reversal. Remember that the foulants are often on the anion exchange membranes' surface and not "inside" the membrane structure. The foulants, which have a little negative charge, are physically "pulled" away from the anion membranes' surface by the polarity reversal. This method effectively removes the foulants as a consequence. Deionized water flowing through chambers that were brine concentrators before polarity reversal removes potential scale-forming ions, making the device more or less "self-cleaning." Moreover, electrodialysis reversal can be carried out in the majority of applications without adding acid or antiscalants to the feedwater.

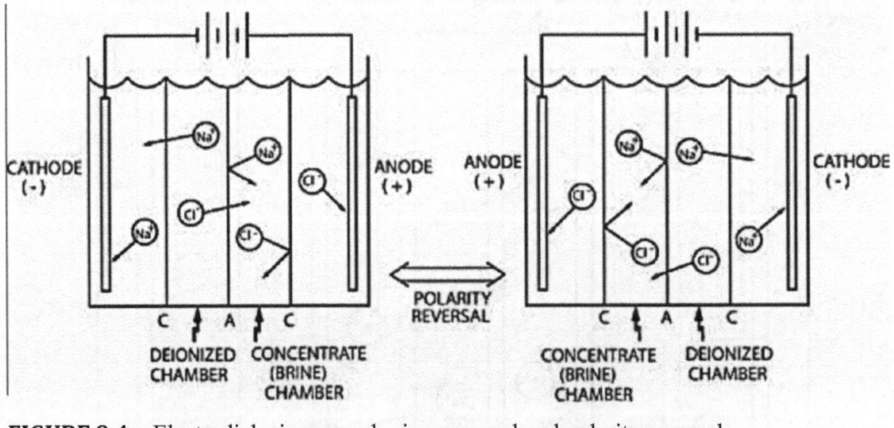

FIGURE 9.4 Electrodialysis reversal—ion removal and polarity reversal.

Without the use of acid or antiscalants, electrodialysis reversal units can function with calcium sulfate concentrations in the brine stream that are 150% higher than the solubility limit.[26] The concentrate (brine) stream can also function with an LSI of 2.1 to 2.2. In general, 40% to 50% of the salt can be removed in one pass.[25]

9.4.3 DESIGN CONSIDERATIONS

CEDI is used to polish the product water from single- and double-pass RO units.

A storage and distribution system is typically fed product water from a single- or double-pass reverse osmosis plant. Product water conductivity at the point of use will not satisfy the USP Purified Water requirements without post-RO unit polishing. Also, according to the literature, one specific CEDI manufacturer will not install the unit without a RO system upstream of the unit.[27] As a result, a single-pass RO unit with feed to a CEDI unit is typically included in system designs along with a suitably constructed pretreatment system. Although a double-pass reverse osmosis system may be used to increase the stack life of a continuous deionization unit, it is argued that the system's capital cost, complexity, operating costs, and maintenance costs do not make them worthwhile.

Although many RO/CEDI systems are built for cyclic operation, continuous flow—even when there is no demand for water—should be taken into consideration, a recommended RO/CEDI system with chemical sanitizers. An RO break tank with a conical bottom is fed with pretreated feedwater. Unpigmented polypropylene or polyethylene is used to build tanks. The tank has a full-diameter cover that fits snugly and is gasketed. The tank volume is sufficient to prevent "short-cycling" of the feedwater makeup valve for pretreated feedwater. The following should be included in the tank's accessories:

- Flanged and gusseted fittings for the introduction port of concentrated liquid sanitizing agents (top), discharge to the downstream repressurization pump (center bottom), RO/CEDI "loop" return (center top), level sensor (top), hydrophobic vent filter (top), pressure relief system (top), and CEDI waste optional recovery stream (top).
- System for hydrophobic vent filtering that uses an O-ring seal to seal the filter to the housing and an "inverted" polypropylene housing.
- Level-monitoring device of the conductance type or a comparable type with analog outlet capabilities.
- A relief device to prevent the tank from imploding or exploding in the event that an attachment malfunctions or fails.
- Steel support stand with a coating that resists corrosion.
- A positive acting pneumatically actuated diaphragm valve (air-toopen, spring-to-close) should be included in the pretreated

feedwater piping or tubing. Just upstream of the valve should be placed a sample valve and a pressure gauge. The piping or tubing should have a minimum physical length and be set up such that it can completely drain.

- The bottom discharge line from the tank to the repressurization pumping system must have an inline full-diameter pump feedwater isolation valve, a zero dead leg-type drain valve, and a compound-type pressure gauge. The compound pressure gauge is necessary to ensure that there is no pump cavitation. To satisfy the pump's net positive suction head, the "low" level tank set point may be modified (NPSH).

The RO/CEDI system with hot water sanitization capacity should incorporate the chemical sanitization provisions mentioned in the text above. Periodic hot water sanitization will not eradicate biofilm. Assuming there are no dead legs or that dead legs are exposed to the peracetic acid and hydrogen peroxide solution, chemical sanitization using a 1% solution of peracetic acid and hydrogen peroxide will eradicate biofilm. Three to six months is the recommended time between chemical sanitizations for non-hot water systems. For hot water sanitizable RO/CEDI systems, annual chemical sanitization is advised.

9.5 SYSTEM DESIGN

9.5.1 THEORY AND OPERATION

Pharmaceutical water systems should incorporate storage systems for a variety of technical reasons. A storage tank might just offer a practical way to repressurize compendial water before a distribution loop in less complicated distribution systems. Storage systems are needed for various uses, such as Water for Injection systems with distillation, in order to maintain hot Water for Injection at a high temperature, typically greater than 808°C, and to allow hot Water for Injection to be circulated. Although condensate receivers can also be used to collect the product water from distillation units, this does not replace the requirement for a storage tank because the product water must be recirculated, usually at a high temperature to ensure microbiological control (FDA, 1986). Product water from a single-pass or double-pass reverse osmosis (RO) plant, with or without polishing continuous electrode ionization (CEDI), may be minimally pressured for use in USP purified water systems (10–30 psig). Without a storage tank for a USP Purified Water system, RO membranes will not be able to withstand abrupt variations in the product water pressure caused by "point-of-use" draw-off, which, unfortunately, leads to a quick loss of RO membrane integrity. Pretreatment, ion removal, and storage and distribution are the three loops that make up a sound system architecture, as was previously mentioned.

Without citing particular instances of the necessary specifications and supporting paperwork, it is inappropriate to talk about the design, installation, validation, operation, and maintenance of pharmaceutical water purification systems. Specifications are a crucial component because they give the manufacturer or system integrator of water purification equipment the means to outline the design and performance criteria. Several manufacturers of water filtration systems concentrate their sales and marketing efforts on "standard items." Some manufacturers show versatility in terms of the design and manufacturing standards for specific components. The "ideal" specification ought to be created in a way that explicitly outlines the particulars of the unit operations needed for a project. Manufacturers of "standard equipment" will either offer "adders" for specified goods that differ from their standard product offering or will object to particular parts of the specification in their proposal. When possible, the specification should provide design information that is easily accessible from the majority of equipment manufacturers. Things that could be needed for a certain component but would unquestionably drive up its price disproportionately to its intended benefits should either be removed or provided by "others" (mechanical contractor, installer, etc.). With an activated carbon unit, for instance, a port for the removal of activated carbon may be specified. Since it is necessary for the routine replacement of activated carbon, this item is crucial. Both manufacturers of "standard" equipment and "custom" products ought to offer it. On the other hand, if the unit is specified with hot water sanitization requirements, it would be advantageous to provide an outside thermal insulation "package" for it. The majority of equipment manufacturers do not often stock the insulating package. After the equipment is put "in-place," it would be more cost-effective to install the insulating package (using a mechanical contractor). In most installations, the insulating package is a requirement due to worker safety. You should not get rid of it. The "Scope of Supply" should be specified in the system specification. A project manager or construction manager can be of great help in ensuring that nonstandard components from providers of water purification equipment are integrated into the finished system.

Subsections under "documentation and specification":

- General requirements
- Drawing requirements
- Instrumentation and control system requirements
- Operation and maintenance requirements
- Quality control/quality assurance requirements

9.5.2 GENERAL REQUIREMENTS

The system should have a basis of design prepared. As suggested, this document offers standards that outline the needs of the water system. The following ought to be at the very least part of this:

- It is necessary to classify the water according to its intended use, such as "drinking water," "softened water," "RO/DI water," "USP/EP/JP purified water," and "USP/EP/JP water for injection."
- Describe the nature of demand at several points of usage, such as production (ingredient), clean-in-place rinsing, and laboratories.
- A description of the physical points of use, such as the heat exchanger, mass flow or totalizing metre, hoses, "transfer panel," and manual diaphragm valve.
- A list of all the points of use, along with their identification numbers and peak demand (volume/unit time), is provided.
- Each point of use is included together with its identification number and the highest daily (or 8-h shift) demand (volume).
- Point-of-use matrix displaying the quantity of continuously used points. Included in this should be times when there will not be any water demand on the distribution system.
- The delivered water temperature requirements for each operation are listed in a point-of-use matrix. If relevant, the flow rate for each application at each temperature should be included in this list.
- Limits for total viable bacteria that cause "Alert" and "Action" for each type of water, including the proposed enumeration technique.
- Analysis, historical statistics, and source information for the facility's feedwater source (public or "private" supply).
- Accessible services (steam, raw feedwater flow rate, electrical, drain, ventilation, etc.).
- Regulations that are local, state, county, provincial, or other.
- List of all nations where products made with water will be used.
- Any further facility specifications.

An Equipment Specification is necessary for every pharmaceutical water filtration system. The requirements specified in the Basis of Design are illustrated in this document. It is a crucial piece of paperwork in the process of ultimately validating or qualifying the system. After considering conceptual and early design possibilities, the specification is created. For the effective water purification equipment supplier, the "submittal" requirements must be spelled out in detail. During the preparation for Installation Qualification (IQ) and Operation Qualification, the Basis of Design, full specifications, and submission materials, along with a "turnover package" from the successful equipment supplier, will be heavily utilized (OQ). Both the IQ and OQ documents are references to the Basis of Design and comprehensive specification. The "submittal package" serves as the foundation for the final handover package, which must adhere to the precise specification and is directly included into the IQ, OQ, and standard operating procedures (SOPs), which are then formatted for the end user.

The development of specification sections for unique water purification unit activities is quite advantageous. To prepare the IQ and "integrate" with all system components to prepare the OQ, this unit operation concept can be expanded. The unit operation strategy has a number of advantages even if some system components, including controls and instrumentation, are common to system operation. A pharmaceutical water purification system's design, operation, maintenance, and validation (or qualification) involve many different technical disciplines, as was covered in prior chapters. It would be incorrect to assume that everyone involved in the process will be "in-depth" knowledgeable about the parts of water purification systems, how they work, and other technical aspects of the process. Thus, the intricate design, operation, maintenance, and validation requirements of a water filtration system can actually overwhelm people. On the other hand, the unit operation technique is a very effective way to reduce the complexity of a component's technical characteristics to a level that is understandable. The "package" for qualification (or validation) should include a training handbook. During training sessions, staff can receive personal copies of this document. The sessions ought to be set up such that they can be presented to people from a variety of disciplines in a setting. The training sessions start with the system's initial component and work their way through each one. Similar to the prior chapters in this book, the training programmed is organized to cover the fundamental theory, application, design concerns, operation, and maintenance issues. It is also important to talk about the validation criteria, particularly the verbatim conformance to SOPs. When finished, a proper training session should outline the function of each component, the rationale behind its configuration with particular accessories, the best practices for operating and maintaining each component, and the importance of the validation process.

The design criteria for each component should be explicitly stated in a properly produced specification. The obvious criterion would be things like column height and diameter. Items that are less important, but they would be essential to a properly constructed specification, feedwater characteristics, product water sampling valves, pressure gauge face diameter, range, construction materials, and type, leveling provisions for column legs, column exterior paint specifications, motor horsepower, motor voltage, anticipated motor amperage requirements, and motor voltage phases. It is appropriate to take into account a description of the component as though it were being assembled while creating the full specification for a component. The specifications should specify the submission of a Bill of Materials with the equipment assembly drawings. If the specification does not explicitly state the amount of complexity, submittal requirements, and documentation, it will unavoidably lead to an "extension of scope" from the producer of water purification equipment, with a corresponding rise in the system's overall cost. With very few exceptions, a correctly written specification can never have too much information. The one thing that should be taken into consideration is specification, to the extent that the individual unit operations are actually "custom components," having details of options not easily accessible from the bulk of equipment makers, as was previously discussed.

Performance criteria for each component as well as the system as a whole should be outlined in addition to the specific design requirements for each component. This is crucial, particularly in light of the performance standards for individual components. The performance of individual components may not be stressed enough in the majority of pharmaceutical water purification system specifications, especially "performance specifications." In most cases, this leads to a system that initially satisfies defined performance standards but ultimately falls short of delivering the intended system performance. This is related to the fact that certain system components, particularly those involved in pretreatment, are not appropriately designed or sufficient for the application.

Each component's specification, including those for accessories, should make it clear that the necessary paperwork must be submitted for approval and/or delivered with the equipment's final turnover package upon delivery or reception.

9.5.3 DRAWING REQUIREMENTS

There should be a process flow diagram (PFD) available. The PFD is a simple diagram that shows the main parts of the system together with their attachments. Typically, components are displayed as “blocks.” Lines with flow direction arrows show the flow path.

Drawings of the instruments and piping (process) should be included in the P&ID. The P&ID(s) specified as part of the specification and those provided by the equipment manufacturer should align. Attachments, equipment, and P&ID(s) format should adhere to criteria established by the American Instrument Society (ISA) (chap. 10). The “legend” for all symbols used on all system P&IDs should be specified in a “lead sheet” (for one or more system P&IDs). At contract award, the manufacturer of the water purification equipment (or general contractor/system integrator) should get this information if the pharmaceutical manufacturer utilizes an established legend for parts and equipment that is not entirely consistent with ISA language. The final P&IDs should explicitly identify equipment labeling, valve number designations, instrument number designations, etc. Although the P&IDs included with the comprehensive specification are likely to have dotted lines (“envelopes”) designating specific components, only the equipment manufacturer or a professional engineer may offer P&IDs with complete instrument and valve numbering references, employing drawings provided by the manufacturer of the water purification equipment, the pharmaceutical manufacturer, or their authorized agent. It is strongly advised that the organization (equipment manufacturer or pharmaceutical business) that establishes and provides the valve and instrument tag numbers and physical tags be defined within the specification.

Drawings of the general organization ought to be offered. The operations of the water purification unit are typically arranged in two dimensions in the physical area that is available for installation in the general arrangement drawings. The drawings are obviously “scaled” in some way. In general, isometric drawings are required for elements like distribution loops with “drops” to specific points of usage. Drawings of general arrangements are a useful tool for locating utilities. Also, before “fixing” the location of support utilities, the general arrangement drawings that were submitted for review and approval aid in resolving space and access difficulties. The rigging of potential equipment can also be handled. Finally, the operational weight of each component should be included in the overall arrangement drawings. Before releasing the equipment for manufacturing, floor loading restrictions might be examined.

There should be generic assembly drawings for the equipment. The drawings typically show all aspects of a single unit's operation, including the valves and instrumentation, from top to bottom, left to right, in front and in back. Information from the general layout drawings is improved by the drawings. All "tie-ins" such as inlet, outlet, and drain are shown in the drawings, which are made to scale.

9.5.4 CONTROL SYSTEM REQUIREMENTS

The following, among others, is a list of the specific documentation that should be supplied for system controls and instrumentation:

- a hardware setup picture
- technical articles
- annotated source code
- a software development program
- a software test procedure
- I/O lists
- screen shots

9.5.5 DOCUMENTATION REQUIREMENTS FOR OPERATION AND MAINTENANCE

- All equipment manufacturers' operating instructions should be made available. It is strongly advised that the pharmaceutical manufacturer, system integrator, or hired specialty organization convert the operational manuals into specially created SOPs.
- According to the project's scope, particularly the volumetric capacity of the system, it could be necessary to create an equipment installation manual.
- A manufacturer's maintenance manual should include advice on how to maintain system parts and accessories. It is strongly advised that the details provided in the manufacturer's maintenance manual(s) be a manufacturer, systems integrator, or contracted specialty organization-prepared maintenance manual.
- All components and accessories should have vendor "cut sheets" available. Throughout the IQ, references to the catalog cut sheets will be made.
- As necessary, data sheets that add to the information in the cut sheets should be provided.

- A list of spare components ought to be provided. The list should be created taking into consideration each component individually. The model number and serial number on the list should be provided to the original manufacturer. If obtained from a supplier, spare parts should not have a "coded" number to the equipment manufacturer.

9.5.6 REQUIREMENTS FOR QUALITY CONTROL/ASSURANCE

Stainless steel and/or stainless-steel welding are mentioned in many of the QC/QA standards listed below. A reference specification for a stainless-steel distribution loop with valves for purified water can be found in Appendix B.

- For the purpose of confirming the integrity of distribution and/or linking tubing/pipes, a hydrostatic or pneumatic test process should be offered. No plastic tubing or piping material should be subjected to pneumatic testing to determine its integrity. The preferred test is a hydraulic pressure "hold" test. The pressure test acceptability criteria should be defined in the test protocol. The test technique should explicitly address temperature management since temperature will have a major impact on the pressure in tubing or pipe that is totally filled with water.
- A report from a hydrostatic or pneumatic test should be supplied. The original test log sheet with the date, time (start and stop), pressure and temperature values, and the signatures of both the test performer and a trained observer should be included in the report.
- Any stainless steel surfaces in contact with water, such as tubing, valve bodies, sanitary pressure gauges, and conductivity cells, should have material test reports available. The test reports ought to include a thorough chemical analysis of the stainless steel, the name of the organization that produced them, the date they were produced, a traceable "heat number," and a signature.
- All instruments should have certificates of calibration. Indicators and sensors are included in this. According to experience, pharmaceutical facilities prefer to commission installed instrumentation by performing a field calibration on it. The necessity to present Certificates of Calibration from the instrument manufacturer is still in place notwithstanding the "field calibration." And finally, the National Institute of Standards and Technology (NIST) must be able to be traced back to a source for Certificates of Calibration.

- For materials, a Certificate of Analysis must be given. These would contain components used for orbital welding stainless steel tube, such as gaskets, valve diaphragms, RO membrane preserver, O-rings, and inert gas.
- Weld test coupons must be offered. The lowest number of coupons needed for orbital welding stainless steel is stated in the literature (ASME BPE, 2009), and a little more rigorous test program is shown in Appendix B. It is advised that test coupons be provided for socket welded and "bead and crevice-free" plastic welds [polypropylene, polyethylene, and/or polyvinylidene fluoride (PVDF)], even if these procedures are not usually carried out.
- Procedures for mechanical polishing and testing stainless steel tubing should be offered. The methods for mechanically polishing stainless steel surfaces and for measuring the finish should be specified in the procedures.
- The mechanical polish and test results for stainless steel tubing should be delivered in strict adherence to the protocols. The test report needs to be both signed and date.
- Passivation instructions ought to be included. The type, intensity, and temperature of the alkaline cleaning process used during prepassivation should all be considered during the passivation procedures. The process should additionally include the temperature, exposure period, chemical concentration, and composition of the passivation solution. All chemicals used should have material safety data sheets (MSDS) and certificates of analysis available.
- Results from Feroxyl Testing or another alternative test method that has been approved by the "owner" should be included in the passivation test report.
- Riboflavin test results proving the required spray ball system "coverage" for storage tanks should be submitted.

9.5.7 COMMERCIAL PERSPECTIVE

The manufacturer of water purification equipment and other parties should have a clear understanding of the Scope of Supply. A multimedia filter and an activated carbon unit, for instance, are typically offered and shipped without media installation. Who is in charge of installing the media should be made very clear in the specification. It is advised that the manufacturer of the water purification equipment install all internal parts for the various activities of the water purification unit. This recommendation is based on possible warranty problems. Although it would seem like a reasonably straightforward task that could be completed by either facility staff or a subcontractor, installing support and filter media for a multimedia filtration machine can provide challenges. The operation of downstream components may be impacted if the installation procedure (done by others) directly damages the lower distributor in the multimedia filtering unit, perhaps "nullifying" any warranties offered by the provider of water-purifying equipment.

Every area of the specification should make clear the equipment and performance warranties for each component. The system warranty and, more significantly, the system performance warranty should be explicitly defined in a descriptive section of the specification that is included in the specifications for each individual component. In general, warranties for individual system parts and the entire system should not be any shorter than 12 months. The system should not be fully installed and operating before this 12-month term starts. Third-party warranties should not be covered by the warranty. This specifically applies to accessories that the maker of water-purifying equipment buys from third parties. A distribution pump or recirculating pump, for instance, would fall under this category. The language in the specification should make it clear that the water purification equipment supplier, and not the pump and/or motor manufacturer, is responsible for providing labor, travel expenses, living expenses, etc. for a service person to fix the problem in the event that the pump and/or motor fails. Once more, this is a very crucial component because, if not specified properly, it would lead to a system with several warranty contacts. Furthermore, if an inappropriate item is installed by a third party, it is advised that this problem could lead to a potential voiding of the guarantee by the equipment manufacturer.

The foundation of a performance or component warranty should also be addressed. Some equipment manufacturers will concur with the warranty conditions spelled out in the technical specification in exchange for paperwork that "qualifies" the warranty's conditions. The relationship between RO and electrodeionization units is a prime illustration of this circumstance. The performance of the unit will be qualified by the manufacturer's post-contract award submittal package or "terms and conditions," which state that it is based on feedwater characteristics to the unit and may not be feasible for the specified system, even though the manufacturer will indicate acceptance of warranty items detailed in a detailed specification. Prior to a RO system, a Silt Density Index (SDI) value is frequently provided. According to the system's (components, membranes, and performance) warranty from the equipment manufacturer, the SDI value of the feedwater must always be between 3 and 5. If SDI readings are measured at lower levels as part of an analytical monitoring program the system warranty may be "null and invalid" more often than stated in the terms and conditions of the equipment makers. A specification's phrasing should be clear and concise to prevent misunderstandings over this and comparable items.

The thorough specification should clearly explain the requirements for system startup. Start-up usually happens following the delivery of the equipment, unpacking, and connecting the necessary utilities (electrical, drain, air, waste, etc.). Startup of the apparatus is frequently delayed. Numerous of these issues are linked to "software" issues for systems using programmable logic controllers and services needed for the components (PLC). Since it is typically not specified in the specification, the amount of start-up offered by the manufacturer of the water purification equipment, in terms of the number of person-days allocated, is a problem. There are frequently "scope expansion" problems connected with "delays" related to the connection of utilities, even when the specification specifies that start-up would be necessary as part of the water purification equipment supplier's scope.

Start-up concerns ought to have the proper "admission criteria." It goes without saying that a definition is needed to establish when system turnover is approved by the pharmaceutical facility and start-up is complete. The release of any financial "retainer" for the project should be based on the acceptance criteria. To put it another way, the conditions of payment to the supplier of water purification equipment should contain a suggested 10% retainer up until the system is in compliance with the requirements specified in the acceptance criteria. Everyone is motivated by the obligation to finish the project as soon as possible while adhering to the specification's precise requirements.

The specification should be quite detailed about the shipping needs. It is strongly advised that up until it is delivered to the facility of the pharmaceutical company, the manufacturer or accessory supplier of the water purification equipment "maintains" ownership of the equipment. Here it is achieved by designating the pharmaceutical plant as the "freight-on-board" (FOB) location. Remember that it is insufficient to just state that the provider is responsible for the freight costs. The equipment belongs to the pharmaceutical producer after it leaves the facility of the equipment manufacturer or supplier, unless the FOB point is stated as indicated. The receiving of materials at the pharmaceutical facility, unloading and unpacking the equipment, checking the equipment for visual damage sustained during transport, and other related tasks are thought to be the most crucial aspects of shipping. When possible, equipment should be transported in special trailers or containers that are used "exclusively." The specification should expressly indicate that upon delivery of the equipment at the facility of the pharmaceutical producer, the manufacturer or component supplier of the water purification equipment shall give an authorized person to certify that shipping damage has not happened. Almost majority shipping companies will not take responsibility for equipment damage discovered more than 24 to 48 h after the goods was received at a location. Although if the water purification equipment could not be installed at a facility for days, weeks, or even months after it is delivered, it is crucial to conduct an initial examination. If a component has damage that necessitates replacement, starting the system up could be severely delayed if the problem is not discovered until the system is being unpacked and installed, possibly much later than when it was ordered.

As was previously said, the system should come with a turnover package that contains the paperwork needed to create the validation standards. Receiving a satisfactory turnover package should be one of the "acceptance criteria" mentioned before. Without a comprehensive evaluation by the competent staff at the pharmaceutical facility, particularly those responsible for creating the validation (or qualification) documentation, this package should not be considered "approved." Prior to the final distribution of any monetary retainer, it is far simpler to obtain "missing" documentation from a water purification equipment seller than it is once it has been released. Yet, the necessary information is more easily accessible at the time of, or immediately after, shipment from the manufacturer of the water purification equipment, supporting their position.

Moreover, "production milestones" for the system's components should be specified in the full specification. The specification should specify that the equipment may be inspected at particular "milestones" by the pharmaceutical producer and/or a designated representative. Regular inspections conducted throughout the manufacturing process can prevent unnecessary delays brought on by little issues at system startup. Installing a different sample valve or pressure gauge, for instance, at the equipment manufacturer's plant is undoubtedly far simpler than having to fix the system while it is starting up at the pharmaceutical manufacturer's facility. Also, routine inspections could turn up things that were unintentionally missed when the full specification was being written. This can involve unintentional relocation of one or more components from the facility's unloading area to the installation area, equipment dimensions that would hinder installation, etc.

A final inspection should be specifically described in the specification in addition to the routine examinations of the equipment parts at the manufacturer's site. The people who will be responsible for unpacking and examining the equipment when it is received at the facility of the pharmaceutical company should carry out this last check. The personnel from the company that manufactures the water purification equipment should be present during the final inspection in order to start the system. A Factory Acceptance Test can be utilized to complete these tasks (FAT).[28]

9.6 OPERATION

Pharmaceuticals uses water in a variety of applications. It can be employed in formulations during the production of pharmaceutical products, as raw material, as a cleaning agent inside a restricted production area, and many more applications. Depending on how the water will be used, different water quality levels may be necessary. Typically, eight different types of water are used in the production of pharmaceutical products. Non-potable, potable (drinkable) water, USP purified water, USP water for injection (WFI), USP sterile water for injection, LUSP sterile water for inhalation, USP bacteriostatic water for injection, and USP sterile water for irrigation are all examples of water that is not potable.[29] Each has distinct specifications or recommendations that are required or advised. Physical procedures, biological processes, and chemical processes can all be used to purify water in various ways. The final water quality criteria, the amount of water needed by the user, the overall life-cycle expenses, the continuity of operational usage taking into account throughout the day, year, and planned downtime should all be taken into account when choosing a water purification technique. The location, potential system temperatures, corrosion resistance, a sanitary design, and many other factors should be considered while designing the equipment, storage, and distribution.[30] Water quality is a serious issue because it is typically carried from a system only when it is needed, which prevents testing and batch or lot release before use. Establishing a preventative maintenance programme will help keep the water system under control. The programme should include instructions for using the system, quality control routines, and calibration processes for crucial devices, apart from a timetable for routine sanitation, preventative upkeep of parts, and management of modifications to the mechanical system and operating circumstances.[31] Monitoring is a critical quality attribute that consists of the frequency of sampling, the requirement for evaluating test results, and the reason for initiating corrective action. Some sampling techniques that should be considered are TOC, conductivity, endotoxin, and microbial growth. The use of alert and action level specifications can assist with avoiding the pass/fail specification. An alert level excursion would lead to alerting personnel involved, quality assurance, and the possibility of additional monitoring.

9.6.1 CONSIDERATIONS FOR OPERATIONS

1. Water system startup and commissioning

It is necessary to create as-built drawings for the interconnecting tubing in a water purification system. Dimensions or a point of reference should be included in the designs, which should also be scaled. As a minimum, the stainless steel tubing after the RO unit needs to make use of orbital weld numbers and "heat numbers," just like the 316 L stainless steel distribution loops used for filtered water or water for injection. It is recommended that utility plumbing such as steam, compressed air, drains, etc. should not be required as constructed drawings. However, it is advised, for reference, that utility plumbing routing be shown on dimensioned "single-line" drawings whenever practicable. Interconnecting tubing and piping should be pressure tested. The prerequisite and recommended test approaches were previously covered in chapter 10. Each component's PLC-based operation sequence should undergo a functional test. Each of the component's typical operations should be covered by this test. As an illustration, normal operation, backwash, settle, post-backwash rinse-to-drain cycles, and automated return to normal operation should all be tested when evaluating a multimedia filtration unit. During cycle testing, the location of each automatic valve should be confirmed. Moreover, testing should be done for support component functionality, including upstream pump speed, flow metre indication, pressure gauge indication, etc. The manual opening and closing of each automatic valve on the multimedia filtering unit using the "Operator Override" display panel.

It is important to check the manual on and manual off modes of multimedia filtering unit accessories such an upstream repressurization/recirculation pump. When I/O testing is finished, it should demonstrate that the PLC is operating each automatic valve and accessory as planned. The suggested testing will be repeated when the Operation Qualification for a validated system is executed in order to address any differences in the PLC programme. It might be utilized to meet the needs of a "qualified" system's "commissioning operational qualification." OQ-related commissioning tests should be checked during the execution of the OQ, even if it is advised that Installation Qualification items can be verified during commissioning for a validated system. OQ-related commissioning tests should be checked during the execution of the OQ, even if it is advised that Installation Qualification items can be verified during commissioning for a validated system. To fulfil validated system OQ test requirements, Factory Acceptance Testing (FAT) and Site Acceptance Testing (SAT) are frequently utilized. The system Standard Operating Procedures (sops) must be created, reviewed, and approved before the OQ for a validated system is executed. The FAT and/or SAT are not conducted using well-accepted sops. The "repetition" of testing that is advised is justified by the significance of sops for the OQ execution of validated systems. The system should undergo an SAT, including the control function. This test ensures that the operations of every system unit are running well. Issues with ant code, components, or accessories need to be addressed and fixed. Execution of the Commissioning Installation Qualification (CIQ) is required. As was already said, documentation and CIQ execution can serve as an IQ for a qualified system. Also, the data may be utilized as a reference for sign-off in the actual IQ when implemented in accordance with standards specified in a validated system IQ (full documentation, execution member requirements). The system should be started formally. All of the system's parts ought to operate normally once the startup process is finished.

The Installation Operation Qualification (IOQ) can be done once the system startup is complete. The IOQ's execution validates a qualified system's correct system operation. Moreover, control verification during IOQ execution should enable approved system OQ execution to be carried out without "deviations." Initial system sanitization can be done following the execution of the IOQ and IOQ, if necessary. Pretreatment elements, a RO/CEDI loop (or a RO/rechargeable canister system), as well as a storage and distribution system, may be included. It is recommended that approximately a week of intra component and point-of-use testing be carried out in line with a timetable in an Installation Performance Qualification before continuing with formal validated system execution as discussed in chapter 13. (IPQ). Testing at this step may reveal a problem that may impair the predicted quality of the water and can be quickly fixed before the execution of a validated system's IQ, OQ, and PQ. Also, this document may be utilized to satisfy the requirements for a qualified system by having an extended execution time period (e.g., 2 to 4 weeks).[32]

A pharmaceutical water purification system's long-term success depends on its installation, startup, and commissioning. The focus of this text has been to correct component design, specification of the component and system, and component documentation. The system as a whole may not function properly if one or more components fail to function properly as a result of poor component installation. Moreover, numerous minor yet significant system operational faults could be discovered during startup. A little but significant modification to the system control logic, for instance, might be discovered during appropriate system startup and could have a big impact on how well the system functions. In addition to confirming the appropriate installation and operation of components throughout all sequences, proper system commissioning also gives a successful validation of water systems that require planned, well-defined, successful, and well-documented commissioning and qualification. The commissioning process should involve setting up the system, configuring the controls, fine-tuning the loop, and capturing all of the system performance metrics. The commissioning work's quality, together with any related data and documentation, must be in line with the validation plan's standards if commissioning data is to be used or referred to during the validation process. A chance to check items in the IQ and OQ before the execution is "official." This can greatly lower or perhaps completely eliminate "deviations" in IQ and OQ execution. Proper installation should address the following:

- Facility considerations
- Utility considerations
- Technical considerations

Each installation item of concern is discussed below.

9.6.2 INSTALLATION OF THE SYSTEM

Facility considerations:

It is necessary to assess the physical characteristics of the physical area used for the installation of the water system, including its dimensions. Since distribution tubing must cross the area to go from the storage tank to the production area using water, the area should not be far from the sites of usage. But, the location should also take into account availability to services such as drains, electrical service, clean, dry, oil-free air, facility steam, chill water, cooling water, and raw water. Wherever possible, the installation area should be enclosed with restricted access. Also, since non-particulate contamination, such as volatile organic compounds, will flow through the hydrophobic vent filtration systems during storage tank drawdown, the region should not be in an airborne contamination environment.

The floor loading for the elements of the water purification system should be carefully examined. Support "pads" are found in numerous water filtration component parts. The floor loading is determined by the size and quantity of the support pads. If the floor loading seems insignificant, the size of the support pad can be specified during equipment specification. The installation area might be changed if the floor loading is insufficient. As an alternative, the area for equipment installation can be excavated and replaced with suitable reinforced concrete to meet the floor loading requirements. This is especially true for older facilities where floor loading is either unavailable or dubious. Tanks and/or equipment may occasionally be placed on mezzanines. An access mezzanine may also be present in a tank. A "support band" may be placed around the center of a tank to evenly distribute the weight along the periphery in order to meet tank mezzanine loading standards. Large volume tanks that cannot be erected on a single floor due to height limits may also employ this technology. The specification for the tank must specify the support band requirements and set them forward. Access for personnel, maintenance tools, and replacement tools to the installation area should also be taken into account. The route from an operator's office to the equipment should be taken into account when planning personnel access. Buildings that are not physically connected should not be crossed if at all possible (outdoor passage). Also, the path should not pass through locations that are clean (contamination-controlled/classified) and where people must don gowns and remove shoes. The installation and removal of RO membranes, activated carbon media, water softening resin, and other components are all part of routine maintenance.

Physical space is frequently limited for the construction of a medicinal water system. The manufacturing space is prioritised, and space for facility support equipment is minimized. Water systems may therefore be built with other facility equipment in physical spaces. Boilers, air compressors, condensate receivers, HVAC equipment, vacuum pumps, waste collection tanks, waste neutralization/treatment equipment, etc. are some examples of this equipment. Modern buildings could contain sections that are specifically designated for installing equipment. There are difficulties when installing pharmaceutical water purification equipment in a "shared area." For instance, vacuum pumps or air compressors may discharge oil into the environment.

A hydrophobic vent filter on a storage tank or condensing unit of a distillation unit will not be able to remove the oil. From a rigorous regulatory/compliance perspective, any oil added to stored purified water, water for injection, or condensed water for injection would be regarded as an "added substance." The official monographs for both purified water and water for injection restrict the addition of any substance; thus, it is necessary to test the presence or absence of oil or find a way to remove oil from the air before applying it to the hydrophobic vent filter. It should be noted that the chemical characteristics of injection water (conductivity, TOC, and nitrates for EP) and purified water (water for injection) may be within specification. But, as mentioned, a monitoring programme is necessary due to the potential presence of oil. The location of the hydrophobic vent filter ("intake") can be changed if oil is found. A thorough plan for rigging components into their installation position should be devised before finalizing the system specifications in this "rigging plan."[33]

Take into account any existing equipment, access, temporary barrier removal (if possible), and the size and kind of rigging equipment needed. It is proposed that building a "model" of the equipment would be acceptable for essential (tight-fitting) rigging applications. The use of cardboard or inexpensive schedule 40 PVC piping models can be quite helpful for rigging that involves movement in many directions through doorways and hallways, according to field experience.

It is important to determine where the eye wash stations and safety showers will be located. These safety features ought to be put in places with a higher "risk," like activated carbon and water softening units where the media will frequently be changed. Both eye wash stations and safety showers need to have raw water "hard piped" to them. Portable eye wash stations may be used for specific applications. The portable units are recommended for higher risk areas where access to raw water is a challenge. For colder raw water or hot water lines, it is advised that tubing and pipe be as well-insulated as feasible. Lines carrying chilled water, raw water, and cooling water will "sweat" in the summer and/or when relative humidity levels are high. A pharmaceutical water filtration system's floor space ought to be dry. This makes it possible to find system leaks. Tubing and pipe carrying hot water or steam should, of course, be insulated to protect people. Experience shows that raw water, chill water, cooling water, and hot water tubing/piping may all be adequately protected using 1-inch-thick fiberglass insulation with a PVC jacket. For all steam tubing and piping, a 2-inch-thick fiberglass insulation with a PVC jacket is advised.

The use of flow direction indicators on color-coded tubing and piping identification labels is recommended. It is possible to order personalized labels. Labels are advised so that staff can identify the tubing/piping section for "interface" people, such as those taking water samples, and also help them understand how the system works. In vertical or horizontal portions of tubing or piping longer than about five feet, at least one label must be visible. Where possible, the operational crew should be able to read the label's font size from floor level. Floor drains might already be there when the system is installed, or they might be added. For proper draining, floor drains need a "pitch" in the floor level. In regions with unsuitable slopes, the floor drain system should be able to reduce "puddles" of water. According to practical experience, many older facility drain systems will need to be "upgraded" in order to accommodate the installation of a modern water purification system. Because the floor area surrounding a drain really slopes away from the drain, an upgrade may be necessary. This seems to be related to the floor settling over time near the drain while the drain's physical height is maintained by the steel drain piping that is attached to it. The capacity of the drain piping should also be assessed by a field test not drawing pipe diameter size. Over time, drain piping may rust, give way, or accumulate debris. This can call for digging and pipe replacement for drains. As an alternative, a collection tank may be selected, purchased, and constructed for high flow rate operations such as multimedia filtering unit backwash. Via an adjustable valve (such as a butterfly type), the gathered backwash and rinse water can be discharged to drain at a flow rate significantly lower than the backwash flow rate. The tank may, however, take up physical space that would otherwise be used by water purification components. The discharge of water is finally regulated. The "combined" waste from a plant may not exceed a maximum temperature of 1408 F. At larger facilities with high water usage overall, this might not be a problem. To lower the temperature of waste water from processes like activated carbon unit or RO/CEDI loop hot water sanitization, smaller facilities would need a waste heat exchanger (chill water cooling). Water should not be diluted with cold raw water as this is against numerous local and state regulations. As previously mentioned, the installation of the water purification system should be done in a way that restricts access to the machinery. Installation of detachable walls and doors may be necessary for this. The "control" of a "component," purified water, and/or water for injection is compromised by unauthorized entry to the water purification system. "Card scan" systems may be used to simplify access for authorized personnel. As part of a facility-wide access system, these systems can also be used to keep track of who enters a specific area as well as the time and date of entry and exit.

The presence of chemical vapors or fumes near the installation site is a concern, as was mentioned earlier in this chapter. The environment may have an impact on both the long-term performance of crucial components as well as the quality of the water (through hydrophobic vent filtration systems). For instance, the majority of pump motors are completely enclosed fan cooled (TEFC). The fan on the motor will allow vapors or dust from the environment to enter, harming the bearings and seals in the process. The cooling air passing through a "vortextube" cooler, which is widely used for power panels with heat-generating components like CEDI transformers, may also become clogged with environmental contaminants. All equipment, especially manual valves, should be visible in the installation area thanks to adequate lighting. Water flows to the floor as a result of routine maintenance tasks like RO membrane "rotation" and activated carbon media replacement. A reduction in the need for "extension cords" when performing maintenance results in increased worker safety. Prior to the equipment's arrival, the installation area should be painted and ready. For floors, a chemically resistant epoxy-type paint should be taken into consideration.

Metal barriers and guiderails may need to be removed when equipment installation is complete to protect the equipment. While "traditional" yellow guiderails may be utilized, a more aesthetically pleasing detachable barrier can be created by welding specially made stainless steel tubing with fittings. There will be an overhead sprinkler system in the installation area. Sprinkler heads may need to be moved if a "fake" ceiling is built. It is also necessary to check the rules governing the separation between sprinkler heads and any "obstructions." Sprinkler head relocation or augmentation may be necessary. Installation of "curbing" and/or "equipment pads" may be taken into consideration once the precise component measurements have been determined. This chapter already said that the ground beneath the equipment should be dry. While preventative maintenance may introduce water to the floor beneath some components, with the right post-maintenance "housekeeping," the water can be trapped or diverted to a drain and removed. Over time, curbing commonly develops "cracks," notably at edges and corners. The height is increased by equipment pads.[33]

System installation: utility considerations:

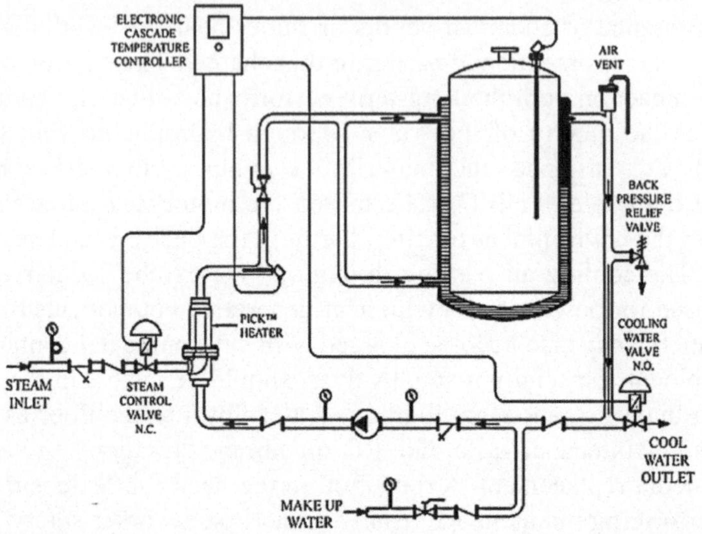

FIGURE 9.5 Steam and condensate piping for storage tank heating jacket.

Image credit: Pick Heaters Inc. https://www.pickheaters.com/Blog/improve-jacketed-heating-systems-with-direct-steam-injection.htm

Field verification should be done to determine the location and state of the steam supply to the installation area (for systems requiring steam). Pressure, flow should all be tested in the field steam with thermodynamic quality. By conducting an outside inspection of the piping and gathering clean steam condensate, the state of the steam piping should be assessed (iron oxides). Connecting steam-requiring components requires the use of the proper accessories. The manual isolation valve, pneumatically operated ball valve, pressure gauge, pressure regulator, steam modulating control valve (fail close), pressure gauge, temperature indicator, and relief valve should all be included in the supply steam pipe to the component shown in Figure 9.6. Between the controlled supply steam and the steam trap, a "drip leg" with an isolation valve, strainer with a drain valve, and steam trap should be attached between the regulated supply steam pressure and the steam condensate line from the component.

As mentioned, the drip leg helps during times of high steam demand to remove condensate from the supply steam. The steam relief valve must match the application that includes computations that are saved in the project file. Regulations for each component using steam should strongly consider having independent pressure release valves. Each relief valve's discharge should ideally be pointed upward and through the roof. The roof's egress point needs to be properly sealed. To prevent debris buildup, the relief line should be terminated with a double elbow and "bird screen" wrapped around the piping. A "drip pan" should be present near the water purification component's outlet of the relief valve to prevent water "backflow" from rain, snow, or ice. The drip pan feeds water to a collection "cup" that could potentially reach the relief valve piping. To deliver water to the drain, a tiny diameter portion of piping can be attached to the cup. A steam trap, temperature gauge, pressure gauge, and isolation valve should be included in the steam condensate piping. For return to the facility's boiler condensate return/feed water system, the line should be directed to a nearby condensate receiver. Condensate return might not be technically necessary for applications that only seldom or infrequently use steam, such as a steam-heated jacket on a hydrophobic vent filter housing or monthly hot water sanitization of a RO/CEDI loop. Technically speaking, condensate return might not be necessary. You could direct the condensate to drain. To protect workers and lower the condensate's temperature (below the previously mentioned temperature discharge limit), raw water should be added to the drain at the same time. In order to insulate steam and condensate piping, 2-inch-thick fiberglass material with a PVC outer cover should be used.

The installation's compressed air needs should take into account both control air for solenoid valves that open and close pneumatically operated valves and higher volume uses like vortex coolers for electronic panels. The compressed air supply should ideally be dry, free of dirt and oil. To eliminate the listed impurities, filters should be used in the compressed air line. At each significant system use point, there should be an air-regulating mechanism in place. A filter and pressure gauge should be included with each regulator. Prior to each regulator, a shutoff valve should be installed to allow for maintenance. The status of the water filtration system in the event of an electrical power outage is a crucial factor to take into account. The storage, distribution, and control parts of a water purification system should ideally include a backup emergency generator power supply. However, a central or dedicated air compressor is a requirement of this emergency electrical supply. The system has electrically powered control valves that are pneumatically operated. Although some valves are designed to fail in a "safe" position, others, like the automatic feed water valve for a RO unit, must be opened when electrical power is present and the RO unit is operating. These valves fail in a safe closed position upon lack of air. In other words, both electrical power and compressed air are necessary for the correct operation of the system.

A pressure switch ought to be installed in the compressed air supply line as a safety measure to prevent catastrophic component failure if electrical power is available but compressed air is not. When air pressure is not detected, the central control panel will receive a distinct control signal that will disable control (similar to pressing the "Emergency Stop" button on the panel's face) until compressed air pressure is once again detected. Ventilation for the installation area, together with humidity and temperature management, is highly recommended. With rising temperatures, bacteria multiply substantially more. Also, the components of the central control panel—the processor, input/output (I/O) cards, etc.—have high temperature working requirements. Other panels' parts, such as the transformers in the CEDI power cabinet, produce heat. Maintaining the ambient temperature in the installation region below 1008 F is highly desired. Panels should be placed with cabinet coolers (such as vortex type) if this criteria cannot be met. Also, it should be thought of using inline heat exchangers to provide microbiological control of stored and recirculated water. Regulations usually outline HVAC needs based on the "volume" of the installation area. Airflow should be distributed uniformly across the water purification system's components. I am taking into account the production of heat from individual components as well as recurring maintenance tasks like hot water sanitization. The installed system's raw feed water pipe needs to have a backflow preventor. Once more, this is a legal necessity.[33]

System installation: technical considerations:

Underwriters Laboratory (UL) certification for electrical components is required. The majority of electrical parts made or assembled in Europe will bear the "Conformite Europeenne" (CE) stamp certifying that the product complies with European Union safety, health, and environmental laws. Regrettably, some state, local, or insurance agency rules in the US restrict the acceptance of CE-stamped components and call for the UL stamp. Both the basic criteria for a project and the component specifications should mention this criterion. A component with a CE stamp could also be UL-inspected and UL-stamped. Yet, the procedure is time-consuming, expensive, and occasionally harmful. Considering past experience, all components that would need a proper UL stamping are suggested. The specification and submittal process ought to have included the determination of the electrical power requirements. The following needs to be taken care of before and during equipment installation:

- Electrical power needs, including motor starters, for each panel. Voltage, single or three phase, cycles, and amperage are all included.
- The closest motor control center (MCC) location for the installation area's electrical power feed.
- An acceptable conduit for electrical wiring for controls.
- This can include threaded galvanized steel conduit that is "water-tight" or flexible conduit.
- Conduit size in l. l the UL specifications that were covered earlier in this chapter.
- The emergency generator provisions covered in this chapter's prior sections.
- Electrical regulations from the county, state/province, and federal governments.

The supply and installation of control and power panels should be done in accordance with a specific National Electrical Manufactures Association (NEMA, 2008) classification. Installations must take into account the surroundings, as well as activity near the panels or cabinets. Cabinet interiors are protected by NEMA designations against contaminants such as dust, water, vapors, corrosive compounds, noncorrosive liquids, steam, and explosive substances. The worst case scenarios should be taken into account while determining the NEMA classification (Type). It is advised that all control wiring be run through conduit to and from cabinets and field devices. In the field device and in the cabinet, the field wire needs to be tagged. Label names should match those on project drawings. According to practical experience, some analogue signals from parts (such as some electrolytic ozone generator power indicators and UV unit intensity metres) need "isolators" to prevent "feedback." Equipment manufacturers frequently fail to mention these components. Since it may take many days to find these issues during system commissioning, each component should have them assessed before installation. Tank rupture discs' discharge locations should be set up to prevent operator exposure to pressured water after pressure relief, especially if the water is hot. This is rapid yet dramatic relief from a "solid" hydraulic system. At the base of the tank, there should be enough room provided in a place with restricted or impossible access. The rupture disc discharge tubing should have a diameter that is one to two sizes larger than the rupture disc. For instance, a rupture disc with a 3" diameter should "exhaust/relieve" into a tubing portion with a 3 12" or 4" diameter. Solenoid valves may be put on the local I/O panels in systems having a central control panel powered by a PLC and distant I/O panels.

I/O panels need to be placed as close as possible to the component(s) they are associated with. There should be a local manual "override" option for each solenoid valve. While the central control panel's "output override" feature should be utilized for "manual" overrides, some maintenance procedures can call for the application of the local override features. It is recommended that the physical method of solenoid valve override use a mechanism requiring a simple tool, such as a screwdriver, rather than a push button-type device. Push buttons could accidentally be pressed (manual override), according to experience. As a result, the central control panel will not be able to identify or display faulty component operation. The type of pneumatic tubing used from individual solenoid valves is the last consideration.

It is important to talk about components that receive a controlled air signal between 3 and 15 psig. It is not advised to utilize "hard-piped" copper or stainless steel tube. Copper tubing oxidizes to generate a blue/green tint, needs constant support to prevent bending with time and component operation, and occasionally needs galvanic isolation due to technical reasons. Also, the way it is physically positioned lacks visual appeal. Stainless steel tubing lacks the necessary flexibility for connecting to solenoid valves and actuators and necessitates extensive installation work to achieve aesthetically pleasing horizontal and vertical "runs." Polyethylene tubing is advised to be taken into consideration. A "spaghetti-like" look should be avoided. The tube can be put in an electrical

conduit that runs through the middle of the "controlled" component's outer surface. When compared to "hard-piped" stainless steel or copper, little visible parts of installed tubing do not obstruct maintenance procedures. Also, the flexibility of the tubing improves connection to solenoid valves and valve actuators. Flexible polyethylene tubing should not be used in applications where there is a chance of coming into direct touch with a hot surface. Some applications call for stainless steel tube. When the RO/CEDI loop is being recirculated, recovering softened or RO wastewater should be taken into consideration. The reclaim system may include collection tank(s) with air breaks and overflow directed to the drain. The reclaimed water can be used for tasks including closed-loop cooling application makeup, facility boiler feed water, and cooling tower makeup. It may be beneficial to use RO waste recovery and treatment (makeup mode) for facilities that have daily volumetric discharge limits. An "industrial-grade" RO unit may be used to process RO waste, with the product being recovered for the aforementioned uses. Conservatively, at least 50% of the RO waste produced during the application of makeup (non-recirculating) may be recovered.[33]

2. CREDENTIAL:

WPU, BPW, BHPW, and BWFI systems are all regarded as direct impact, quality-critical systems that should be qualified 2012 WHO Expert Committee on Requirements for Pharmaceutical Preparations Technical Report Series No. 970. The validation convention for the 46th report should be design review or design qualification (DQ), IQ, OQ, and PQ. This recommendation focuses on the specific PQ technique that should be used for WPU systems to demonstrate their consistent and dependable performance rather than the normal requirements for the traditional qualification stages DQ, IQ, and OQ. To achieve the goal of demonstrating the system's resilience and dependability over an extended period of time, a three-phase strategy should be adopted. The validation procedure must include tests on the source water, and such tests must continue as part of the regular monitoring. The validation procedure must include tests on the source water, and such tests must continue as part of the regular monitoring. It should adhere to any internal specifications for drinking water specification. The specification and submittal process ought to have included the determination of the electrical power requirements. The following needs to be taken care of before and during equipment.

Phase 2:
Verify the quality of the incoming feed water by taking daily samples or monitoring it continually:

Two weeks should pass during the test period when the system is closely watched. The system must function without interruption or performance variations during this time.

1. Do tests for chemicals and microorganisms in accordance with a define plan.
2. To the quality of the incoming feed-water every day, sample it or continuously monitor it.
3. Samples or ongoing monitoring after each procedure step "process of cation."
4. At every site of usage and in other areas, sample or continually monitor at each sample point.
5. Create suitable operating parameters.
6. Create and finalize your maintenance, cleaning, and sanitizing processes.
7. Show that product water is produced and delivered in the required quantity and quality.

8. Usage and reuse "follow the standard operating procedures (SOPs) for use, upkeep, sanitization, and troubleshooting."
9. Check temporary alert levels.
10. Create and revise refine process for test failure.

Phase 2:

Following the successful completion of phase 1, a further test period of 2 weeks should be spent doing further rigorous monitoring while implementing all the modified SOPs. The sampling scheme should generally be the same as in phase 1. Providing both commissioning and phase 1 data confirms suitable water quality and the practise is authorized by QA, using the water for FPP manufacturing during this phase may be permitted.

1. show consistency in operation within predetermined limits
2. show consistency in the production and distribution of water that meets the required standards for quantity and quality when the system is run in accordance with the SOPs.

Phase 3:

Phase 3 normally lasts for a year after phase 2 has been satisfactorily completed. During this phase, which has the following goals, water may be used for FFP manufacturing purposes:

- To show consistent performance over an extended period of time.
- To make sure seasonal differences are included.
- Based on established protocols that have been demonstrated throughout stages 1 and 2, all sample locations, sampling frequencies, and tests should be reduced to the usual routine pattern.

9.6.3 CONSTANT SYSTEM OBSERVATION

After phase 3 of the qualification programme for the WPU system, a system evaluation needs to be done. Based on the findings of phase 3, a routine monitoring strategy should be formed after this assessment. Monitoring should also incorporate monitoring with online tools (with suitably qualified sample testing for physical, chemical, and microbiological characteristics. Ed alert systems) of parameters such as pressure, temperature, conductivity, and total organic carbon. When sampling from points of use is not possible, one sample should be taken from a specific sample point or a point of use. 100% water.

- Tests should be performed to check that the company specification and approved pharmacopoeial have been satisfied.
- If necessary, it may also provide the water's microbial quality. Trend analysis should be performed on monitoring data (trending should normally be within 2 sigma). Based on previously provided data, appropriate warning and action levels should be determined.
- Any pattern of repeatedly exceeding alert limitations should be the basis for a full assessment of the fundamental cause and the implementation of the necessary corrective measures.[32]

9.6.4 MAINTENANCE OF WATER SYSTEMS

A controlled, documented maintenance programme that takes into account the following should be used to maintain WPU systems:

- Defined frequency for system elements.
- The calibration programme.
- The sops for particular jobs.
- The control of allowed spares.
- The issuance of a clear maintenance plan and instructions.
- The review and approval of systems for use once work is completed, and the recording and review of issues and faults during maintenance.[32]

9.6.5 SYSTEM EVALUATIONS

WPU (BPW, BHPW, and BWFI) systems should be assessed on a regular basis and by a team that includes members of engineering, quality assurance, microbiology, operations, and maintenance. A review should take into account things like the following:

- modifications made since the previous review
- system performance
- reliability
- quality trends
- failure events
- investigations
- results of monitoring that fall outside of specifications
- modifications to the installation
- updated installation documentation

- log books
- status of the current SOP list.

The following should also be examined for new systems or systems that exhibit instability or unreliability:

- the need for an investigation
- the CAPAs (corrective and preventative activities);
- The qualification (DQ, factory acceptance test (FAT), IQ, site acceptance test (SAT), OQ, and PQ) or comparable verification papers, and monitoring phases of the system.[32]

9.6.6 PRETREATMENT TECHNIQUES

9.6.6.1 OPERATING AND MAINTENANCE CONSIDERATIONS

1. Chemical injection systems

With both seasonal and climatic changes in the raw feed water supply, the amount of disinfectant needed to achieve a predetermined residual disinfecting concentration in the feed water to the pretreatment system will change. It goes without saying that this is especially true for groundwater sources that are influenced by surface sources or feed fluids from those sources. It is significant to remember that a lot of municipal systems in the US that use chloramines for microbial control in the distribution system switch the residual disinfectant for a set amount of time every year. During a month out of every year, the majority of these systems use residual chlorine to manage microbes. The operating features of the municipal treatment facility (long and short terms) should be evaluated and documented by those in charge of the pharmaceutical water purification systems at a facility, even if this technique is typically used in a spring month like April or May.

The analytical method used to evaluate the residual disinfectant should take into account the possibility of influence from very small amounts of specific contaminants. The literature provides an overview of methods for calculating residual chlorine and residual chloramine concentrations. In addition to residual disinfectant, samples taken downstream of the residual disinfectant injection port should also be examined for total viable bacteria and total coliform bacteria. The analytical techniques utilized should be in line with the techniques employed by the municipality and described in (34). The residual disinfectant storage tank should have features that enable operational staff to gauge the volume (or weight) of residual disinfectant strictly for private use. By using a cheap "load cell," you may determine this by level or, more precisely, by weight. The residual disinfectant injection pump has to have routine maintenance. Even with the best choice of materials and components, the "pulsing character" associated with pump operation causes "wear and tear" on surfaces and components.[34]

2. Injection of reducing agent for removal of residual disinfectant agent

The discussion of suitable operating and maintenance issues is pertinent to the injection of residual disinfection. The concentration of the reducing agent solution may change over time as a result of a possible reaction between dissolved oxygen and stored reducing agent. To ascertain the concentration, standard operating procedures (SOPs) should be defined, specifically stating the frequency of sampling as well as the "shelf life" of the reducing agent solution. Protocols should be devised to reduce the entry of bacteria from the tubing used to store and transmit reducing agents. This may considerably raise the bacteria levels in the water used to feed a RO system. It is advised to prepare the reducing agent solution frequently (daily to bimonthly). One option is to combine an "internal" chemical monitoring programme with a Certificate of Analysis for the reducing agent. The volume and pace of the reducing agent injection should alter in accordance with seasonal and climatic variations in the raw water supply's residual disinfectant content. This would entail modifying the injection rate to account for predicted annual variations in the type of residual disinfectant used by some towns (chlorine vs. chloramines), as was previously described.[35]

3. Sodium hydroxide injection in systems using reverse osmosis

Consider the following when operating and maintaining caustic injection systems.

- Regular post-caustic injection monitoring must be carried out, either through in-line analysis using a calibrated system or sampling and laboratory analysis. The measuring electrode must be regularly maintained for inline pH monitoring systems, as well as frequent calibration with "buffer" solutions. It is crucial to ascertain whether even minute amounts of insoluble hydroxides are forming. As previously mentioned, these insoluble hydroxides will impair the long-term effectiveness of RO systems. When "off-site" cleaning is done, the scaling of the "tail" membranes in the RO array can be calculated by looking at the RO membrane precleaning data. Loss in product flow that may be verified by removing and doing an "autopsy" on a tail may indicate scaling of RO units cleaned in place. As previously mentioned, even little changes in the pH of feed water can have a big impact on whether sodium hydroxide injection prior to RO units produces the required outcomes.

- It is crucial to keep in mind that pH is a logarithmic representation of the concentration of hydronium ions. One unit of pH change corresponds to a 10-fold change in the measured hydronium ion concentration. But more crucially, as was mentioned in chapter 2, the presence of weakly ionized material "buffers" the pH of the raw water supply. As sodium hydroxide is required to achieve the optimum pH in the feed water, changes in raw water pH values as minor as 0.1 to 0.3 units can indicate changes in the amount of sodium hydroxide needed required to increase the feed water's pH by a factor of 1 to 10 or more in order to achieve the desired pH in the RO unit.[35]

4. *Thermal Blending Systems*

Before the principal ion removal unit operates, many systems have recirculation features in the pretreatment portion (e.g., reverse osmosis). Valve for thermal mixing is commonly utilized in systems with smaller capacities. Water is provided continuously for 8 h per day, 5 days a week, using numerous lesser capacity systems. The pretreatment system's recirculating features, which are covered in more detail later in this chapter, will use the recirculation pump to introduce mechanical (Joule's) heat. This will raise the water's temperature inside the pretreatment equipment, speeding up the growth of bacteria.

The thermal blending valve can be manually set to a colder temperature at the end of each day or before the weekend to avoid this situation. By doing this, operating personnel can flush water through the pretreatment system and subsequently avoid the higher pretreatment system operating temperatures and associated bacterial growth. By employing the "cold water" bypass valve and a "drain valve" that have been installed in the system, it is obvious that the goal of this operation can be accomplished with just minor design changes. There would be a "normal" operation under this arrangement. By only cycling the valve briefly, the thermal blending valve's functionality should be routinely checked. By performing this action, the valve's position would be changed to a warmer or cooler temperature and check to see if the temperature rises or falls as shown by the downstream temperature indicator. Periodically checking the hot water shutdown and cold-water bypass solenoid valves should also be done. By purposefully raising the temperature setting on the thermal blending valve and making sure the hot water flow ceases, the operation of the temperature switch and hot water solenoid actuation should be checked. Periodically calibrating the temperature switch and temperature indication is recommended. The thermal blending valve will need to be adjusted on a regular basis to maintain the desired temperature for the specified intended service. This goal can be accomplished by including a data input "box" for the effluent temperature from the thermal blending valve in the system's operational log. According to seasonal/climatic conditions, it is completely anticipated that periodic valve adjustment will be necessary around twice a month to maintain a product water temperature value between 208 and 258°C.[36]

5. *Heat Exchangers*

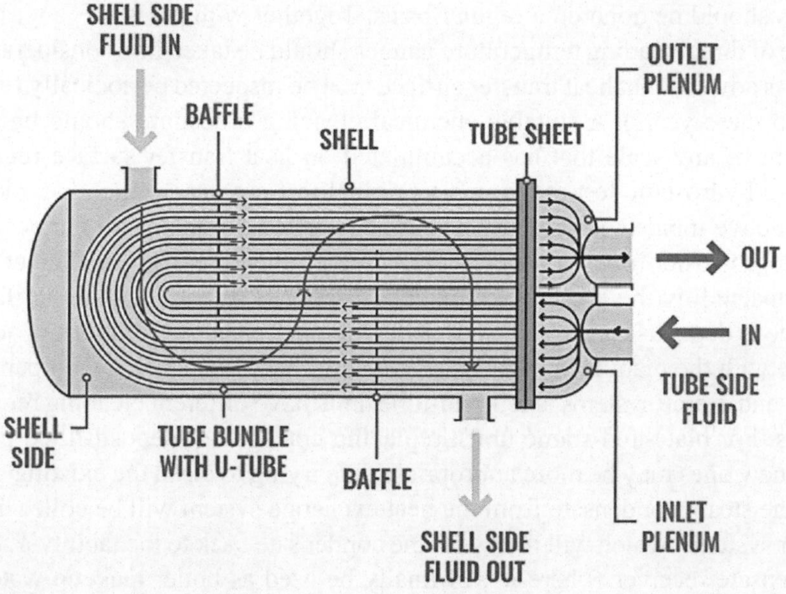

FIGURE 9.6 Heat exchanger.

Image credit: From Powerblanket. https://www.powerblanket.com/blog/using-electric-heaters-as-heat-exchangers/

The facility's modulating steam supply valve's role in the heat exchanger's operation is crucial. The output (current or amperage) associated with the temperature control system should be calibrated on a regular basis (6–12 months). It is important to check the current-to-pneumatic converter's pneumatic output as well. The thermodynamic quality of the steam is crucial since it influences the overall properties of the heat exchange system, as was mentioned earlier in this section. Periodically, the steam's quality should be assessed. While the traditional "drip leg" (bypass with strainer, trap, and isolation valve) should always be used in supply steam lines, the use of a "bypass" steam line with a valve and steam trap should be taken into consideration for heat exchangers installed in "branch" sections of steam piping where condensation will occur. Regularly checking the quality of the steam can easily confirm the need for this bypass device. It is crucial to maintain the steam trap, including checking that it is working properly.

As the design of the heat exchanger assumes that the full heat transfer surface area is capable of transmitting latent heat, rather than sensible heat, from steam condensate, the steam trap should not allow liquid to build up in the heat exchanger. The calibration of all the heat exchanger-related instruments should be done on a regular basis. Together with pressure gauges, the usage of direct reading temperature gauges should be taken into consideration.

It is advised that heat transfer surface area be inspected periodically (every one to three years). A suitable chemical cleaning procedure should be used to remove any scale that has accumulated on heat transfer surface regions. Also, a hydrostatic test at the unit's original test pressure, which is typically 50% above the design pressure, should be carried out to ensure that leakage due to corrosion does not occur. The overall heat transfer coefficient will be impacted by the presence of siliceous deposits on heat transfer surfaces. Siliceous deposits can be removed using the right cleaning solutions in accordance with the manufacturers' instructions for heat exchangers. Compared to plate-and-frame systems, shell-and-tube units have different cleaning requirements. For plate-and-frame units, replacing corroded or deposit-filled plates with new ones may be more appropriate than trying to clean the existing ones.

The steam condensate from the heat exchange system will be collected by larger systems, which will then send the condensate back to the facility's steam condensate receiver, where it will finally be used as boiler makeup water. A "receiver" tank and a pump should be part of the condensate return system. To guarantee that the upstream steam trap operates correctly and to prevent flooding of the heat exchanger's heat transfer surfaces, components for the condensate return system must undergo routine maintenance. Systems with low condensate flow rates and infrequent use should be concerned about this.[37]

6. *Particulate Removal Filters*

The differential pressure drop through a unit should be used as the foundation for determining the backwash frequency. The frequency of the backwash operation will change depending on the season and the weather, especially for raw supply water from a surface source. As necessary, the backwash flow rate should be adjusted periodically to account for variations in raw water temperature. The backwash procedure should go long enough to get rid of any trapped particle matter. The length of time may also vary depending on the season and the climate. To assess the efficacy of the particulate removal filter, operating employees should set up an analytical monitoring programme. The concentration of total suspended solids is a superior analytical indicator of filter efficacy (TSS). Both input water and output water should have their TSS concentrations routinely measured. Both feed water and product water samples should be frequently tested for TSS concentration. Throughout the unit's operational "cycle," the monitoring programme should periodically collect multiple samples (feed water and product water) (between backwash cycles). If accessible, sight glasses mounted on the straight side of the particulate removal vessel should be used to visually monitor bed expansion. Moreover, particulate matter or water color should be checked in the transparent piece of tubing in the waste line. It is also possible to utilize sight glasses in the product water pipework of a particulate removal unit to confirm that filter media are not "passing through" the column as a result of a lower distributor failure.

The interior of the apartment should be checked once a year. The integrity of the coating should be first and foremost confirmed by a visual examination of "interior-coated" steel columns. The integrity of the PVC-sprayed and baked or rubber-lined steel columns can be checked using a dielectric testing tool like the teslacoil. Visual examination of the distributor integrity should be part of the inspection (feed water). After column access, manway and handhole gasket replacement should be done. If air scouring is utilized to improve the backwash process, the air's purity needs to be checked routinely using the right analytical monitoring techniques. Any study of the scouring air, for instance, should confirm that it does not contain any materials that would be considered "foreign substances and impurities" according to the USP General Notices section, such as oil or other contaminants.

Although it is a physically demanding process, filter and support media should be changed approximately every five years for steel column units and once every three to five years for fiberglass column units. Depending on the quality of the feed water and product water, the replacement frequency may need to be changed. Since access to the lower distributors necessitates removal of filter and support media, which typically necessitates system downtime, it is strongly advised that filter lower distributors be changed each time media are replaced. Although a particulate removal filter's supply water typically has some residual disinfection, slime and/or algae can still grow inside the unit, especially on the gravel that serves as the base of the filter. The TSS value in the feed and product water may not be the sole factor to consider when deciding whether to replace the filter media and support gravel. An excessive amount of microorganisms may be proliferating, for instance, if the total viable bacteria levels in the product water from the unit have increased over time and cannot be lowered by backwashing or sanitizing. It is necessary to record operating parameters every day, including backwash frequency, pressure drop through the unit, feed water flow rate, and feed water and product water pressure.

Regular calibrations should be performed on all instruments. The maintenance of valves should be proactive. It is advised to keep the full spare valve assembly as "spare components" for fiberglass column systems that use multiport-type valves. Consider a one- to two-year diaphragm replacement cycle for devices with manual or automatic diaphragm valves. Every one to two years, a complete repair of the valve seals and seats is required for bigger machines with manual or automatic butterfly valves.[38]

7. *Activated Carbon Units*

The activated carbon unit needs to be periodically backwashed (at room temperature). The foundation of this backwash procedure ought to be microbial effluent level management. However, accumulation of particulate matter with a corresponding rise in pressure drop may necessitate increasing the backwash frequency for the unit if a particle removal filter is not positioned before the activated carbon unit. However, excessive backwash is undesirable because it can lead to the premature breakthrough of organic material and residual disinfectant (assuming that the residual disinfectant agent is chloramines). This is because heavier, denser activated carbon media containing organic material literally "transfers" to the lower part of the unit (at the effluent). Hot water should be sanitized periodically. Based on trends in product water bacterial levels from the activated carbon unit, particularly samples taken following ambient backwash of the unit, the criteria for sanitization should be developed.

With seasonal and climatic variations, it will be necessary to adjust the backwash at ambient temperature and hot water sanitization frequency. As the temperature of the raw feed water to the system rises, it is expected that the backwash frequency would need to be increased. Assuming there is sufficient freeboard space in the activated carbon vessel, it is recommended to alter the backwash flow rate in response to changes in the raw feed water temperature in an effort to reduce the creation of activated carbon fines. Hence, the efficiency of the backwash operation and by using cooler, more viscous water, the activated carbon bed will expand more quickly. The backwash operation should last long enough to achieve the appropriate bed expansion, specifically to get rid of activated carbon fines and lower the bacterial counts in the postbackwash product water. The duration should often be altered based on operating experience because it depends on the particular conditions for a unit.

To measure the levels of TOC and total viable bacteria, as well as the concentration of residual disinfectant in product water, an analytical monitoring programme needs to be designed. In order to assess the performance of the unit, particularly with regard to lowering the TOC level, it is also advised to do periodic monitoring of the feed water. Last, but not the least, analytical monitoring should include TSS testing in feed and product water to confirm that activated carbon fines are not being produced or going through a lower distribution system that is broken. To make sure that media are not being removed from the column, the level of activated carbon media in the column should be frequently monitored and logged. If excessive volumes of activated carbon fines are being created during ambient temperature backwash, hot water sanitization, or regular operation (due to improper lower distributor design), this logging process will show it. As stated, activated carbon media needs to be changed approximately every six months.

Suspended activated carbon can be taken out via the active carbon sluice port on the column's side. It is crucial to make sure that the activated carbon media used as replacement have been thoroughly hydrated, acid washed, and rinsed. Activated carbon needs between 24 and 36 h of "soak time" to be properly hydrated (i.e., to ensure that water has entirely permeated the carbon "pores"). By purchasing prehydrated activated carbon, this procedure and the ensuing "downtime" can be avoided. Activated carbon that has been acid washed and prebackwashed (rinsed) is also crucial. Common natural impurities found in commercially accessible products including barium, aluminum, and strontium are removed via acid washing. Because these ions cause the production of very insoluble precipitates, their presence will have a negative impact on the performance of RO units placed downstream of the activated carbon unit. Because of their strong bonds with the cation resin, these heavy-molecular-weight multivalent cations are also unfavorable in deionization systems. Lastly, silica, sodium carbonate, and potassium carbonate are all common components of activated carbon sources. Because sodium carbonate and potassium carbonate are "basic salts," their presence causes product water to have a higher pH than usual. The contaminants will be eliminated, as well as any possible issues with downstream components, by acid washing and pre-rinsing. It is advised to rinse the interior of the column with water and a chemical disinfectant after removing the used activated carbon media and before replacing it with new media. The bacteria in a "biofilm" on the inside wall of the column can be effectively removed using a 1% solution of peracidic acid and hydrogen peroxide.

In the event that the typical operational flow rate is either too high or too low, channeling via the activated carbon unit may occur. Even under recirculating conditions, operating flow rates should be provided through the unit to provide the requisite face velocity of 3 gpm/ft2 over the cross-sectional bed area.[39]

8. *Water-Softening Units*

It is crucial to maintain microbial control in water-softening systems. A water softener regenerates using a non-bacteria-destroying salt solution and runs at a neutral pH, in contrast to deionization systems, which employ bacteria-destroying acid and caustic regenerant chemicals and function in the hydronium and hydroxyl form. The backwash process may help eliminate some of the bacteria in the carbonaceous cation bed during the regeneration cycle of a water softener. Sadly, the amount of bacteria in product water will rise over time.

Moreover, regenerant solutions introduce microorganisms, especially for systems that use wet brine storage. Periodic hot water sanitization at 908°C for 2 h or, more effectively, chemical sanitization using a 1% solution of hydrogen peroxide and peracidic acid can both efficiently suppress bacteria. Thereafter, it has the capacity to eliminate biofilm using the proper sanitization method. It is feasible to perform effective sanitization using a 1% solution of peracidic acid and hydrogen peroxide if an old water-softening system has a carbon steel vessel with an internal lining (or coating) that cannot tolerate hot water sanitization temperatures. It is possible to administer the chemical cleaning solution concurrently (downward through the cation resin bed). Effective chemical sanitization necessitates both dynamic and stationary exposure conditions.

The cation resin will oxidize if the supply water to the water softener includes any remaining disinfectant. Depending on the kind and concentration of the residual disinfectant, this oxidation process will remove some of the residual disinfectant from the feed water supply. Monochloramine will not oxidize the cation resin as quickly as chlorine, for instance. The cation resin will decrosslink as a result of the residual disinfectant oxidizing it, which will cause a slow reduction in capacity and the elution of compounds that break down the resin. While the capacity of the resin will often decrease slightly over time as a result of the oxidation process, the concerns related to the elution of organic material are more significant because they cannot be monitored for their existence or composition.

The presence of trace amounts of multivalent cations, such as aluminum, barium, and strontium, in the feed water to the softening system is crucial for water-softening units used as a pretreatment to single- or double-pass RO units. Establishing an analytical monitoring programme that makes use of "supplemental" analysis (in addition to that mentioned in the USP Purified Water Official Monograph) is recommended. The salt dose should be increased if even minute amounts of these contaminants are found in the water-softening unit's effluent. As a result, the sodium content rises, improving the regeneration process' efficiency (the sodium ion replaces the heavier multivalent cation). As was previously mentioned, regenerant salt purity is crucial, especially when the water-softening unit is utilized as a pretreatment step in a membrane process like reverse osmosis. It is necessary to include a Certificate of Analysis with the salt. The pharmaceutical facility (the "end user") should do routine salt analysis. The analysis should confirm that there are no detectable amounts of extremely undesirable contaminants. In an effort to preserve iron in a soluble form, some towns may add proprietary compounds throughout the treatment process. It is undesirable for particle iron to emerge and stain household sinks, basins, and other surfaces. However, the iron produced by these unique compounds is often in a colloidal form. This substance typically causes scaling of downstream RO membranes as it passes through the water softener. It is advised to conduct a thorough examination of the municipal water treatment system and the raw water supply.

Moreover, a connection ought to be made at the municipal treatment facility. The performance of the pharmaceutical water purification system must be ensured that treatment methods have not changed by conducting an annual tour with this contact. It is important to keep an eye out for resin particles in the product water of the water softening unit's stainless steel resin traps. Operating staff should take note of the presence of resin and/or resin fines on RO prefilters if resin traps are not being used. The overproduction of resin fines is a symptom that the system is not working properly (or design). This may be related to a design element (such as insufficient freeboard area) or an operating variable (such as an excessive backwash rate). The presence of resin fines in the backwash water will likely be linked to the observation of resin fines, which will eventually lead to a drop in the volume of cation resin. In order to make up for this, it might be required to occasionally add cation resin to a functioning unit. It has been previously explored how long cation resins can last in systems that use feed water that contains iron. Nonetheless, it is crucial to collect yearly "core" samples from the cation bed for systems where iron is not present in the feed water in low amounts. To ascertain the state of the resin, these samples should be examined for important factors. Important criteria include oxidation by residual disinfectant and the percentage of exchange sites transferred from sodium to calcium and magnesium sites during regeneration. The dilution water pressure must be established and maintained at a constant amount during the regeneration process in water-softening systems that use eductors to introduce regenerant brine. The eductor needs to be examined. It is likely that a tiny buildup of particle matter will have a big impact on how well the eductor functions. Samples of the feed water and product water should be collected and examined on a regular basis. Total hardness, calcium and magnesium concentrations, TSS, total iron, and trace amounts of multivalent cation contaminants should all be included in the study (aluminum, barium, strontium, etc.).

The end product water from two units operating in series should have calcium and magnesium concentrations of less than 1 mg/L as calcium carbonate. As part of the system's data trending, feed water and product water data should be plotted as a function of time. In addition, analysis of USP Water for Injection and low-endotoxin USP Purified Water systems should be done for total viable bacteria and bacterial endotoxins. To control bacteria inside, periodic sanitization (semiannual or annual) using hot water (where appropriate) or a chemical sanitizer should be done to regulate the microbial population in the carbonaceous cation bed. As was previously mentioned, the action limit for total viable bacteria in product water indicated by a water-softening unit is 500 cfu/mL. It is crucial to carry out a preoperational "rinse-to-drain" cycle on water-softening units that offer regeneration for primary mixed-bed deionization units.

This procedure guarantees that calcium and magnesium, which may have moved to the lower part of the resin bed, are removed before the regeneration process and removes bacteria from the stagnant bed. During the mixed-bed deionization unit's regeneration, the multivalent cations would cause precipitates of calcium and magnesium hydroxide that are intractable. The brine tank should undergo routine chemical sanitization on a yearly or biennial basis. Sodium hypochlorite is a potent sanitizer. In an effort to reduce microbial growth inside the salt storage tank, sodium hypochlorite may also be intermittently supplied. To ascertain whether the resin capacity is degrading over time, regeneration cycles as well as the outcomes of periodic feed water tests for total hardness, calcium, and magnesium should be considered.[35]

9. *Organic Scavenging Resins*

Particularly for macroporous anion resins, the production of resin particles and concomitant resin attrition are of concern. With time, irreversible organic resin fouling will develop. To evaluate the resin's physical state on a regular basis (every six months), core samples of the resin must be analyzed. The success of the regeneration of the anion organic scavenging resins depends on the concentration of regenerant sodium chloride and sodium hydroxide. Also important is temperature. Both the temperature and the amount of regenerant introduced should be periodically checked.

The main factor controlling the regeneration cycle is sulfate breakthrough via the anion exchange resin. The system design should specify a breakthrough mechanism that will make the frequency of the regeneration step clear. Chemical analysis or inline monitoring methods designed specifically for the sulfate ion may be used to determine this. Depending on the resin chosen for the application and the "life cycle" of the anion organic scavenging resin, the predicted reduction in TOC will range from 50% to 80%. When flow through the unit is restarted after using macroporous resins and stopping the flow rate through the anion bed, it is possible to see a discernible drop in the product water's TOC level. Although it would seem advantageous, what is actually happening is that traded organic material is being diffused toward the center of the anion resin beads. This greatly reduces the chance that the transferred organic material will be removed throughout the regeneration process. Thus, it is advised to maintain continuous flows, especially for systems using a macroporous anion organic scavenging resin. The regeneration time period needed for anion organic scavenging resins is substantially more than that needed to remove the inorganic ions, as stated. The sodium hydroxide and regenerant salt should be added to the anion bed over a period of 10 to 16 h to "completely regenerate" it.

The TOC readings in feed water and product water should be assessed often. Around 0.04 to 0.05 mg of TOC/mL of resin/min should be eluted from the anion organic scavenging resin. Resin replacement and/or prolonged regeneration should be taken into consideration if calculations show that TOC elution is much greater than the allowed values. It is strongly advised that frequent SDI measurements be made in addition to the periodic measurement of TOC values from samples acquired from the unit in order to confirm the effectiveness of the latter. The amount of contaminants in the feed water that can foul a RO membrane can be accurately determined using SDI measurements (chap. 4). The operation and regeneration cycles for the unit should be chosen based on the TOC, SDI, and product water sulfate levels.

It may be advantageous, especially for larger flow rate systems, to consider online TOC monitoring from the product water of the unit since elution of organic material may occur, particularly if the inorganic anion concentration increases. Given that this is an ion exchange process, a number of "breakthrough curves" will be produced by the numerous organic compounds (NOM) that pass through the unit. While these curves appear to favor the early passage of organic material with a lower molecular weight, then the passage is necessary to determine the precise properties of dynamic anion organic scavenging resin unit performance for certain water sources of inorganic material, followed by heavier-molecular weight organic material.

The use of the technology as a pretreatment unit operation is highly restricted, mostly to small capacity rechargeable canister ion exchange-based applications, due to the clear maintenance-intensive requirements for organic scavengers. The use of this technology for pharmaceutical water systems, where the TOC product water specification is orders of magnitude higher than that for other applications like those in the semiconductor industry, is limited by the continuously improving performance characteristics of RO membranes combined with the declining cost of replacing membranes.[40]

TABLE 9.3 TOC Reduction by Various Anion Resins.[40]

Resin "skeleton"	Pore structure	Mean pore radius (nm)	Surface area (m2/g)	% TOC reduction
Styrene	*Gel*	*N/A*	*0.1*	*15*
Styrene	*Porous*	*10*	*409*	*50*
Acrylic	*Gel*	*35*	*60*	*60*
Acrylic	*Porous*	*N/A*	*0.1*	*60*
Styrene	*Porous*	*100*	*<5*	*80*
Styrene	*Porous*	*3500*	*7*	*88*

10. Inline Ultraviolet Unit

Researchers looking at how lamp life affects ultraviolet exposure hypothesize that there is a 10% reduction in ultraviolet radiation dose within the sanitizing chamber for every 1000 h of operation (Zinnbauer, 1985). Hence, when the unit's effluent bacteria levels go above a predetermined value, the lamp needs to be replaced. This circumstance might arise before the inline ultraviolet unit manufacturer's suggested 8000–9000 h working life has run out. It is strongly advised that cautious inline UV unit maintenance be taken into consideration due to the effects of insufficient microorganism inactivation. Given their high cost, it is advised that, as part of a preventative maintenance programme, when replacing lamps and sleeves, it is crucial to keep your eyes away from ultraviolet radiation because it might cause cataracts (Wieringa, 2006). As part of a microbiological monitoring programme for an inline ultraviolet sanitization unit, the total viable bacteria levels in feed water and product water should be assessed. The inactivation of bacteria is a concern, even if the enumeration technique used for routine samples should be the same as that for samples obtained from other pretreatment components.

Thus, it is advised that feed water and product water samples be taken prior to the advised six-month lamp and sleeve replacement frequency and measured using R2A media at a temperature of 208 to 228°C for a 10- to 15-day incubation period. The microbiological concentrations in product water should be displayed as a function of time. Once more, if microbiological levels rise above a certain level, lamp replacement, sleeve replacement, and/or an examination of unit functioning should be done. An indication of the strength of ultraviolet radiation is among the most crucial operating parameters for an inline ultraviolet unit. To calibrate the intensity metre, use a recognized UV radiation source. This should ideally be carried out with a source that has been calibrated to a National Institute for Standards and Technology (NIST) standard.

Sadly, this rarely occurs since the source must be compatible with the monitor and is frequently either not offered by makers of inline ultraviolet units or is only offered at a premium price. As previously stated, measurements of the total live bacteria present in feed and product water must be used to confirm that an inline ultraviolet sanitization effectively inactivates microorganisms. Most calibration instructions offered with operating instructions for equipment manufacturers are based on determining a "100% transmittance" result for newly installed sleeves and lamps on the unit. An appropriate supply voltage that meets the requirements of the equipment manufacturer has a significant impact on an inline ultraviolet unit's performance. A slight decline in supply voltage, such as those experienced during "brownout" periods in major cities during the summer, can dramatically reduce the actual UV strength. An increase in temperature inside the sanitization chamber has a substantial impact on the functioning of an inline ultraviolet unit. Several systems are available for purchase with internal alarms or "cutoff" switches that will prevent electricity from reaching the unit if the temperature rises above a predetermined level. Excessive temperatures will destroy the ballast and reduce the effectiveness of the device's ability to kill microorganisms (s). An external temperature switch placed in the feed water line to the unit should be provided if the device does not have a high-temperature sensor, alert, or shutoff during conditions of high water temperature. The switch must be electrically connected to the inline UV unit's power source in order to shut off the unit's power when temperatures are too high and to provide an audible and/or visual alarm indication. It is advised that units having access to the sanitizing chamber remove the flange and gasket approximately every two to three years, depending on the application, to check the interior of the chamber. After the inspection, the access flange should receive a replacement gasket. Cleaning surfaces with the proper material should be part of this inside examination as well. Passivation or derouging of the sanitizing chamber should be taken into consideration if there is significant rouging of the stainless steel.

Determinement of the periodic feed water particle matter should be a component of the sampling procedure for monitoring unit performance. TSS and turbidity measurements will typically offer the necessary data to assess if UV light will be absorbed. Any apparent leaks come from either of the quartz sleeves. Alternatively, the connections between the flanges in the sanitizing chamber (or any other penetrations) need to be fixed. Bacteria may be introduced by leaks. Furthermore, the leaks could "aspirate" air into the sanitizing chamber, resulting in a "air bubble" and impairing the unit's functionality (41).

11. Recirculation and Repressurization Pumps

Pump discharge pressure and flow rate entries should be made in the operating logs. Pump feed water pressure and any "strange" noise made by the pump and/or pump motor should both be recorded in the log. Pump seals should be replaced on an indicated annual basis as a preventive maintenance item. Every year, the interior of the pump casing should be checked. Moreover, the shaft's "alignment" needs to be examined during seal replacement.

If possible and consistent with system design, every effort should be made to run a pump constantly. In comparison to operation in a cyclic mode, a pump will function for a longer amount of time with less maintenance when it is operated constantly. Two pumps should not be used, and switching between them once a week is not recommended. When the pump motor is first installed, when it is starting up, and whenever the electrical supply to the pump motor is changed, operating employees should check the rotation direction for the pump motor. If a pump feed water pressure switch is used, its proper operation should be checked once a year by purposefully restricting flow to the pump's suction side, which results in a drop in pressure. This will check to see if the pump's functioning ends at or near the minimum feed water pressure that has been specified. The temperature of the recirculating water for the particular pretreatment component(s) being recirculated should be recorded in operational logs.

To ensure that mechanical heat input from the pump is not raising the temperature of the recirculating water and, consequently, the water in the pretreatment component(s), which would increase the rate of bacterial proliferation within the component, this value should be plotted as a function of time (s).[35]

12. Cartridge Filtration

Taking into account all of the circumstances surrounding the usage of cartridge filters, frequent feed water and product water sampling and accompanying microbiological level analyses should be conducted. If cartridge filters are used as "traps" (activated carbon fines or ion exchange resin), they should be replaced based on the amount of installation time that has passed rather than the amount of pressure reduction. In general, accumulating activated carbon fines or resin fines would not be expected to result in a significant pressure drop. Nonetheless, routine replacement should be done in order to reduce microbial buildup inside the units.

Setting a maximum pressure drop through the filtration system is crucial for cartridge filters used to filter raw water. Based on the fact that there has not been a material breakthrough, the maximum pressure decrease should be applied. The removal of cartridge filters during "change-out" should therefore be followed by regular "autopsies." The filter cartridge should be "sliced" perpendicular to its height during the autopsy to ensure that no particles have made it to the "clean" side of the cartridge. Since the kind and nature of particulate matter present in the raw water supply will fluctuate due to seasonal and climatic changes, it could be essential to modify the DP value calculated for cartridge filter replacement.

As filter cartridges are taken out, they should be examined for any signs of visible microbial growth, particularly slime, algae, and specific types of mould. Both the clean and dirty sides of the cartridge should be examined during this check. Also, it is advised that "swaps" or "swipes" of the filter cartridges be purchased if odor and/or color are seen so that the microorganisms can be recognized. The frequency and programme of pretreatment system sanitization are significantly influenced by the presence of pathogens.

For the TSS measurement, samples of feed water and product water should be taken periodically. Filter discs having a micrometre rating equivalent to the rated pore size of the cartridge filter element should be used for this measurement. The frequency of cartridge filter element replacement should also be determined using this measurement. Microbial development on the cartridge filters is a worry, as was already mentioned. Filter discs having a micrometre rating equivalent to the rated pore size of the cartridge filter element should be used for this measurement. The frequency of cartridge filter element replacement should also be determined using this measurement.

Microbial development on the cartridge filters is a worry, as was already mentioned. The filter housing should be routinely cleansed with a liquid sanitizing agent to reduce the impact of this circumstance. Input of bacteria from a cartridge filtration system will be reduced as a result. Every year, the gasket connecting the filter housing's "base" and "bell" should be changed. For circumstances requiring non-routine repair, spare gaskets should be kept on hand. Isopropyl alcohol (IPA), which can provide an effective cleaning solution, should be frequently removed from the gasket and the physical space beneath the gasket. IPA was used to clean the surface because it can serve as a breeding ground for microorganisms.

It is not advisable to keep filter cartridges in a place where they will be exposed to dust, fumes, or other airborne contaminants. When fresh replacement cartridges are received from the cartridge filter supplier, it is advised that they be placed in airtight bags. It is advised that a label with the cartridge filter lot number, serial number, manufacturer, and date of installation be affixed to the filter housing in order to give a record of cartridge filter change as well as other information. Moreover, this data needs to be entered in the system logbook.[35]

9.7 VALIDATION OF WATER SYSTEMS

FIGURE 9.7 Water system validation.

Image credit: Auriga Research. https://aurigaresearch.com/pharmaceutical-testing/water-system-validation/

Every pharmaceutical product contains water, so it is important to evaluate the water system to ensure the reliable production of high-quality water. The pharmaceutical business places a great priority on the quality of water used in completed product manufacture, intermediate reagent preparation, and analytical operations, particularly in the case of parenteral medicines where the quality of water must be in accordance with Pharmacopoeia. It is crucial to plan the water system correctly and have staff members that are knowledgeable about all the physical, chemical, engineering, and microbial challenges related to water.[43]

1. Obtain and maintain compliance with pharmacopoeia requirements.
2. Use a proper sampling system from the right spots at the relevant frequency.
3. Solve typical contamination issues.
4. Continually generate water that satisfies industrial standards for quality.

Currently, a good pharmaceutical water system must be in place in order to preserve the quality of pharmaceutical water. This system must be qualified in terms of system design and must pay attention to regularly updated regulatory requirements. A clear legislative necessity and a significant financial burden on staff and resources, regular monitoring of the water system is required.[44,45]

In the production, processing, and formulation of prescription items, water is mostly employed as a substance, raw material, or ingredient. Controlling the microbiological quality of this water is crucial since the distribution, storage, and purification of this storage substance may result in the rapid growth of germs that are common to water. These microbes or their metabolic byproducts could possibly have negative effects if water is employed in the final product. The National Primary Drinking Water Regulations (NPDWR) (40 CFR 141) established by the Environmental Protection Agency must be complied with in order for water to be used in the initial stages of the production of drugs and as the source or feed water for the preparation of various types of purified water (EPA). It is permitted to use comparable drinking water rules from the European Union or Japan. These conditions guarantee the absence of coliforms, which, if found to be fecal in origin, may foreshadow or imply the existence of other fecal microbes, including viruses that may be harmful for humans. On the other hand, if these National Primary Drinking Water Standards were met, it would not preclude the inclusion of additional bacteria, which while not being a major public health problem, could, if present, pose a risk or be deemed undesirable in a drug substance or manufactured product. Pharmaceutical water comes in a wide range of grades as a result. Organic scavengers, activated carbon beds, chemical additives, and filtering selected unit processes with validation issues include deionizations, electrodeionizations, electrodialysis, reverse osmosis, and ultra-filtration. The goal is to draw attention to problems with installation and design. Criteria for operation, maintenance, and monitoring aid in water system validation.[46]

Purified water is one of the most essential ingredients used in the production of all pharmaceutical products. To make sure that the desired water quality is consistently generated, the water system needs to be validated. Water system certification is unquestionably essential for the production of your pharmaceutical products if you are a pharmaceutical firm.

As water is a crucial component of all pharmaceutical products, its validity must be ensured in order to guarantee the reliable manufacture of high-quality water. The most crucial factor to take into account in the pharmaceutical sector is the quality of the water utilized in the creation of finished goods. It is essential for the manufacture of intermediate reagents and analytical procedures, particularly when it comes to parenteral goods where the water quality must meet pharmacopoeial standards.[47]

The pharmaceutical water system's purity should satisfy all legal standards. It is undeniable that any negligence in the correct monitoring of the water system can put a financial strain on your manufacturing company by forcing a product recall and wasting time and resources. To understand all the physical, chemical, engineering, and microbial challenges associated with water production and distribution, water system validation is crucial.

Significance of the water system in pharmaceuticals

- Pharmaceutical product manufacturing requires high-quality water.
- In the production of pharmaceuticals, water is the most often used raw resource.
- Water is used in the manufacture of pharmaceuticals, either directly or indirectly, for cleaning manufacturing equipment and as a key ingredient in injectable medications.
- In GMP and in validating the production process, water is a crucial raw element.

Why is it necessary to validate water systems?

- The water quality must be specific to the quality of the product. Deterioration, pollution, product loss, and financial gain can all result from poor water quality.

- Water typically has microbiological contamination, organic and inorganic pollutants, endotoxins, and particle contamination that can be inappropriate for the production of pharmaceutical goods. Also, it may significantly affect a product's effectiveness and quality.
- Manufacturing pharmaceutical items requires a lot of water. Contamination may also result from the distribution procedure. Because of this, a crucial component of pharmaceutical manufacturing plants is the validation of the water system.
- Validating water systems for conformity with regulations is a need for factories.
- For the purpose of ensuring dependable, regular water output and quality.
- To avoid unacceptably high levels of microbiological, chemical, and physical contamination during manufacturing, storage, and delivery.

Requirements for pharmaceutical water system validation

- It must continue to adhere to pharmacopoeia regulations.
- A good sample procedure using the right spots and the right frequency.
- Identifying and fixing frequent contamination issues.
- To continuously provide water that meets the requirements for quality set by the industry.

A brief examination of a few aspects for pharmaceutical water system validation maintenance and control

- In order to maintain control over the water system, a systematic maintenance programme needs to be implemented.
- The operation of the water system, regular maintenance, and remedial action should all be done according to established procedures, which should specifically specify when action is necessary.
- Operating parameters should be recorded and monitored, and a thorough monitoring programme is a crucial quality aspect. The programme may combine manual documentation of operating parameters (like the pressure drop in the carbon filter) with laboratory testing, in-line sensors or recorders (like a conductivity metre and recorder), and other elements.
- To keep the system under microbiological control, a regular sanitization procedure must be carried out in accordance with the system design and the chosen units of operation.

- To determine what preventive maintenance is to be carried out, the frequency of maintenance work, and how the work should be documented, a preventive maintenance programme is important.[48–50]

Several methods for treating water

- De-chlorination (sodium bisulphate carbon filter)
- Filtration
- Ultra-filtration
- Softening
- Demineralization
- Reverse osmosis
- UV treatment
- Deionization
- Ozonization

Concept of validation

It is critical to show that the system or process reliably generates the required quantity and quality when run and maintained in accordance with detailed stated operating and maintenance procedures.

Validation requires proof, which is essential

- Construction engineering
- Operational methods and permitted control parameter ranges
- Its maintenance process to make it happen
- The system must be thoroughly planned, installed, and tested throughout the construction process, afterward, and under all operational circumstances.

Lifecycle of water system validation

Identifying quality characteristics

- To clearly comprehend the water quality that is needed and the purpose for which it will be put.
- It should be decided before the validation process begins.
- It is crucial to identify the necessary quality criteria before establishing validation processes.[51–53]

Protocol for validation

- A documented strategy outlining the validation process and specifying quality acceptance standards.

- A manufacturing process's protocol specifies
- Crucial process variables
- Process equipment
- Sampling
- Product characteristics
- Number of runs for validation
- Test data to be collected
- Satisfactory test results

Validation steps

- Defining systems and their components
- Establishing standards for quality attributes.
- Designing machinery, controls, and monitoring systems.
- Establishing standards for operating procedures.
- Establishing an OQ and IQ stage.
- Setting the alert and action levels.
- Developing and future PQ stage.
- Completing procedures and recording each action.

System of water for pharmaceutical use

The specifications of its use determine the qualities of water that are appropriate for a given use. Schematic figure illustrates the sequential processing stages used to treat water for various pharmacological uses.

9.7.1 INSTALLATION, CONSTRUCTION MATERIAL, AND COMPONENT SELECTION

Installation: Methods are significant because they have the potential to impact the system's mechanical, corrosive, and sanitary integrity. Gravity drainage should be promoted during valve installation. Pipe supports should have the proper drainage slopes and be built to adequately sustain the piping in the worst possible heat conditions. To avoid potential issues, connecting techniques for system parts such operating units, tanks, and distribution pipelines must be carefully considered.

Welds made of stainless-steel ought to produce sturdy joins with no internal corrosion. To guarantee satisfactory weld quality, low carbon stainless steel, compatible wire filters, when necessary, inert gas, automatic welding machines, routine inspections, and documentation are all helpful. To get rid of pollution and corrosion products and restore the passive

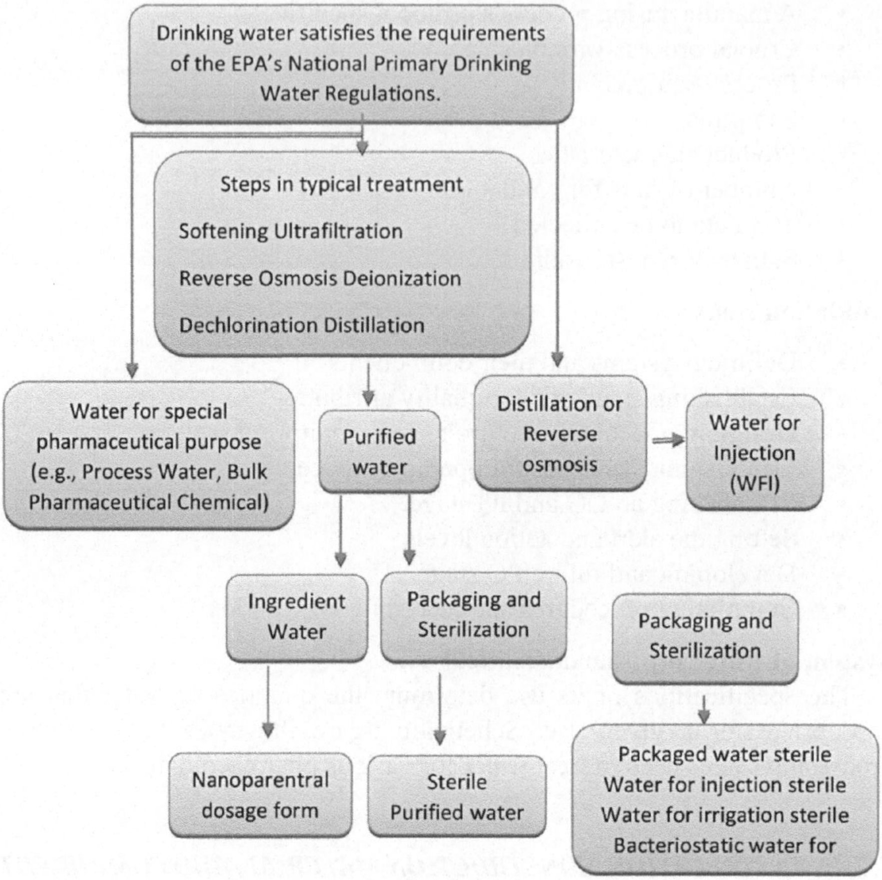

FIGURE 9.8 System of water for pharmaceutical use.

corrosion-resistant surface, follow-up cleaning and passivation are crucial. In some circumstances, plastic materials can be fused (welded), and they also need internally smooth, uniform surfaces. The possibility of cavities and chemical reactions makes adhesives to be avoided. When using mechanical connecting techniques like flange fittings, caution must be taken to prevent the development of offsets, gaps, penetrations, and voids. Good alignment, appropriately sized gaskets, suitable spacing, homogeneous sealing force, and avoiding threaded fittings are a few examples of control procedures.

Construction materials must be chosen so that they work well with preventative actions including sanitizing, cleaning, and passivating. Because surfaces may need to withstand high operating and sanitization temperatures, temperature rating is a crucial consideration when selecting the right materials.

Materials should be able to withstand turbulent flow and high velocities without wearing down the corrosive barrier, such as the stainless-steel surface with its linked chromium oxide passivation.[54–56]

9.7.1.1 WATER PURIFICATION, STORAGE, AND DISTRIBUTION SYSTEMS: VALIDATION AND QUALIFICATION

Pharmaceutical water purification, storage, and distribution systems must be reliable for a sufficient amount of time before being put into use. The chemical purity of purified water and water for injection is typically kept under control with few issues. It is more challenging to regularly achieve defined microbiological quality norms, nevertheless. After operational criteria have been set for each sample point, a typical programme entails intense daily sampling and testing of main process points for at least one month.

The process of obtaining and documenting evidence to a high level of certainty that a particular process will consistently produce a product adhering to a predetermined set of quality features is known as validation. The validation specifies the important process parameters and their permissible limits. A validation programme certifies the design, installation, operation, and performance of the equipment. It starts with the definition of the system and progresses through the following stages: Design Qualification (DQ), Installation Qualification (IQ), Operational Qualification (OQ), and Performance Qualification (PQ). Figure 9.3 illustrates a graphic depiction of a typical water system validation life cycle.

The following actions are often included in a validation strategy for a water system:

1. Setting criteria for operational parameters and quality qualities.
2. Creating systems and subsystems that would use the available source water to create the needed quality qualities.
3. Choose hardware, software, and monitoring methods.
4. Creating an IQ stage that includes instrument calibration, inspection to ensure that the drawings accurately portray the water system's as-built configuration, and, when needed, extra tests to ensure that the installation complies with design specifications.
5. Creating an OQ stage that consists of testing and inspection to ensure that the controls, equipment, and system alerts are running dependably and that the proper alert and action levels are set. There may be certain components of the next stage that overlap with this qualification phase.

6. Creating a hypothetical PQ stage to validate that the operating ranges for crucial parameters are adequate. To show system repeatability over a suitable time frame, a contemporaneous or retrospective PQ is carried out. Alert and action levels for important quality features and operating parameters are checked during this phase of validation.
7. Adding to a validation maintenance programme (also known as a continuous validation life cycle), which organizes and carries out planned preventive maintenance, including instrument recalibration, as well as a mechanism to manage changes to the water system. A monitoring programme for crucial process metrics as well as a remedial action programme are also included in validation maintenance.

9.7.2 SEQUENCE OF VALIDATION

A. Design qualification (DQ)

Priority should be given to the equipment design that makes up the water purification system. It comes from the specifications for the process of purifying water. It is important to identify the process capacities, including total volume required per hour or day, average consumption, peak demand requirements, reverse capacity, minimum circulation requirements, and whether or not elevated temperature storage is required. All suitable organizations, such as engineering design, production operation, quality assurance, analytical services, etc., should participate in the design qualification. The complexity of the project makes a team approach necessary. There are many factors at play, including the choice of materials, the suitability of the equipment, operational controls, construction methods, cleaning and sanitizing techniques, component compatibility, preventive maintenance, sterilization programmes, sampling, and regulatory requirements. The actions required for the reliable production of the specified grade of water are listed in the design qualification. Additionally, it offers the calibration of important devices. The design qualification provides sampling plans and ports for chemical and microbiological testing, the microbial action and alarm limits, the sanitizing techniques, and the procedure for data analysis and plotting.

The following should be part of the fundamental design package:

1. For the ease of use, flow schematics for the planned water system that illustrate every piece of instrumentation, control, valve, and component should be numbered.

2. A thorough explanation of the system's attributes and capabilities. To ensure that manufacturing and quality assurance staff, who might not be familiar with technical jargon, completely comprehend how the system is to be developed, built, operated, monitored, and sterilized, this is crucial.
3. Detailed specifications for the pretreatment and water treatment equipment that will be employed.
4. Detailed specifications for all other system parts, including pipelines, pumps, valves, heat exchangers, and storage tanks.
5. Exact specifications and explanations of how sanitary system controls work.
6. Requirements for building methods to be used in situations when quality is crucial.
7. Steps for cleaning the system both immediately after installation and on a regular basis.
8. Provisional operating, sampling, and sterilization standard operating procedures (SOPs). Cross references between these methods and the valve and component numbers on the system schematics will be made.
9. Draft SOPs for filter maintenance, integrity testing, and replacement.
10. First sampling techniques to check the equipment's performance and the purity of the water.
11. Methods for preliminary system certification.
12. Procedures for preliminary preventative maintenance.

B. Installation qualification (IQ)

This is the initial qualifying document, and it will include a section on procedures after the description of the system. The installation and assembly of the various pieces of equipment must be confirmed to be correct. Each piece of equipment ordered and reviewed must be carefully examined for resemblance before being recorded. It makes sure that every component of the unit is placed in accordance with the design drawing and according to the requirements. Construction compliance with set specifications is verified by IQ. Instrument connections, a review of the instrumentation designs, a review and verification of the MOC, an examination and documentation of the welds, a check for dead legs and pipe slopes, a confirmation of the passivation of the stainless steel, and any other information are also included in this.

IQ guarantees the appropriateness of the finished system and conforms the "As-Built" drawing. Leaks should also be tested for absence. The water filtration system should be fully described in the IQ along with the why and how it works. This stage requires identifying the feed water.

List the primary system components, such as the pump, filters, UV lamps, controls, valves, drains, and control system, and make sure they all meet the design requirements. Create a list of the instruments and controls, and ensure that the calibration of each can be traced back to regional, national, and worldwide standards. Instrument calibrations can be completed at the conclusion of the IQ process and documented as a part of IQ or at the start of the operational qualification.

C. Operational qualification (OQ)

The system's OQ is possible when the IQ has been completed successfully. To reduce the possibility of corrosion and contamination, the system needs to be thoroughly cleaned and all construction derbies removed. Equipment should be started up and carefully checked for appropriate operation when cleaning is finished. OQ tests a processing unit's ability to function properly within operational constraints. OQ involves taking into account the system capacity, temperature management, and flow rates. During OQ, concentrate on the crucial elements and criteria. Limits for pressure differential alarm control for utilities like steam pressure (high/low) shall be examined. Each limit's calibration requirements are established. The system should present obstacles with respect to the minimum and maximum inputs and outputs for processes. Findings must be examined and must fall within acceptable bounds. Operation, maintenance, and preventative measures SOP must be completed along with real use as well as operation and maintenance manuals. After the SOP is completed, OQ must include training for the technical personnel of the SOP. Via a change control mechanism and approvals, each change may be addressed. Each variation must be authorized and documented. Complete the OQ documentation and confirm that all functional and operational characteristics meet the acceptance criteria. Evaluate and accept the OQ report and protocol.

The system is prepared for Validation or Performance Qualification (PQ).

D. Performance qualification (PQ)

PQ's goal is to do thorough testing to show how well the entire integrated process works and can be replicated. To achieve the goal of ensuring the system's reliability and resilience over an extended length of time, three phase approaches must be implemented. The regulatory expectation is a three-phase validation.

1. PHASE 1

Two to four weeks of testing, or at least 14 days, are required for thorough system monitoring. The system should function continuously during this time without failure or performance deviance. The testing strategy should take the following into account.

- Conduct tests for bacteria and chemicals in line with a predetermined plan.
- To check the quality of the incoming feed water, sample it every day.
- Daily samples should be taken after each purification process.
- Create suitable operating ranges.
- Create and complete working, cleaning, sanitizing, and upkeep processes.
- Show that product water is produced and delivered in the required quantity and quality.
- Check the action levels and provisional alerts.
- Apply and improve the SOP for cleaning, sanitizing, and shouting problems.
- Develop and refine test—failure procedure.

2. PHASE 2

After phase 1 has been well completed, a second test period of 2–4 weeks (30 days) should be used to conduct additional rigorous monitoring and develop all the revised SOPs. Generally speaking, the sample plan should be the same as in phase 1. During this phase, water can be used for manufacturing purposes. When the system is run in accordance with the SOP, the approach should also demonstrate

- Consistent operation within predetermined ranges; and
- Consistent production and supply of water in the requisite quantity and quality.

3. PHASE 3

Phase 3 normally lasts one year after phase 2 has been satisfactorily completed. During this phase, which has the following goals and characteristics, water may be used for production purposes:

- Show extended, dependable performance.
- Ensure that seasonal differences are considered.
- Based on established practises validated during phases 1 and 2, the sample sites, sampling frequency, and tests should be reduced to the typical routine pattern.
- Based on the outcomes of phase 3 of the water system qualification programme, a regular plan should be prepared after phase 3 is finished.

9.7.3 REVALIDATION

Revalidation should only be carried out when the system or the operational parameters have undergone a significant change. The same conditions that prevailed during the initial validation will apply to routine monitoring and inspection. There should be a precise written protocol for routine maintenance and replacement of parts, and it must have been validated when it was first validated.[57–60]

9.7.3.1 SANITIZATION

Sanitation procedures are the main means of achieving microbial control in water systems. System sanitization techniques include thermal and chemical methods. Moreover, the water in the system can be continually sanitized using in-line UV light with a 254 nm wavelength.

A wider range of construction materials can be treated with chemical procedures when they are compatible. These procedures frequently make use of oxidizing substances such as halogenated substances, hydrogen peroxide, ozone, or peracetic acid. Halogenated substances are efficient sanitizers, but they are challenging to remove from the body and frequently leave biofilms intact. By creating reactive peroxides and free radicals, substances like hydrogen peroxide, ozone, and peracetic acid oxidize bacteria and biofilms (notably hydroxyl radicals). Ozone, in particular, may need to be supplied repeatedly during the sanitization process due to its short half-life. Ozone, hydrogen peroxide, and peracetic acid all quickly break down into water and oxygen in the presence of UV radiation.

9.7.4 LEVELS OF ALERT AND ACTION

Specific microbiological limits are not listed in the various monographs for purified water and water for injection. They were purposely left out because the majority of the currently used microbiological procedures need at least 48 h to get conclusive results. By then, the industrial process has already used the water from which the sample was collected. Rejecting the product lot in question would be necessary if it did not meet a compendial specification. And this is not what an alert or action guideline is intended to do. It is necessary to create quantitative microbiological guidelines for water used for pharmaceutical purposes because they will specify what must be done in the event that major excursions beyond these limits take place.

Alert levels are thresholds or ranges that, when surpassed, show that a process may have deviated from its intended state. Levels of alertness are a warning but may not always call for a response.

Action levels are thresholds or ranges that, when surpassed, show a process has strayed outside of its typical operating range. If the process exceeds the action level, corrective action should be done to return it to its regular operating range.

9.7.5 CONTROL, MAINTENANCE, AND OPERATION

In order to maintain control over the water system, a preventative maintenance programme should be implemented.

Operation procedure—Written operating, maintenance, and corrective action procedures for the water system should be created. These procedures should also specify when action is necessary. The method should be thoroughly documented, specify the purpose of each task, designate the person in charge of carrying it out, and provide instructions on how it should be carried out.

Monitoring programme—It is important to record and keep track of operating metrics as well as critical quality aspects. The software may combine manual documenting of operational characteristics (such as carbon filter pressure drop) and laboratory tests with in-line sensors or recorders (such as a conductivity metre and recorder).

Sanitization—To keep the system in a state of microbiological control, regular periodic sanitization may be required, depending on the system design and the units of operation chosen. The aforementioned descriptions of sanitization technologies.

Preventative maintenance—It should be carried out according to a plan. The programme should specify the types of preventative maintenance to be carried out, the frequency of maintenance work, and the methods for documenting the work.

Change control—It is necessary for the mechanical setup and state of operation. The effects of proposed changes on the overall system should be assessed. After changes are made, it should be decided whether the system needs to be requalified. Drawings, manuals, and procedures that will be impacted by a water system modification should be updated.[34,61–64]

9.7.6 TYPES OF VALIDATION

- **Retrospective Validation**

  It may be necessary to "redesignate" a system as a purified water system if it has been in operation for some time but is not yet certified as a validated system, such as a "qualified" high purity water system. This kind of thing happens frequently. Sometimes a pharmaceutical manufacturing company will decide that a particular application needs purified water rather than drinking water or deionized water. Nonetheless, the particular manufacturer has been using this water purification system to supply water for the application for a number of years. The purified water delivery system is regularly maintained and complies with all chemical requirements, including internally determined Alert and Action Limits for total viable bacteria. Documents for proper validation can be generated. It may require a lot of work to submit the necessary paperwork for the IQ and OQ. It is necessary to produce a basis of design, specifications, and suitable drawings. Through the PQ Summary Report, the validation procedure described must be finished. As the validation for this specific application is being carried out after the system has been deployed and is in use, it is known as a "retrospective" validation. Product water at points of use can be suitably labeled as purified water after validation, presuming that the Summary Report clearly confirms functionality as a purified water system. It is not the intention of this technique to imply that water produced before the validation process was purified water. To put it another way, a system is not retroactively designated as a verified Purified Water system.

- **Prospective Validation**

  The best way to validate a pharmaceutical water system is to carry out each of the actions outlined previously in this chapter, from creating a basis of design to compiling and creating a PQ Summary Report. Before using water as purified water or water for injection, validation would be finished. Because all validation steps, including confirming that water quality at points of use meets Purified Water or Water for Injection requirements, were established prior to the water being used as Purified Water or Water for Injection for any applications at the facility, this method of validating a system is referred to as "prospective." It is crucial to produce "simulated" water demand conditions when the PQ is being executed for prospective validation. To put it another way, water needs to be drained from the system at the appropriate flow rate, anticipated volumetric demand, and frequency. This is best done at the points of consumption.

- **Concurrent Validation**

  It is occasionally necessary to validate water systems using the final approach. Water from the system is used as Purified Water or Water for Injection during a "concurrent" validation before the PQ execution is fully finished and the PQ Summary Report is created. Water used before the validation process is complete is utilized "at risk." It is extremely advised to isolate any product created during this process. All papers, from the Basis of Design through the OQ, including SOPs, the maintenance manual, and the training manual, have been prepared using this validation process. Executions of the IQ and OQ procedures have been made. The PQ protocol is ready, has been accepted, and is being partially used. The pharmaceutical manufacturing facility will typically wait until at least two complete weeks of PQ data are available before using water from individual points of use when concurrent validation is carried out. Water may be used, at risk, as purified water or water for injection if it has been established that the quality of the product water at the points of use complies with all applicable chemical, bacterial, and bacterial endotoxin standards. This form of validation comes with danger, as was already mentioned. By holding the product in quarantine until PQ execution is finished in full, this danger can be reduced. The PQ is kept running for the recommended 30-day time to complete concurrent validation, and the PQ Summary Report is created. Purified water systems are typically the only ones where it is decided to do concurrent validation. A USP Purified Water system where the stored and recirculated water is continuously kept at a temperature >808°C is another circumstance that could be taken into account for concurrent validation.

- **Revalidation**

  System revalidation is the last topic that needs to be briefly covered. The quantity and scope of Change Control documents produced for a particular system affect how frequently a system needs to be revalidated. Drawing updates and markups to IQs, OQs, SOPs, and maintenance manuals eventually get confusing and difficult to manage. Indeed, the Change Control process broadens the type of documentation to some extent. A significant modification to a significant component of the system may necessitate system revalidation, even though there is no "fixed" time period designated for this. For instance, in the last five to ten years, continuous electrodeionization has been "upgraded" into many older purified water systems. The use of PLCs has been updated in numerous control systems over this time, allowing for flexibility in system monitoring and "operator-friendly" control features. Installation of continuous electrodeionization systems and/or controllers necessitates significant adjustments. Due to the nature of the Change Control process, these modifications will necessitate system revalidation. As technology for water purification, control, storage, and distribution systems advances, revalidation will therefore rise.[65,66]

9.7.7 PROTOCOL FOR INSTALLATION QUALIFICATION: USP WATER STORAGE TANK WITH ACCESSORIES

Experience suggests that using an actual water purification system as an example is the best way to illustrate an IQ's contents. Each year, seminars and workshops are held to examine the verification of water filtration systems. This method gives staff members a thorough understanding of the validation procedure, although it generally does not particularly demonstrate an actual IQ. Despite the fact that "template" validation should never be used, it is nevertheless appropriate to go over a step-by-step guide for a particular pharmaceutical water system component. In chapter 11's examples for writing specifications, purified water system components were covered. This chapter will concentrate on a storage tank with accessories, a part of USP Water for Injection.

9.7.7.1 INSTALLATION CERTIFICATE

A Certificate of Installation, comprising signatures from representatives from technical services, facilities engineering, and quality control, should be included on the first page of each part of the IQ document. Other disciplines covered earlier in this chapter can be added to this list. It should be noted that each IQ component should have its own "header" as a general rule. Moreover, a "footer" is appropriate, especially for indicating the document's most recent change. The Certificate of Installation has a notation about Change Control and potential requalification/revalidation of the system, which should be mentioned in closing.

9.7.7.2 PURPOSE

There should be a broad one-paragraph statement that explains the goal of each IQ segment. The structure of the IQ parts is outlined in this.

9.7.7.3 EQUIPMENT RATIONALE

The purpose of each component is described in the system's Basis of Design. The equipment justification gives a summary of the element that is discussed in the IQ. The Water for Injection storage tank and accessories in this instance, according to the equipment logic, offer a depressurized place for the collection of water from the upstream multiple-effect distillation unit. The tank also serves as a site for distributing water to a recirculation pump, a distribution loop, and a reservoir of water to support manufacturing activities. A hydrophobic vent filter, compound-type rupture disc assembly, spray ball, back pressure regulating valve, level control system, and temperature control system are just a few of the included accessories on the tank. A succinct yet useful part merging the design criteria (determined by thorough specifications) and the validation requirements can be found in the equipment rationale.

9.7.7.4 ACCEPTANCE OF PROTOCOL FORMAT

The IQ document's approval section includes a "sign-off" requirement that certifies those in charge of overseeing the validation of the Water for Injection system have read the IQ procedure before proceeding. Prior to execution, it is crucial that the right people sign both the cover sheet for the Certificate of Installation and the protocol format approval. The lead disciplines have only reviewed execution of the document by signing the first page of the IQ for the Water for Injection storage tank. By approving the protocol format, the lead disciplines attest that they have read the document carefully, reviewed it, and agree that it is sufficient as an IQ for the water for injection storage tank.

9.7.7.5 MCONCLUSION

The tank chosen complies with the system design parameters, equipment requirements, and design criteria outlined in the Basis of Design for the system, according to the conclusion section. This conclusion "weaves" the validation process from the Basis of Design to the IQ through the project.[67,68]

9.7.7.6 ACCESSORIES AND PARTS—INSPECTION ITEMS

The primary items that will be included, in detail, with descriptions in the IQ are tabulated in the "Components and Accessories" section. For instance, the detailed component inspection for the tank may need to include numerous line items for item 14 in the sample IQ, "tank fittings size and placement." The inspection list provides an overview of the several issues that must be resolved as part of the IQ for the storage tank for water for injection.

9.7.7.7 COMPONENT EXAMINATION

The Water for Injection storage tank is the component that is covered in this portion of the IQ for the project. The "Component Inspection" section outlines the various components related to this component in depth. The exact object to be inspected is listed in the "Parameter" column, which is the first column of the table. The "Test Method," which will be "visual" for most items, is indicated in the table's second column. "Real Results" is written in the table's last column. Any variations from the precise data in the Parameter column should be made very apparent in the Actual Results column.

This section exemplifies the wide range of variation that was encountered during document production. For instance, the sample IQ's second line item instructs the person completing it to "check, using tank nameplate information, that the tank is serial no. 12345." It is stated that the only way to properly prepare an IQ for any component is to thoroughly inspect it as you prepare the document. In fact, the person or people who are creating the document should act out its implementation as they are creating it. The person working in the pharmaceutical facility, who is expected to be familiar with the water purification system, should confirm the precise information in the IQ.

There are several approaches to creating validation documents, as was previously stated. The actual serial number could be left blank, which would force the person who executes the document to fill in the blanks. This is a less preferable method of producing the IQ. This might be seen literally as the Intelligence being prepared as the task is being carried out. Without ever visiting the facility, inspecting the water purification system and supporting components, or making sure the information they received from the equipment manufacturers and suppliers is accurate, many organizations will prepare an IQ based solely on the documentation in the turnover package they provide to customers. The creation of validation documents, particularly the IQ and OQ documents, should be a hands-on process. Only an on-site visual evaluation of thc numcrous components can determine the essential elements determining the nature of the components and accessories.

9.7.7.8 ACCESSORY INSPECTION

An inventory of tank accessories is provided in the "Accessory Inspection" part of the sample IQ. The hydrophobic vent filter (housing and membrane filter), the compound-type rupture disc, the vent tubing from the compound rupture disc, the divert valve on the supply line on the upstream multiple-effect distillation unit, the feed water tubing from the upstream multiple-effect distillation unit (with indicated slope), the back pressure regulating system, which includes a pressure sensor installed on the return tubing to the storage tank from the distribution unit are some of the accessories.

9.7.7.9 INSPECTION OF THE CONTROL SYSTEM

The specific components that interface with the sensors for temperature, level, pressure, etc., addressed in the "Accessory Inspection" section are covered in detail in the "Control System Inspection" section. The level-indicating transmitter, temperature-indicating transmitter, return tubing pressure transmitter, current-to-pneumatic converter for the back pressure regulating valve, facility steam modulating supply valve to the tank heating jacket, divert-to-waste valve in the feed water line to the tank from the upstream multiple-effect distillation unit, and the solenoid valves to the pump are all included in this section for the Water for Injection storage tank.

9.7.7.10 INSPECTION OF INSTRUMENTATION

The Water for Injection storage system's sensors and transmitters are all identified in the instrumentation examination section. The calibration reference standard's identification should also be included in this section. Also, it should make a distinction between "critical" and "noncritical" instruments that are both necessary for routine system monitoring. The transmitters, indicating controllers, and temperature or level sensor components are a few examples of important instrumentation. Pressure and temperature gauges fitted in the supply steam line to the tank's heating jacket are an example of noncritical instrumentation. The tank differential level-indicating transmitter, the tank temperature sensor/transmitter, the tank level sensor/transmitter (differential with two sensors), and the distribution loop return tubing pressure sensor/transmitter are only a few of the items covered in this area.

9.7.7.11 UTILITY INSPECTION

The utilities inspection section lists the tank's utilities, such as the supply of steam for the heating jacket. The specific components covered in this section include, but are not limited to, the facility steam modulating supply valve to the heating jacket, the condensate return from the heating jacket, the regulated [2–3 pounds per square inch gauge (psig)] supply steam to the heating jacket for the hydrophobic vent filter, the condensate return from the hydrophobic vent filter heating jacket, the waste line for the divert-to-waste valve from the upstream multiple-effect distillation.

9.7.7.12 VERIFICATION OF DOCUMENTS

The "Documentation Verification" section lists every supporting document, manufacturer information, and accessory "cut sheets" that were included in the "turnover" package from the component supplier. To ensure that the IQ for the Water for Injection storage tank is, as previously discussed, the tank's encyclopaedia, more information should be gathered as necessary. The title pages of the documents included in this section should each relate to one of the line item designations listed under "document or drawing reference." While being many pages longer than the actual IQ for the tank, enclosures offer a single source in a convenient location for supporting long-term operation, maintenance, troubleshooting, and replacement of items for the Water for Injection storage tank and its accessories. Instructions for manual diaphragm control valves, a manual for actuator installation and operation maintenance on control valves, a manual for hand-wheel operated diaphragm valve weir and straightway and diaphragm valve maintenance, a manual for an electronic sanitary temperature transmitter, a manual for an electronic sanitary pressure transmitter, a service manual for an air-actuated diaphragm valve, and other items are presented in this section of the IQ.[69]

9.7.8 OPERATIONAL QUALIFICATION PROTOCOL: USP WATER FOR INJECTION STORAGE TANK WITH ACCESSORIES

The OQ for a specific unit operation shows, as was mentioned previously in this chapter, that the system runs in a regular mode, including routine maintenance procedures such as multimedia filtering unit backwash, and is able to react to undesirable transients. An OQ will typically interface with upstream and downstream components as well as, more crucially, the system's central control panel or numerous control panels, unlike an IQ, which can be specifically built for a particular unit operation. The best way to demonstrate the suggested criteria of an OQ is to present an actual OQ, similar to the IQ for the Water for Injection storage tank. For conformity with other validation processes for a certain pharmaceutical manufacturing business, the format of the OQ may be changed as necessary. But as was already mentioned, the manufacturer's facility should be where the paperwork is prepared. It is very likely that the set points and limitations created for the control and instrumentation documents provided by the equipment supplier/installer won't match the actual values experienced during the execution of the OQ. The OQ offers a verification of system monitoring and control capacity beyond that carried out during system startup, more so than any other validation document. This is a list of the OQ sections for the USP Water for Injection storage tank.

9.7.8.1 *OPERATION CERTIFICATE*

Similar to the Certificate of Installation is the Certificate of Operation. By signing and dating the Certificate of Operation, the designated parties confirm that the information inside the document has been properly implemented. This sign-off format ensures that this has happened. In addition, if major changes to the system's operational characteristics are introduced, the Certificate of Operation makes reference to Change Control and the probable need for requalification/revalidation of the system.

9.7.8.2 *PURPOSE*

The OQ for the water storage tank for injection has a specific purpose. It states unequivocally that the OQ will confirm that the system functions as intended, sound particular alarms when the system fails, and alerts the user or takes action in the event of an undesirable excursion while the system is in operation.

9.7.8.3 *EQUIPMENT JUSTIFICATION*

The purpose of the specific unit operation, in this example the storage tank for water for injection, inside the system, is defined in the "Equipment Rationale" section. This equipment justification is the same as the justification for Intelligence.

9.7.8.4 *ACCEPTANCE OF PROTOCOL FORMAT*

The acceptance of the protocol format in the OQ document performs the same function as it does in the IQ document. It guarantees that those appointed to check that the OQ has been correctly produced have read it before it is executed, are aware with its contents, and agree with its content.

9.7.8.5 *CONCLUSION*

The purpose of the OQ document is to ensure that components and support accessories operate in an integrated manner from a controls standpoint during normal operation and unacceptable transients or abnormal conditions that generate an alarm warning operating personnel that an unacceptable condition exists, according to the conclusion, which reiterates this point. This OQ ensures that a temporary condition in one or more unit operations will not impair the quality of water at points of use when combined with other OQs for the overall system.

9.7.8.6 LIST OF COMPONENTS AND EQUIPMENT DESCRIPTION

In the section titled, "Equipment Description and List of Components," references are made to essential papers that will be needed throughout the OQ execution. In particular, the P&ID for the storage and distribution system is mentioned. The electrical diagram for the control panel would also be used for executing OQ. The manufacturer support materials are also mentioned, which have to be in the system's IQ, SOPs, and maintenance guide.

9.7.8.7 CRITICAL COMPONENTS

The essential tools needed to complete the OQ for the water storage tank correctly are highlighted. Critical components might include, for instance, the individual indicating lamps on the central control panel (or display panel icons), the temperature control system for the storage and distribution system, the level control system for the storage tank, the alarm system on the central control panel, and the appropriate system electrical interlocks located within the central control panel for the system. As was previously stressed, when compared to the organized individual unit operation character of the "example" IQ, reference to these crucial components shows the integration of individual OQs.

9.7.8.8 CRITICAL PARAMETERS AND DATA SHEETS

The various steps necessary to create a transient condition are presented in the "Critical Parameters" section to insure that the system responds in an appropriate fashion. The body of this section is presented as a three-column table. The first column, Parameter, represents a critical step in creating the unacceptable transient condition or a description of how the transient condition will ultimately be noted by operating personnel. The second column,

Test Method, describes the method for creating the transient condition. The final column, Actual Results, would be initialed by the individual(s) from the pharmaceutical manufacturing firm executing the OQ for the storage tank. Initials (preferably a signature) and date are required in this column. Each of the steps required to create the transient condition must be verified within the Actual Results column. The Actual Results column must contain verification for each of the procedures needed to produce the transitory condition.

OQ's primary goal is to check the tank's level control system's functionality. A four-point level control system is installed in the storage tank for water for injection. "High-high," "high," "low," and "low-low" are used to denote the four level set points that run from the top to the bottom of the tank. The dedicated control panel for the multiple-effect distillation unit is included in the device's "package." The "manual/auto" switch (display screen indicator) on the face of the multiple-effect distillation unit control panel is highly beneficial for OQ execution because it allows the "auto" control function to be "defeated." The multiple-effect distillation unit's switch is in the "manual" position when the tank level is raised to the highest setting. The high level scenario has been encountered, as evidenced by an indication light on the symbol on the central control panel or display screen. The level is raised even further to a high-high level. The OQ requires confirmation that when the high-high tank level is reached, an audible alert, an indication light/display screen icon on the central control panel, and a red-colored strobe light are all engaged. After this procedure, a valve at the tank's base is used to purposefully drain (but not totally) the stored Water for Injection. Individual OQ line items are presented during the draining process to confirm that the high-high and high level indications, as well as the relevant alarm circumstances, are "clear." Up till the low level is reached, the tank is being drained. An OQ line item confirms that when this level is achieved, an indicator light or display screen icon is activated. Once low-low level—an excursion condition—is reached, the tank draining process continues. An indicator light or display screen icon on the central control panel, a red strobe light, and an auditory alert are all activated by this condition. The downstream Water for Injection distribution pump is automatically turned off at a low-low tank level, according to a line item of OQ. The multiple-effect distillation device can now fill the tank after closing the tank drain valve. Alarm circumstances should end as the level rises above the low-low set point. The low-low level-indicating light and display screen icon should also be turned off as the water level rises, and the Water for Injection distribution pump should be turned back on.

The following section of the OQ adjusts the set points to create alert conditions in order to verify the appropriate operation of the temperature-indicating control system. The high temperature set point in this example is 958°C, which is extremely close to boiling conditions, despite the fact that it may seem unreasonable to alter set points to obtain alarm circumstances. Allowing the tank to reach boiling point would be totally unacceptable. The effectiveness of the temperature control system can be tested by making a tiny adjustment to the set point, to a value just below 958°C. A temperature recorder or data logger records the temperature excursions. The central control panel's control light and display screen icon, a red strobe light, and an auditory alert are all activated in response to the high tank temperature condition. The low distribution loop return temperature is verified using a similar process. This value is determined by a temperature probe that is inserted into the return distribution tubing upstream of the back pressure regulating valve.

By manually disconnecting the electrical line from the central control panel to the "mating connection" from the compound rupture disc, the OQ ensures the integrity strip (electrical continuity strip) within the compound rupture disc assembly is functioning properly. The electrical connector being disconnected simulates disc failure and sounds an alarm. The central control panel's light/display screen icon, a red strobe light, and an auditory alarm all turn on to signal the situation. The transient conditions mentioned in the OQ for the Water for Injection storage tank show that when an unsatisfactory system transient situation occurs, a visual indicating light/display screen icon, an auditory alarm, and a red-colored strobe light are all activated. The OQ for this component has this as its goal.[70]

9.7.9 PERFORMANCE QUALIFICATION PROTOCOL: USP WATER FOR INJECTION STORAGE TANK WITH ACCESSORIES

A PQ for a USP water for injection storage and distribution system with few sites of use is shown in Appendix C. The appendix's content makes an effort to illustrate the PQ's format, sample criteria, analytical criteria, data logging, and timetable preparation. It should be noted that just the first week of a 4-week sampling schedule is included in Appendix C. The remainder of the sampling schedule has not been included due to the small number of sample points. The following list contains important considerations that should be made when preparing the PQ.

9.7.9.1 ACCEPTANCE OF PROTOCOL FORMAT

The approval of the protocol format for PQ is the same as that for IQ and OQ for the USP Water for Injection storage tank. Reviewing the PQ and confirming that the suggested format is acceptable to the technical services, facility engineering, and quality control disciplines is the responsibility of designated persons from the pharmaceutical manufacturing company.

9.7.9.2 SUMMARY

A summary of the sampling programme can be found in the "Summary" part of the PQ for the storage tank for water for injection. The initiative, according to the statement, would include a rigorous monitoring period of 4 weeks. As there are only three places of use in the distribution loop, samples will be taken from each one every day. Samples are only collected from Monday through Friday, during business hours. Moreover, samples will be taken from both the supply and return sides of the distribution loop. Samples of the water used to feed the multiple-effect distillation unit and the water used as its output will be taken once a week. A 10 cfu/100 mL total viable bacteria Action Limit is mentioned in the summary as being applicable to all Water for Injection samples. Also, it is mentioned that any microorganisms found in samples of Water for Injection would be subcultured and speciated. The "WFI" samples are those that were taken from points of use inside the Water for Injection distribution loop. Samples taken from the multiple-effect distillation unit's feed water, including the distillation unit's product water, will be referred to as "WGS" samples.

9.7.9.3 TESTING FOR CHEMICALS—WFI SAMPLES

According to the standards for Water for Injection systems that were previously covered in the text, chemical testing is described. TOC, conductivity, and nitrates are the chemical variables that are measured. Each points-of-use sample is subjected to laboratory chemical analyses. Online TOC measurement was not used due to the small number of points of usage. A conductivity sensor and a non-temperature compensated conductivity metre are both present in the return line tubing from the recirculating loop. Individual conductivity measurements made at the PQ's conclusion were contrasted with constantly recorded, non-temperature-adjusted conductivity measurements made at the Water for Injection return tube conductivity sensing device.

9.7.9.4 TESTING FOR BACTERIA—WFI SAMPLES

"Bacteria Testing—WFI Samples" describes the bacteria testing sampling process, shipment requirements, analysis format, and necessary documentation. The sampling process outlines how a sample should be collected correctly. As it stands, this PQ does not take into account flushing the outsides of point-of-use valves or other potential sample locations, including the ends of hoses. It should be emphasized that running water must drain for at least a certain amount of time before a sample can be taken. There is simply a 10-s time limit given. This brief window of time is actually intended to give sampling staff time to collect the sample. It makes no inference that a "representative" sample must be obtained through flushing. If water is not flushed for the same length of time from each point of use as specified in the PQ sampling process, extensive flushing prior to sample collection may be contested by regulatory inspectors. Flushing prior to sampling is also discouraged because it suggests that the system is flawed fundamentally. Samples that will not be tested within a suitable amount of time, recommended as 30 min, are to be chilled (dedicated room) at 2 to 4°C, it is specified. Moreover, all methods for counting bacteria must be started within 16 h of the sampling. When using a contract laboratory, bacteria enumeration should not start more than 24 h following sampling. The requirements for mailing the samples are listed. The samples should be transported in a transportable cooler with "ice packs" to keep the temperature between 2 and 4°C. 108°C is the maximum temperature that the sample should be at during shipping. The sample temperature should also not drop below 1°C. Membrane filtration of a 100-mL sample with incubation in R2A culture media for 72 h at 228°C 28°C is the bacteria monitoring approach recommended for the PQ of the USP Water for Injection system. According to the protocol, at least one sample should be examined each week in order to produce data seen after 5-, 8-, and 10-day incubation times. Additionally, the protocol specifies that any bacteria colonies found should be subculture and speciated.

In order to appropriately identify the samples with a sequential logbook number, as was previously discussed, a chain of custody must be completed if samples for bacteria detection are delivered to a contract laboratory. A typical sample bottle label can be seen in Data Sheet B. This label includes the necessary details, including the sample's sequential logbook number, the date and time of the sampling, the location of the sample, and the initials or signature of the person who took the sample. Findings from the contract laboratory should be presented in a formal report, along with the necessary supporting paperwork and the signature of the designated person in charge of overseeing the bacteria determination method at the contract laboratory. Getting an official report from the contract laboratory could take 2 to 3 weeks. Therefore, it is strongly advised that the contract laboratory send results when they become available as part of their "Scope of Supply." The pharmaceutical manufacturing company should receive a copy of the findings electronically or by facsimile machine. The contract laboratory should also be given the Alert and Action Limits for the total viable bacteria. The designated person at the pharmaceutical production facility should be contacted right once by the contract laboratory if a sample shows the presence of germs above the Alert Limit. Rarely, bacteria colonies can be seen after an incubation time of up to 72 h—in some cases, even just 24 h. The contract laboratory should regularly check on the bacteria colonies without affecting the 72-h results. A thorough assessment of the matter should be prompted by the pharmaceutical manufacturing company's discovery of significant bacterial growth before the 72-h incubation time limit. If bacteria monitoring is carried out immediately at the pharmaceutical manufacturing site, it is evident that this straightforward yet extremely responsive programme should be used. This programme should be used for samples that are obtained following PQ execution and should adhere to SOPs.

If a contract laboratory is employed, a qualified designated representative from the pharmaceutical manufacturing company should visit the facility on a regular basis. The inspection should, at a minimum, assess the protocols for receiving samples, the means of ensuring proper chain of custody, the techniques for monitoring bacteria, the separation of samples with high anticipated bacterial levels (from other clients), the qualification programme for the technician performing the analysis, the laboratory supervisor, the administrative personnel qualification, SOPs, and the established quality assurance/quality control programme. The contract laboratory should undergo annual inspections going forward.

9.7.9.5 TESTING FOR BACTERIAL ENDOTOXINS

The "Bacterial Endotoxin Testing" section contains information on the sampling process, transportation requirements (if any), analytical method, and necessary documentation. The only difference between the sampling methods and the methods used to collect bacteria samples is the size of the sample container. When using a contract laboratory, the shipment requirements are similar to those for bacteria samples. Depending on the specific preferences of the pharmaceutical production facility or, in the case of a contract laboratory, the technique available, the method of analysis may change. With a minimum sensitivity at least one order of magnitude lower than the 0.25 endotoxin unit/mL specification specified in the Official Monograph for Water for Injection, the bacterial endotoxin detection technique should be able to produce a quantitative result. It is strongly advised against using a technique that yields a "pass" or "fail" result at the 0.25 endotoxin unit/mL criterion since it does not offer the quantitative value needed to analyze data trends and assess long-term system performance. The paperwork and requirements for the contract laboratory are the same as for testing germs.

9.7.9.6 TESTS ON CHEMICALS—WGS SAMPLES

The "Chemical Testing" part of the PQ for the Water for Injection storage tank outlines the requirements for the chemical testing of WGS samples, the feed water, and the product water to and from the multiple-effect distillation unit. The weekly sample collection frequency, which is determined during PQ execution, is indicated in the first section. A number of characteristics should be examined in the feed water to the pretreatment system feeding the multiple-effect distillation unit. Some of the characteristics are crucial in determining the feed water quality by recognizing the needs to raise the quality of the water, even though they are not always included in the NPDWR as defined by the EPA. Furthermore, a system resembling a standard purified water system has been used to treat the feed water to the multiple-effect distillation unit. Elements that might result in potential scale-forming precipitates have been included in the extensive list of analysis. These would specifically include lead, calcium, iron, magnesium, barium, iron, and other metals. Due to its volatility and potential for transfer to the steam produced by each separate "action" of the multiple-effect distillation unit, reactive silica must be analyzed. The analytical requirements for the distillation unit's product water for chemical constituents are listed in addition to the analysis needed for the feed water to the multiple-effect distillation (TOC and conductivity).

9.7.9.7 TESTING FOR BACTERIA—WGS SAMPLES

The sampling techniques used for point-of-use WFI samples and WGS samples are fairly similar. One of the minute variations has to do with the "flush-to-drain" period before taking a sample. The allowed time for WGS samples is between 5 and 10 min. The additional flush time was chosen because it accurately depicts how the multiple-effect distillation unit operates. For instance, a signal from the USP Water for Injection storage tank is received by the control panel during normal operation of the distillation unit. Prior to flowing to the tank, the product water from the multiple-effect distillation unit will be diverted to waste for a predetermined amount of time, often 5 to 10 min. Thus, it is appropriate to specify a "rinse-to-drain" duration of 5 to 10 min before sample collection. WGS samples must be transported and shipped in a manner that is comparable to that of WFI samples. The analytical portion is comparable as well. The process, however, restricts subculturing and speciation of any detected bacteria to samples taken from the multiple-effect distillation unit's product water and not its feed water. It is crucial to remember that the water used in the multiple-effect distillation unit has undergone pretreatment using a technique that maintains microbial control. A heterotrophic plate count, rather than membrane filtration of a 100-mL sample with R2A culture media at 228°C for 72 h, would often be used to monitor the presence of bacteria in feed water samples to a multiple-effect distillation unit. Using PCA media, an incubation temperature range of 308–35°C, and an incubation period of 48 h, a heterotrophic plate count can be carried out. For this specific Purpose, the ability to "trend" the levels and the predicted bacteria levels are used to determine how to monitor the bacteria in the feed water to the multiple-effect distillation unit. The paperwork needed for this testing, which includes the chemical, bacterial, and bacterial endotoxin testing for point-of-use WFI samples, is comparable to the documentation needed for other portions of the PQ.

9.7.9.8 A SAMPLE LOGBOOK PAGE, LOG SHEET, AND SAMPLE BOTTLE LABEL

The PQ protocol's enclosures contain illustrative chunks of support data. A further sample and monitoring regimen is shown. We offer a sample log sheet for WFI samples with sequential numbers. It should be noted that the actual logbook would be a hardbound book with string binding and sequential page numbering to offer a permanent record of all samples gathered with their results. The structure of the book makes it very evident that no pages have been edited or removed in any way. Just drawing a line through an entry, initialling and dating the line, allows for corrections. For the WGS samples, a comparable logbook page is also included. The conductivity log for the multiple-effect distillation unit's product water conductivity monitoring is on Data Sheet A. A continuous recorder was included in this system's conductivity monitoring system for the Water for Injection storage and distribution loop. Nevertheless, the multiple-effect distillation unit's conductivity metre for the product water lacked a recorder. Thus, it is advised to record conductivity every day. Label for a representative sample bottle is shown in Data Sheet B.

9.7.9.9 SAMPLE LOCATION

A list of sample locations for the distillation unit feed water and product water samples, as well as the Water for Injection distribution loop samples, should be included in the PQ. For the Water for Injection distribution loop, a list of four sampling locations is given. Either "direct" or "sample valve" is referred to as the "kind" of sample. In this instance, the sample points WFI304 and WFI-310 have access to a point-of-use valve and the distribution loop supply valve, respectively, for direct water collection. Yet, two places of use for production vessels are "hard piped" right into the vessel. For each manufacturing point of use, a sample valve is then installed from the point-of-use valves. The "sterile access port type" is the name given to these valves. The water supplied to production vessels 1 and 2, respectively, is represented by the samples taken from sample valves WFI-305 and WFI-307. Last, but not the least, details are given for two WGS samples, "WGS-201" and "WGS-202," which represent the feed water and product water going into and coming out of the multiple-effect distillation unit, respectively.

9.7.9.10 SCHEDULE FOR SAMPLING

The sample schedule for the first week is on the last page of Appendix C. As was said earlier, it should be emphasized that samples are taken everyday from various places of use. Each of the 5 working days is used to gather samples because the operation lasts 5 days (Monday through Friday). The sampling programme is 4 weeks long and includes about 20 sample days. Every week, the WGS samples are taken.[71]

9.8 CONCLUSION

A number of topics pertaining to water filtration methods have been covered in this page. There has been much discussion on the design specifications for a new purified water system. In order to satisfy installation and operational needs, a number of actions were taken into consideration, with a focus on some crucial factors. Before we began the validation operations, we discussed the steps that are essential to a successful process. The validation of water systems, which is necessary to meet regulatory criteria, happens after the installation and operational requirements have been carefully examined and established and have been completed successfully. The design, installation, operating, and GMP specifications for ideal filtered water were examined. Also taken into account was preventive maintenance, a topic that receives little attention but has a significant impact on the quality qualities. We concluded by taking a quick look at cGMP compliance challenges. The type of filtration system used can be significantly influenced by the quality of the raw water used as a starting point. The ultimate design outcome is greatly influenced by an early assessment of the various water qualities that are available and the future water production requirements. The pretreatment techniques given the quality of the feed water and the ratios of the various water qualities to be generated are relevant considerations. System validation, documentation, and capital and operational costs have all been discussed. In pharmaceutical facilities, various types of filtered water systems are used. Each system is specifically created for a particular purpose, even if the majority of them share common features. It takes thorough attention to detail and a solid grasp of system operation to create an effective design. Regardless of how much money is invested, simply adhering to common sense guidelines does not always result in a reliable system. On the other hand, with adequate knowledge, it is frequently possible to design, construct, and validate a viable purified water system with a smaller initial outlay and reduced ongoing expenses.

ACKNOWLEDGMENT

The authors are thankful to the Sandip Institute of Pharmaceutical Sciences, Nashik for providing the necessary facilities to prepare this article.

AUTHORS' CONTRIBUTIONS

All the authors have contributed equally.

CONFLICT OF INTERESTS

There is no conflict of interest from all the authors.

KEYWORDS

- **reliable**
- **resilient**
- **sustainability**
- **water management**
- **pharmaceutical**
- **raw materials**

REFERENCES

1. USP (a). United States Pharmacopeia 33—National Formulary 28, Reissue, Official Monographs, The United States Pharmacopeial Convention: Rockville, 2010; pp 3870–3873.
2. USP (b). United States Pharmacopeia 33—National Formulary 28, Reissue, General Information Chapter, "Water for Pharmaceutical Purposes," The United States Pharmacopeial Convention, Rockville, 2010; p 740.
3. USP (c). United States Pharmacopeia 33—National Formulary 28, Reissue, Official Monographs, The United States Pharmacopeial Convention, Rockville, 2010; p 741.
4. United States EPA. United States Environmental Protection Agency, "Drinking Water Contaminants—National Primary Drinking Water Regulations," 2010. EP 2010 (.
5. European Pharmacopeia, Vol. 2, 7th ed. 2010. pp 3221. EP (.
6. European Pharmacopeia, Vol. 2, 7th ed. 2010. pp 3219–**3220,** 3224–3225.
7. Faroog, S.; Chian, E. S. K.; Engelbrecht, R. S. Basic Concepts in Disinfection with Ozone. *J. Water Pollut. Control Fed.* **1977,** *50*, 1818–1831.
8. Frank, C. A. Destruction of Volatile Organic Contaminants in Drinking Water by Ozone Treatment. *Ozone Sci. Eng.* **1987,** *9* (3), 265–288.
9. Gurol, M. C. Factors Controlling the Removal of Organic Pollutants in an Ozone Reactor. In: *Proceedings of the American Water Works Associates Annual Conference*, Dallas, TX, June 10–14, 1984.
10. Legube, B.; Crove′, J. P.; DeLaat, J. et al. Ozonation of an Extracted Aquatic Fulvic Acid: Theoretical and Practical Aspects. *Ozone Sci. Eng.* **1989,** *11* (1), 69–92.
11. Siddigui, M. S.; Amy, G. L. Factor Affecting DBP Formation During Ozone-Bromide Reactions. *J. Am. Water Works Assoc.* **1993,** *85* (1), 63–72.
12. Westerhoff, P; Song, R.; Amy, G. et al. NOM's Role in Bromine and Bromate Formation During Ozonation. *J. Am. Water Works Assoc.* **1998,** *90* (2), 82–94.
13. Applegate, L. E. Membrane Separation Processes. *Chem. Eng.* **1984,** *91*, 64–89.
14. Kronmiller, D. RO Permeate Water Flux Enhancement. *Ultrapure Water* **1993,** *10* (2), 37–40.
15. Harfst, W. F. Types of Water Treatment Membranes. Ultrapure Water **1995,** *12* (7), 34–38.

16. McClellan, S. A. Membrane Process Technology Basics—Nanofiltration. *Ultrapure Water* **1995,** *12* (7), 39–46.
17. Amjad, Z.; Pugh, J.; Zuhl, R. W. Reverse Osmosis Element Cleaning. *Ultrapure Water* **1996,** *13* (7), 27–32.
18. Collentro, A. W.; Barnett, S. M. Predicting the Performance of Reverse Osmosis Membranes for the Production of High-Purity Water. *Ultrapure Water* **1996,** *13* (7), 40–46.
19. Singh R. A Review of Membrane Technologies: Reverse Osmosis, Nanofiltration, and Ultrafiltration. *Ultrapure Water* **1997,** *14* (3), 21–29.
20. Loeb, S.; Sourirajan, S. Sea Water Demineralization by Means of an Osmotic Membrane. *Saline Water Conversion II: AdvChem* **1963,** *38*, 117–132.
21. Harfst, W. F. Types of Water Treatment Membranes. *Ultrapure Water* **1995,** *12* (7), 34–38.
22. Parise, P. L. RO System Troubleshooting: Diagnoses and Remedies—Part 2. *Ultrapure Water* **1996,** *13* (9), 54–60.
23. Rohm and Haas Co. *Ion Exchange ... Industrial Water Treatment; Technical Brochure IE-247*; Rohm and Haas Co.: Philadelphia, 1978; p 15.
24. McGarvey, F. X. *Introduction to Industrial Ion Exchange; Technical Brochure No. NDJ*; Sybron Chemical Company: Birmingham, 1983.
25. Siwak, L. R. Here's How Electrodialysis Reverses ... and Why EDR Works. Ionics Incorporated Bulletin. *Int. Desalin. Water Reuse Quart.* **1993,** *2* (4).
26. von Gottberg, A. J. M.; Siwak, L. R. Re-engineering of the Electrodialysis Reversal Process. Ionics Incorporated Bulletin. *Int. Desalin. Water Reuse Quart.* **1997**.
27. Hernon, B. P.; Zanapalidou, R. H.; Zhang, L. I. et al. Applications of Electrodeionization in UltraPure Water Production: Performance and Theory. Ionics Bulletin No. TP-371. Presented at the 55th Annual Meeting of the International Water Conference, 30 October–2 November in Pittsburgh, 1994.
28. ASME BPE. *Bioprocessing Equipment, Section MJ-7.2.3*; The American Society of Mechanical Engineers: New York, October 20, 2009; p 116.
29. FDA. Water for Pharmaceutical Use, 2014 Aug. https://www.fda.gov/inspections-compliance-enforcement-and-criminal-investigations/inspection-technical-guides/water-pharmacuetical-use (accessed 21 Feb 2022).
30. WHO. WHO Good Manufacturing Practices: Water for Pharmaceutical Use. https://www.who.int/docs/default-source/medicines/norms-and-standards/guidelines/production/trs970-annex2-gmp-wate-pharmaceutical-use.pdf?sfvrsn=39eb16b8_0 (accessed 21 Feb 2022).
31. Pharmacopeia. <1231> Water for Pharmaceutical Purposes, 2006. from http://www.uspbpep.com/usp29/v29240/usp29nf24s0_c1231.html (accessed 22 Feb 2022).
32. Annex 2. WHO Good Manufacturing Practices: Water for Pharmaceutical Use, pp 83–87.
33. NEMA, C. *250-Enclosures for Electrical Equipment (1000 Volts Maximum)*; National Electrical Manufacturers Association, 2008.
34. Eaton, A. D.; Clesceri, L. S.; Greenberg, A. E. *Standard Methods for the Examination of Water and Wastewater*, 21st edn.; American Public Health Association, American Water Works Association, and Water Environment Federation: Washington, DC, 2005.
35. Collentro, W. V. *Pharmaceutical Water: System Design, Operation, and Validation*, 2nd edn., pp 37–40.

36. Collentro, W. V. *Multicomponent Adsorption in Fixed Beds*; Master's Thesis. Worcester Polytechnic Institute: Worcester.
37. Weber, H. C.; Meissner, H. P. *Thermodynamics for Chemical Engineers*, 2nd edn.; John Wiley & Sons, Inc.: New York, 1958; p 47.
38. Cleasby, J. L.; Hilmoe, D. L.; Dimitracopoulos, J. Slow sand and Direct Inline Filtration of a Surface Water. *J. Am. Water Works Assoc.* **1984,** *76* (12), 44–56.
39. Ghash, K.; Schnitzner, M. UV and Visible Adsorption Spectroscopic Investigations in Relation to Macromolecular Characteristics of Humic Substances. *J. Soil Sci.* **1979,** *30*, 735.
40. Gottlieb, M. C. The Reversible Removal of Naturally Occurring Organics Using Resins Regenerated with Sodium Chloride. *Ultrapure Water* **1996,** *13* (8), 53–57.
41. Zinnbauer, F. E. Ultraviolet Water Disinfection Comes of Age. *Ultrapure Water* **1985,** *2* (1), 27–29.
42. Applegate, L. E. Membrane Separation Processes. *Chem Eng.* **1984,** *91*, 64–89.
43. Jadhav, V. M.; Gholve, S. B.; Kadam, V. J. Validation of Pharmaceutical Water System: A Review. *J. Pharm. Res*. **2009,** *2* (5), 948–52.
44. Swarbrich, J.; Boylan, J. C. *Encyclopedia of Pharmaceutical Technology*, Vol. 16; Marcel Dekker INC: New York, pp 211–247, 293–306.
45. Collentro, W. V. *Pharmaceutical Water System Design, Operation and Validation*; Interpharm/CRC, pp 2, 5–8, 114, 271–274, 437–439, 566–574.
46. Nash, R. A. Wachter, A. H. *Pharmaceutical Process Validation, a International*, 3rd edn., Revised and Expanded; Marcel Dekker; pp 401–422.
47. Sawant, D. A. *Pharma Pathway*, 4th edn., April **2007**; pp 2.235–2.326
48. Carleton, F. J.; Agallow, J. P. *Validation of Aseptic Pharmaceutical Process, Water System Validation*; Marcel Dekker: New York; pp 212–244.
49. United States Pharmacopeia, 23rd edn.; 1984; pp **1637**.
50. Quality Assurance Guide, Volume I, 3rd edn.; Organization of Pharmaceutical Producer of India (OPPI), 1996.
51. Haider, S. I. *Validation Standard Operating Procedures, A Step by Step guide for Achieving Compliance in the Pharmaceutical*; Medical Device & Biotech Industries, 1st edn.
52. Gupta, M. M.; Saini, T. R. Present Industrial Scenario of Pharmaceutical Water System: Preparation of Water, Cleaning and Sanitization of Water System. *Pharma Review*Oct–Nov 2007.
53. Haider, S. I. *Pharmaceutical Master Validation Plan, A Ultimate Guide to FDA*, 1st edn.; GMP & GLP Compliance, **2006**; pp 10–21.
54. Berry, I. R.; Harpaz, D. *Validation of Active Pharmaceutical Ingredient*, 2nd edn.
55. Niazi, S. K. *Handbook of Pharmaceutical Manufacturing Sterile Products*, Vol. 6.
56. Pharmaceutical Engineering Guide. *A Guide for New Facilities, Volume 4: Water and Steam Guide* Dec 2000.
57. WHO. Supplementary Guidelines on Good Manufacturing Practices: Validation. QAS/03.055/Rev.2, Oct 2005.
58. HSA, Singapore Guides, Guidance Notes on Water Systems for Manufacturers of Non-Sterile Products, Ref. No.: GUIDEMQA-010-005, Effective Date: 06 SEP 2004.
59. Bedi, A. S.; Bedi, A. S. *Encyclopedia of Environment and Pollution Laws*, 1st edn.; Orient Law House: New Delhi.

60. Collentro, W. V. Proper Validation of a Water Purification System: Am Inherently Flawed Process? *J. Validation Technol.* **1995,** *1* (3), 17–23.
61. Collentro, W. V.; Angelucci, L. Coordinating Validation Requirements for Pharmaceutical Water Purification Systems. *Pharm. Technol.* **1992,** *16* (9), 68–78.
62. Eaton, A. D.; Clesceri, L. S.; Greenberg, A. E. *Standard Methods for the Examination of Water and Wastewater*, 19th edn.; American Public Health Association, American Water Works Association, and Water Environment Federation: Washington, DC, 1995.
63. Elms, B.; Green, C. Water Systems: The Basics, Part 1: Design as a Prelude to Validation. *J. Validation Technol.* **1995,** *1* (2), 36–42.
64. Elms, B.; Green, C. Water Systems: The Basics, Part 2: Validation and Maintenance. *J. Validation Technol.* **1995,** *1* (3), 716.
65. FDA. *Guide to Inspection of High Purity Water Systems*; Food and Drug Administration, Office of Regulatory Affairs, Office of Regional Operations, Division of Field Investigations: Rockville, 1993.
66. Fessenden, B. A Guide to Water for the Pharmaceutical Industry: Part 3—Purification and Validation Techniques. *J. Validation Technol.* **1996,** *2* (2), 132–148.
67. King, J. A Practical Approach to Equipment Validation. *J. Validation Technol.* **1997,** *4* (1), 84–89.
68. Neal, C., Jr. Validation 101: Back to Basics. *J. Validation Technol.* **1997,** *3* (13), 281–283.
69. Riley, B. S. Rapid Microbiology Methods in the Pharmaceutical Industry. *Am. Pharma. Rev.* **2004,** *7* (2), 28–31.
70. Roczniak, B. J. A Guide to Validating a Purified Water or Water for Injection System. *J. Validation Technol.* **1995,** *1* (3), 24–27.
71. Sutton, S. How Do You Decide Which Microbial Identification Is Best. *Pharm. Microbiol. Forum* **2007,** *13* (1), 4–12.

CHAPTER 10

Rare Earth-Activated Phosphors for Solid-State Lighting Applications: A Case Study on Analyzing the Role of Ce^{3+} as an Efficient Sensitizer in Nanophosphors

S. SISIRA[1], ANN ROSE ABRAHAM[2], KAMAL P. MANI[3], LINJU ANN JACOB[4], CYRIAC JOSEPH[5], and M. KAILASNATH[3]

[1]*Department of Physics, Sree Narayana College, Alathur, Palakkad, Kerala, India*

[2]*Department of Physics, Sacred Heart College (Autonomous), Thevara, Ernakulam, Kerala, India*

[3]*Inter National School of Photonics, CUSAT, Ernakulam, Kerala, India*

[4]*Department of Physics, Baselius College, Kottayam, Kerala, India*

[5]*School of Pure and Applied Physics, Mahatma Gandhi University, Kottayam, India*

ABSTRACT

During this decade, solid-state lighting (SSL) has advanced rapidly for general lighting applications. The currently used standard architecture of SSL is the most popular phosphor-converted light-emitting diode (pc-LED), in which phosphor is an inevitable part. Mostly, rare earth-based phosphors are used in pc-LED because of their extreme stability and often narrow emission lines. Despite these advantages, rare earths have restricted aptness, owing to their weak absorption outside the UV spectral range. A most widely accepted solution to overcome this is incorporating a

Resilience and Sustainability for Energy and Water: New Advancements, Opportunities, and Framework. Cristóbal Noé Aguilar, A. K. Haghi, Amritanshu Shukla, Ann Rose Abraham, Swati Gokul Talele (Eds.)

sensitizer into the matrix so that via a resonant energy transfer mechanism, activator ions can be activated. In this regard, Ce^{3+} has been considered to be the most promising material owing to its advantages over others. Ce^{3+} exhibits strong and broad absorption and emission bands in n-UV and UV regions that originate from 4f to 5d transitions. Because of the emission spectra overlap between the emission band of Ce^{3+} and the excitation spectra of most of the rare earth ions, such as Tb^{3+}, Dy^{3+}, Eu^{3+}, and Sm^{3+}, Ce^{3+} can easily act as an efficient sensitizer to these rare earth activators. So, they could easily produce abundant and tunable emission colors, including white, when co-doped into proper hosts by adjusting doped ion concentrations. Moreover, due to the large ionic radii of Ce^{3+}, co-doping is easily possible in Ce^{3+}-based phosphors, which is an added advantage in achieving tunable color emission. This chapter describes the role of Ce^{3+} as an effective sensitizer in different phosphor host materials, its resonant energy transfer, and color tunability via energy transfer and ion substitution methods for the potential application in pc-LEDs.

10.1 INTRODUCTION

Light-emitting diodes (LEDs) are one of the most accepted and extensively used lighting sources due to their many advantages as compared to their traditional incandescent and fluorescent lamps in terms of power consumption, size, and lifetime.[2] Among various visible LEDs that are mainly used for illumination purposes and applications in optical communication and optical switches, invisible LEDs are preferred.[3] Display applications are mostly fulfilled by organic LEDs; and whenever long-life, better compatibility, and integrated circuits are required, inorganic LEDs are used.[4–5]

Lighting applications using organic, inorganic, or polymer LEDs are commonly referred to as solid-state lighting (SSL).[6] The rapid improvement of the efficiency of LEDs generates new opportunities in the illumination industry.[7–8] The commercial need for LED technology is thereby rapidly increasing as it is projected as an intelligent lighting solution where intelligent lighting indicates its energy-saving capacity. The advantages of SSL based on white LEDs (WLEDs) over older traditional light sources, such as longer lifetime, high color-rendering index, and remarkable efficiency, are abundant. [9–10] Thus, SSL based on LEDs is rising as an extremely competent field, as well as a substitutive method, in the existing lighting technologies.

10.1.1 PHOSPHOR-CONVERTED WLEDS

In SSL, InGaN and AlInGaP semiconductors are utilized for making LEDs that emit red, green, or blue colors.[11] To overcome the limitations and disadvantages of the aforementioned LEDs, down-converted phosphors are preferred to yield white LEDs with high efficiency.[12] The phosphor-converted WLEDs (pc-WLED) are the promising and most-investigated method by which one can produce white light from LEDs.[13]

10.1.1.1 FUNDAMENTAL ASPECTS OF PC-WLEDS: PHOTOLUMINESCENCE

Luminescence is the emission of electromagnetic radiation that occurs when an excited atom returns to the ground state. The process of energizing the atoms is called excitation and the relaxation of atoms from excited state to ground state is called emission.[14–15] The luminescence can be produced by using different modes of excitations and, based on that, luminescence is categorized as photoluminescence, cathode-luminescence, electro-luminescence, bio-luminescence, chemiluminescence, and radio-luminescence.[14–16] Phosphors are the materials in which photons are used for energizing the atoms and in this chapter, we focused on the luminescence from phosphors.[16–17]

10.1.1.2 PHOSPHORS: A BASIC UNDERSTANDING ABOUT ITS COMPOSITION

A luminescent material is often called a phosphor, which means "light bearer" that absorbs radiant energy of a certain shorter wavelength range and reradiates at longer wavelengths. Phosphors are mainly inorganic solids that exhibit a host lattice and luminescent centers, namely, activator ions which are doped in small amounts.[18–19] In phosphors, the exciting radiation absorbed by the activator ions is raised to higher energy states, and the activators, in the excited state, are relaxed to the ground state emitting excess energy as photons having the energy difference between the excitation state and ground state.[20] In some materials, the activators are not able to absorb excitation radiation; so, in such cases, activators are often energized by a donor/sensitizer atom via an energy transfer mechanism.[20–22] In many cases, the host lattice or ions in the host lattice can act as donors and the efficiency of those phosphors is unambiguously decided on the interaction between donor and acceptor.[22–24] The extent of spectral overlap plays an important role in the energy transfer between donor and acceptor and it could be said that there would be no energy transfer if there is no spectral overlap.[25–26] The energy transfer process is described as follows. Initially, a donor (D) absorbs the excitation radiation and transfers the excitation energy to a nearby acceptor (A). The energy transfer process can be schematically expressed as in Figure 10.1.

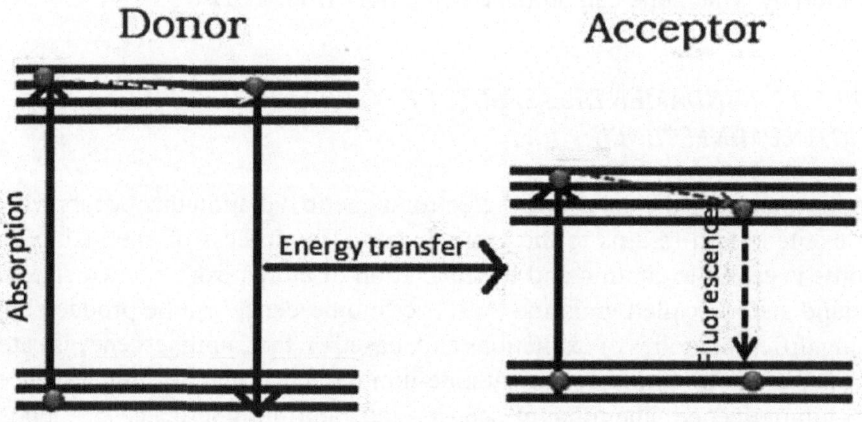

FIGURE 10.1 Schematic representation of energy transfer process.

The relaxation of the activator to the ground state from the excited state can take place radiatively and nonradiatively. If the radiative transitions dominate, a phosphor can be termed efficient.[27]

10.1.2 IMPORTANCE OF ENERGY TRANSFER MECHANISM

As it is already mentioned, the energy transfer mechanism is applicable when the activator ions fail to absorb the excitation radiation. Therefore, the energy transfer mechanism deserves attention since in many phosphors the luminescence is derived from the aforementioned energy transfer mechanism either from the host lattice or an added sensitizer in the host lattice. Several literatures are available on host lattice sensitization that leads to intense narrow emission and some of them are oxalates, molybdates, oxides, and vanadates.[28–33] But it is not always possible to obtain emission via sensitization from the host lattice since not all host lattices are capable of sensitizing activator ions. In such cases, the only solution is incorporating an appropriate impurity ion as a donor to the host lattice. Hence, to design an efficient phosphor, two main factors are to be considered. One is the selection of host and donor and the other is the choice of activator ions. Herein, we briefly describe the choice of activators and concentrate more on the choice of donors that are capable of transferring energy to activators to yield photoluminescence.

10.1.3 CHOICE OF ACTIVATORS

The selection of an appropriate activator is very important while designing a phosphor. Transition elements and rare earth ions are considered to be optically active ions in luminescent materials. Some of the commonly used transition elements are Ti^{3+}, Cr^{4+}, Cr^{3+}, Cr^{2+}, Mn^{5+}, Mn^{4+}, Mn^{3+}, Mn^{2+}, Fe^{3+}, Fe^{2+}, Fe^{+}, V^{4+}, V^{3+}, V^{2+}, etc.[34] Even though transition elements are efficient activators, rare earth ions with high luminescence efficiency and abundant emission colors offering color tenability are preferred in this work.

10.1.4 CHOICE OF SENSITIZERS

10.1.4.1 RARE EARTH ELEMENTS

The main focus of this chapter is the energy transfer mechanism via sensitizers that are used in phosphors. The most commonly used sensitizers in the field of phosphors are rare earths and are considered to be most efficient in yielding intense emissions with high quantum yield.[35,36]

Rare earth elements also known as lanthanides play a significant role in the modern lighting field as active constituents in many optical materials. Their emission spectra include ultraviolet (UV), visible, and near-infrared (NIR) regions, hence, color tuning is easily possible for lighting applications.[37,38] Each lanthanide ion is popular for its sharp and intense emission originating from uniquely characterized energy levels which are relatively independent of the host material.[39,40] The characteristic $^{2S+1}L_J$ energy levels of lanthanide ions doped into $LaCl_3$ were measured by Deshazer et al. and are often used as a reference to the energy levels of the $^{2S+1}L_J$ multiplet manifolds of rare earth ions embedded in any host lattice.[41]

Among different rare earths, Ce^{3+}, Tb^{3+}, La^{3+}, and Gd^{3+} are often used as sensitizers.[42–46] The energy transfer from Tb^{3+} to Eu^{3+} and Sm^{3+} is extensively studied by many scientists. The energy transfer from La^{3+} and Gd^{3+} is also investigated and reported in many literatures. But in this chapter, we focused on a sensitizer capable of transferring its energy to most of the activators that are popular in the field of phosphors nowadays irrespective of host lattices. In the case of Tb^{3+}, Eu^{3+}, La^{3+} Gd^{3+}, etc., the mentioned possibility is limited in number, and in this scenario, Ce^{3+} is mostly investigated as an efficient sensitizer and became the subject matter of this chapter.

10.1.5 THE ROLE OF CERIUM (III) AS A SENSITIZER

Cerium (III) ions exhibit broad and strong absorption and emission bands attributed to the 4f–5d transitions.[46,47] The absorption band assigned to the 4f–5d transition of Ce^{3+} in $CePO_4$ and $ZnA_{12}O_{19}$ are depicted in Figure 10.2(a) and (b), respectively.[48,49] The absorption band observed for all these three matrices is broad but seen to be split. In the case of $CePO_4$, splitting is more evident than that of $ZnAl_2O_{19}$. The difference like band splitting is due to the change in symmetry of the crystal lattice. If the crystal field effect is strong enough, splitting of the 5d level occurs completely and five distinct peaks would be observed.[50,51] In the case of $CePO_4$, crystal field splitting is found to be effective and the broad band appears to split into five peaks. Owing to the symmetry difference, the splitting is not as effective in $ZnA_{12}O_{19}$ as in $CePO_4$ samples.[48]

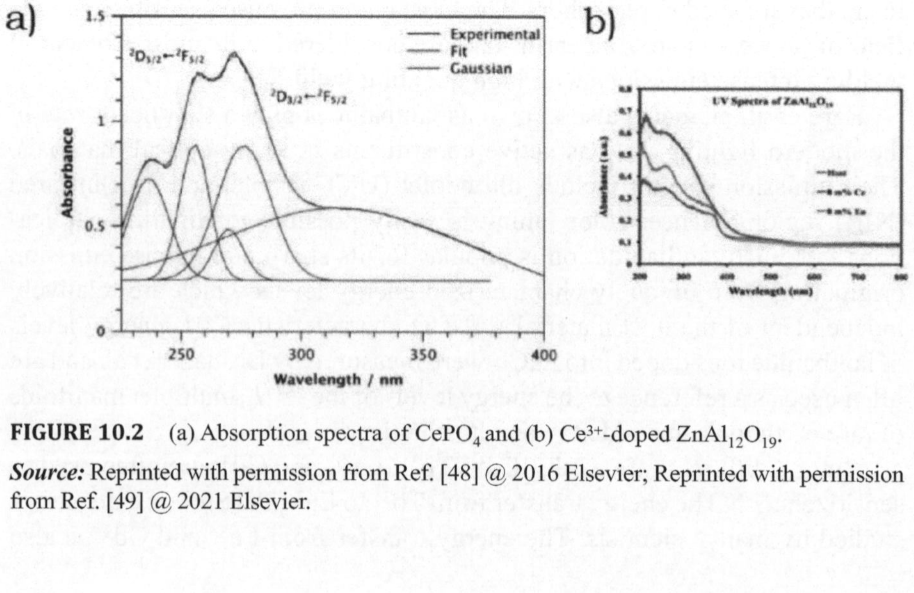

FIGURE 10.2 (a) Absorption spectra of $CePO_4$ and (b) Ce^{3+}-doped $ZnAl_{12}O_{19}$.

Source: Reprinted with permission from Ref. [48] @ 2016 Elsevier; Reprinted with permission from Ref. [49] @ 2021 Elsevier.

Owing to the intense absorption/emission band, Ce^{3+} ions can transfer energy to other ions and are often used as sensitizers rather than as activators. Besides, the large ionic radius of Ce^{3+} is large such that substitutional doping becomes much easier in cerium-based hosts and they are relatively cheaper as compared to other rare earths. Much literature is available on ascertaining the ability of Ce^{3+} as a sensitizer. Yang et al. reported the doping of Ce^{3+} as a sensitizer in $BaGdF_3$ crystal in combination with Sm^{3+}, Dy^{3+}, Eu^{3+}, and Tb^{3+} as corresponding activators and illustrated the potential applications in multiple biolabels and displays.[52] In the following section, we tried to explain the energy transfer mechanism behind Ce^{3+} to certain rare earths.

10.1.5.1 ENERGY TRANSFER FROM Ce^{3+} to Tb^{3+}

Terbium (III) ions are extensively investigated by many researchers owing to their prominent intense green emission during the past few years. They exhibit very strong and narrow excitation peaks that fall in deep UV or Vis-UV regions attributed to the f–f transitions. The emission spectra attributed to Tb^{3+} in $Ca_2Ga_2SiO_7$ is given in Figure 10.3.

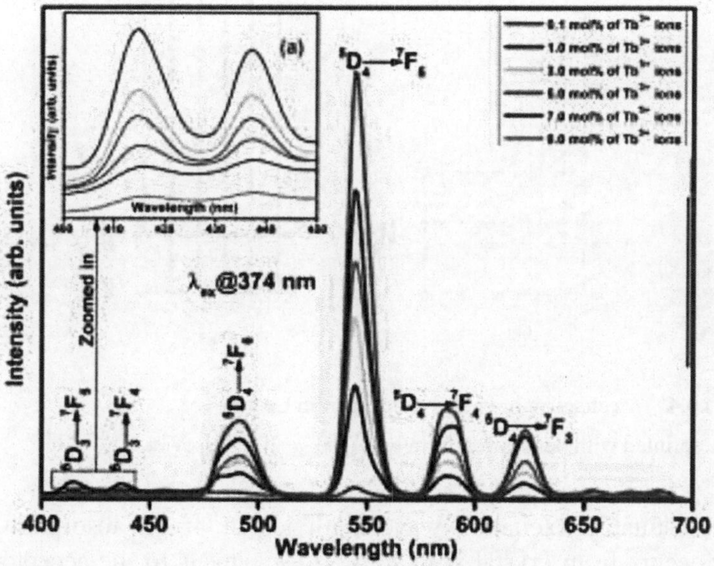

FIGURE 10.3 The emission spectra attributed to Tb^{3+} in $Ca_2Ga_2SiO_7$.

Source: Reprinted with permission from Ref. [53] @ 2020 Elsevier.

In some instances, f–f transitions are observed to be very weak due to their forbidden nature, and energy transfer from an appropriate sensitizer is used to enhance the Tb^{3+} emission. In addition to the green emission originating from its $^5D_4$ energy level, terbium is also known for its blue emission arising from the $^5D_3 \rightarrow ^7F_J$ transitions.[54] The intensity of emission lines from the $^5D_3$ level is dependent on Tb^{3+} concentration and is found to be quenched in most of the hosts at higher concentrations of Tb^{3+} due to the cross-relaxation mechanism.[54,55] The energy transfer mechanism from Ce^{3+} to Tb^{3+} is as follows.

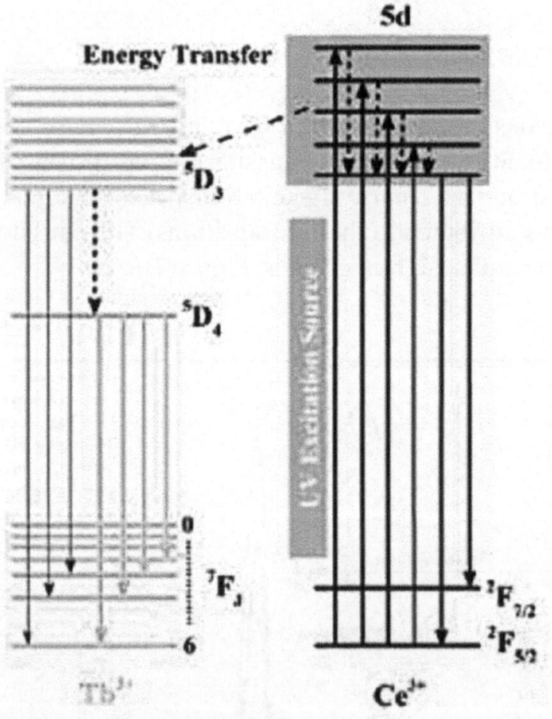

FIGURE 10.4 Mechanism of energy transfer from Ce^{3+} to Tb^{3+}.
Source: Reprinted with permission from Ref. [56] @ 2015 Elsevier.

Under suitable excitation wavelength ie., at 4f–5d absorption, energy transfer occurs from crystal field split 5d Ce^{3+} levels to the acceptor energy state of Tb^{3+}. Nonradiative relaxation occurs to lower $^5D_4$ and $^5D_3$ states of $Tb^{3+}$from excited Tb^{3+} state and is followed by a radiative decay to lower ^7F_J

levels. Two sets of emission peaks can be observed: one is assigned to the $^5D_3$ to ^7F_J transitions and the other is attributed to the $^5D_4$ to ^7F_J transitions of Tb^{3+}.[56] The energy transfer mechanism between Ce^{3+} and Tb^{3+} is investigated in many literatures and some of them are listed in Table 10.1 and shown in Figure 10.4.

TABLE 10.1 Phosphors Giving Tb^{3+} Emission via Ce^{3+} to Tb^{3+} Energy Transfer.

Sl. no.	Name of host	Crystal structure	Excitation wavelength (nm)	References
1	BaF_2:Ce^{3+},Tb^{3+}	Cubic	287	[57]
2	SrB_2O_4:Ce^{3+},Tb^{3+},Na^+	Orthorhombic	319	[58]
3	$GdPO_4$: Ce^{3+},Tb^{3+}	Monoclinic	273	[59]
4	SrAl F_5: Ce^{3+},Tb^{3+}	–	254	[60]
5	$NaBaPO_4$: Ce^{3+},Tb^{3+}	–	301	[61]

10.1.5.2 ENERGY TRANSFER FROM Ce^{3+} to Dy^{3+}

Dysprosium (III) ions are much more popular for intense blue and yellow emissions assigned to the $^4F_{9/2} \rightarrow {}^6H_J$ transitions.[62] Emission spectra of Dy^{3+} doped is shown in Figure 10.5. Blue emission arises from a magnetic dipole transition while yellow emission is originated from an electric dipole transition.[62,63] Therefore, depending upon the symmetry of the host, the luminescence color of the entire material may be blue or yellow. It is also possible to tune the emission color from blue to yellow including white by varying doping concentrations.

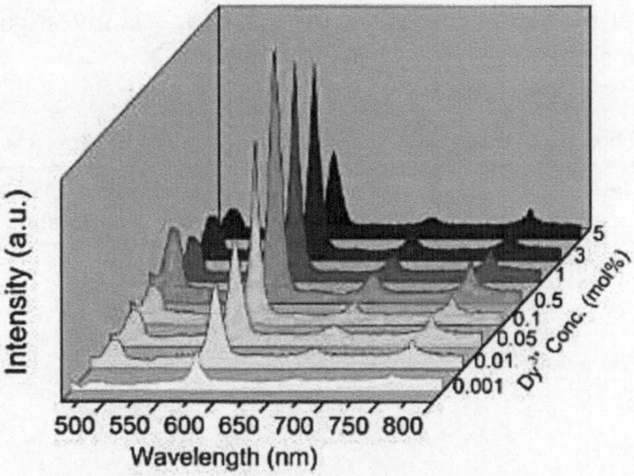

FIGURE 10.5 Emission spectra of Dy^{3+} in $SrBPO_5$.

Source: Reprinted with permission from Ref. [64] @ 2021 Elsevier.

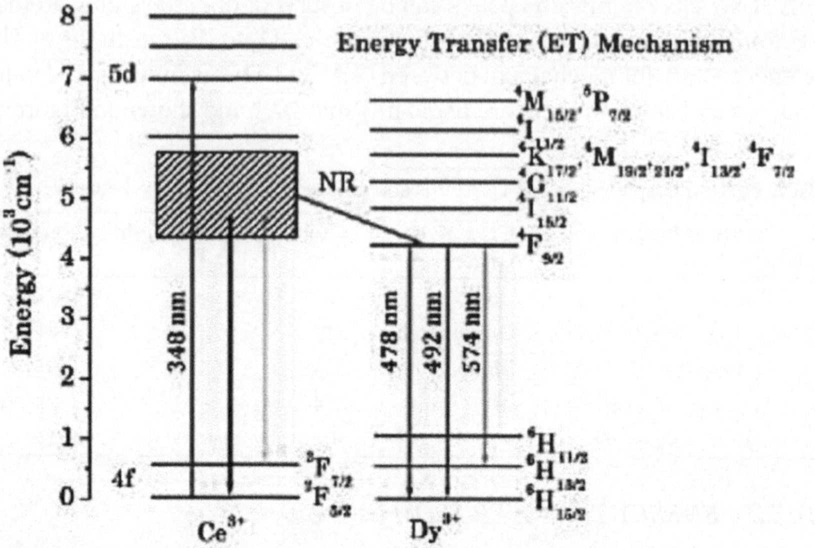

FIGURE 10.6 The energy transfer from Ce^{3+} to Dy^{3+} in CAS: Ce^{3+}, Dy^{3+}.
Source: Reprinted with permission from Ref. [65] @ 2016 Elsevier.

The energy transfer from Ce^{3+} to Dy^{3+} in CAS: Ce^{3+}, Dy^{3+}phosphor is illustrated in Figure 10.6. Under excitation at 350 nm, Ce^{3+} atoms are excited and a part of its energy is transferred nonradiatively to the $^4F_{9/2}$ excitation state of the Dy^{3+} ion. Subsequently, radiative relaxation of this excited level occurs to the ^6H_J ground state of Dy^{3+} ion to generate the emission.[65] The energy transfer mechanism between Ce^{3+} and Dy^{3+} is investigated in many literatures and some of them are listed in Table 10.2.

TABLE 10.2 Phosphors Giving Dy^{3+} Emission via Ce^{3+} to Dy^{3+} Energy Transfer

Sl. no.	Name of host	Crystal structure	Excitation wavelength (nm)	References
1	$Ca_{20}Mg_3Al_{26}Si_3O_{68}$:$Ce^{3+}$,$Dy^{3+}$	Cubic	348	[65]
2	$Ca_3Mg_3(PO_4)_4$:Ce^{3+},Dy^{3+},Li^+	Monoclinic	295	[66]
3	$LiY_9(SiO_4)_6O_2$: Ce^{3+},Dy^{3+}	Hexagonal	310	[67]
4	$Sr_2A_{l2}SiO_7$: Ce^{3+},Dy^{3+}	Tetragonal	335	[68]
5	Al_2O_3: Ce^{3+},Dy^{3+}		355	[69]
6	Sr_3SiO_5: Ce^{3+},Dy^{3+}	Tetragonal	360	[70]

10.1.5.3 *ENERGY TRANSFER FROM Ce^{3+} to Sm^{3+}*

Samarium (III) ions are well known for their intense orange-red emission making them a promising candidate to generate red-emitting material for WLEDs.[71] It exhibits several absorption bands arising from the intra-configurational f-f transitions in the wavelength range of 300 to 500 nm.[71] The emission bands originating from the $^4G_{5/2}$ to ^6H_J transitions are in the wavelength range of 550–700 nm. The Sm^{3+} ion offers high luminous efficiency due to the narrow and strong emission bands. The emission spectra corresponding to Sm^{3+}-doped barium gallium borosilicate glass is shown in Figure 10.7.

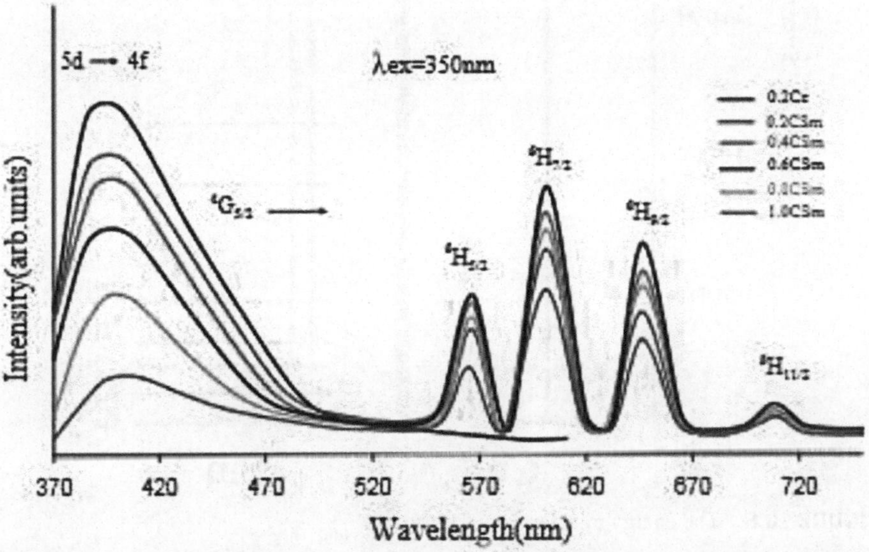

FIGURE 10.7 The emission spectra corresponding to Sm^{3+}-doped barium gallium borosilicate glass.

Source: Reprinted with permission from Ref. [72] @ 2019 Elsevier.

The energy transfer from Ce^{3+} to Sm^{3+} is explained with the help of Figure 10.8. Under the UV excitation, energy transfer takes place from the $^2D_{3/2}$ state of Ce^{3+} to the $^4G_{5/2}$ energy state of Sm^{3+} which decay radiatively to the lower ^6H_J levels giving Sm^{3+} emission lines.[73] The energy transfer mechanism between Ce^{3+} and Sm^{3+} is investigated in many literatures and some of them are listed in Table 10.3.

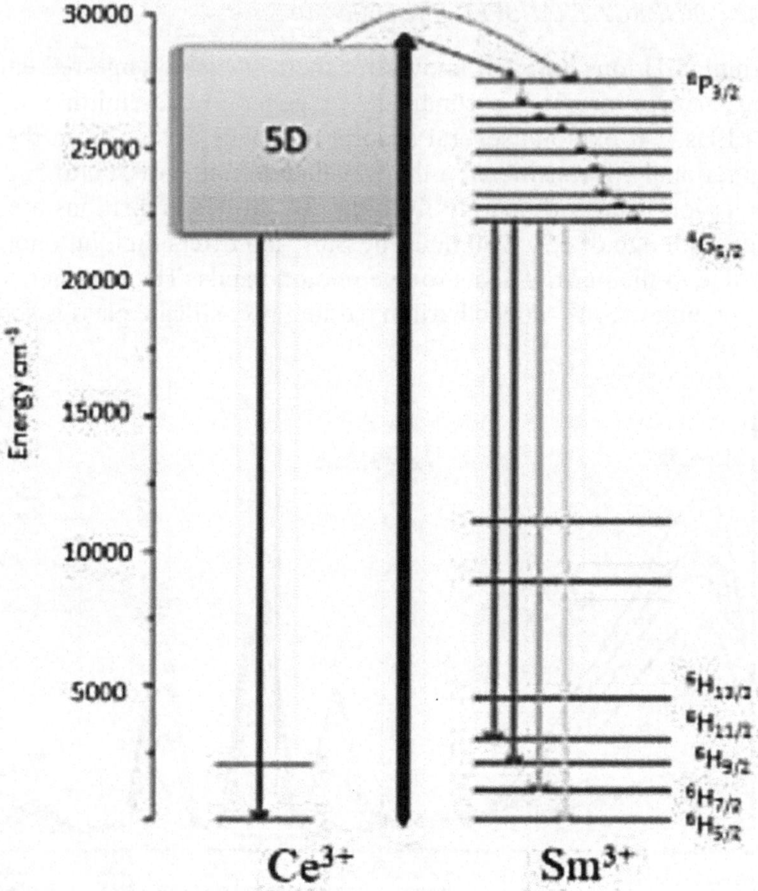

FIGURE 10.8 The energy transfer from Ce^{3+} to Sm^{3+}.

TABLE 10.3 Phosphors Giving Sm^{3+} Emission via Ce^{3+} to Sm^{3+} Energy Transfer.

Sl. no.	Name of host	Crystal structure	Excitation wavelength (nm)	References
1	$CePO_4$:Sm^{3+}	Monoclinic	284	[74]
2	$LiY_{0.99-x}Ce_{0.01}\ Sm_xSiO_4$	-	348	[75]
3	$LaSi_3N_5$	Cubic	354	[76]
4	$Lu_3Mg_2GaSi_2O_{12}$: Ce^{3+}, Sm^{3+}	Cubic	450	[77]
5	$Y_3Al_5O_{12}$: Ce^{3+} and Sm^{3+}-doped zinc–germanate–tellurite glass	Cubic	340	[78]
6	Y_2SiO_5 :Ce^{3+}, Sm^{3+}	Monoclinic	404	[79]
7	$NaLa(WO_4)_2$:Sm^{3+}:Ce^{3+}	Scheelite	–	[80]

10.1.5.4 ENERGY TRANSFER FROM Ce^{3+} to Eu^{3+}

Eu^{3+} is well known for its orange and red emission lines due to the $^5D_0 \rightarrow {}^7F_{i=0,1,2,3,4}$ transitions.[81] The magnetic dipole transition ($^5D_0 \rightarrow {}^7F_1$) of Eu^{3+} can be seen at 590 nm and the electric dipole transition ($^5D_0 \rightarrow {}^7F_2$) is often observed at 615 nm.[81] The emission spectrum exhibiting Eu^{3+} emission lines is shown in Figure 10.9. The relative intensity between these peaks can be used to investigate the nature of the symmetry of the lattice since the electric dipole transition is very sensitive to the site that Eu^{3+} ions occupy.[82,83]

The energy transfer from Ce^{3+} to Eu^{3+} is explained with the help of Figure 10.10. Under the UV excitation, energy transfer takes place from the crystal-field split 5d state of Ce^{3+} to the $^5D_3$ energy state of Sm^{3+} which decays nonradiatively to the lower $^5D_0$ level and is followed by radiative emission to lower ^7F_J levels giving Eu^{3+} emission lines. The energy transfer mechanism between Ce^{3+} and Eu^{3+} is investigated in many literatures and some of them are listed in Table 10.4.

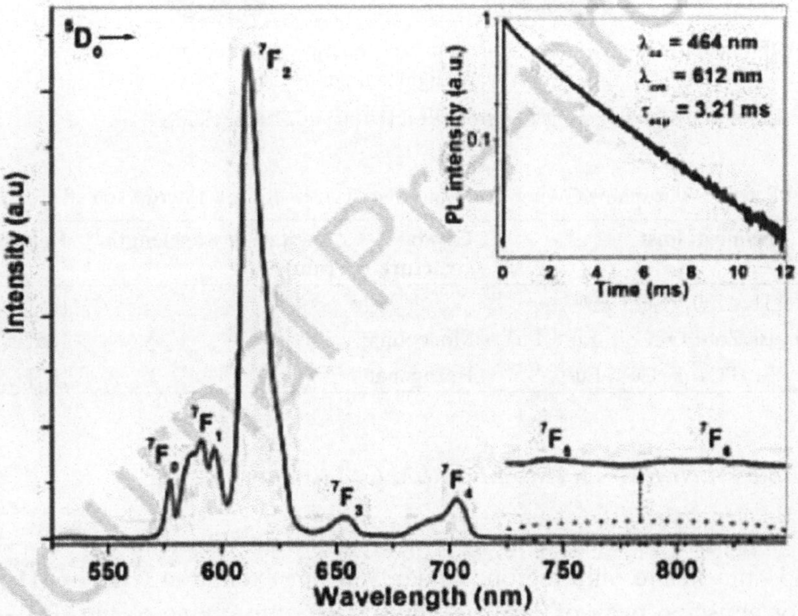

FIGURE 10.9 The emission spectrum of sodium-zinc-lead-borate glass doped with Eu^{3+}.
Source: Reprinted with permission from Ref. [83] @ 2021 Elsevier.

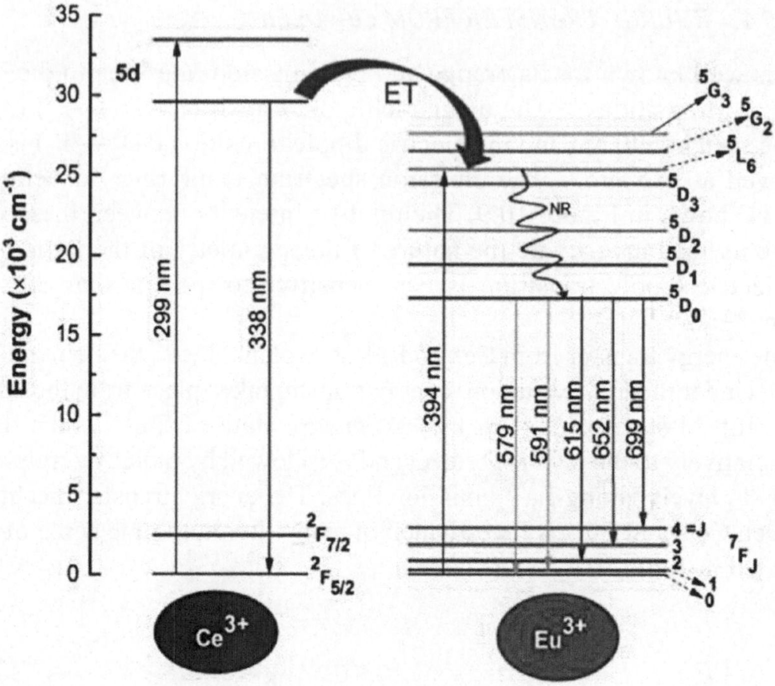

FIGURE 10.10 Intense red emission via energy transfer from Ce^{3+} to Eu^{3+} in P_2O_5+NaF+CaF_2+AlF_3 glasses for warm light sources.

Source: Reprinted with permission from Ref. [84] @ 2021 Elsevier.

TABLE 10.4 Phosphors Giving Eu^{3+} Emission via Ce^{3+} to Eu^{3+} Energy Transfer.

Sl. No.	Name of host	Crystal structure	Excitation wavelength (nm)	References
1	$Tb_3Al_5O_{12}$:Ce^{3+}, Eu^{3+}			[85]
2	$Ba_2ZnSi_2O_7$:Ce^{3+}, Eu^{3+}, Eu^{2+}	Monoclinic	330	[86]
3	$Sr_5(PO_4)_3F$: Ce^{3+}, Eu^{3+}	Hexagonal	354	[87]

10.1.5.5 ENERGY TRANSFER FROM Ce^{3+} to Tm^{3+}

Tm^{3+} ions can produce a sharp blue emission attributed to the $^1D_2 \rightarrow {}^3F_4$ at 454 nm in the blue region. Under 358 nm excitation ($^3H_6 \rightarrow {}^1D_2$), the major emission peak of Tm^{3+} is often at 457 nm owing to the $^1D_2 \rightarrow {}^3F_4$ transition. A weak peak can also be observed at 475 nm corresponding to the transition $^1G_4$–$^3H_6$. These spectral features are not affected by the crystal field.[88,89] The emission spectrum showing Tm^{3+} is depicted in Figure 10.11.

The energy transfer mechanism from Ce^{3+} to Tm^{3+} is explained with the help of a schematic energy level diagram given in Figure 10.12. On exciting at 280 nm, the Ce^{3+} ions are excited and transfer a part of the energy to the excited energy level of Tm^{3+} ($^1D_2$) via nonradiative decay. The excited Tm^{3+} ions released their absorbed energy as radiative emissions and went to the lower $^1G_4$ to ^3H_J energy levels.

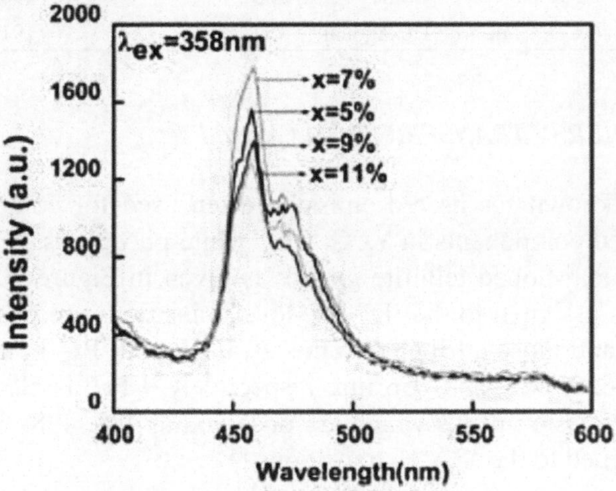

FIGURE 10.11 Emission spectra of $Sr_{2-2x}B_2O_5$: xTm^{3+}, xNa^+.
Source: Reprinted with permission from Ref. [88] @ 2014 Elsevier.

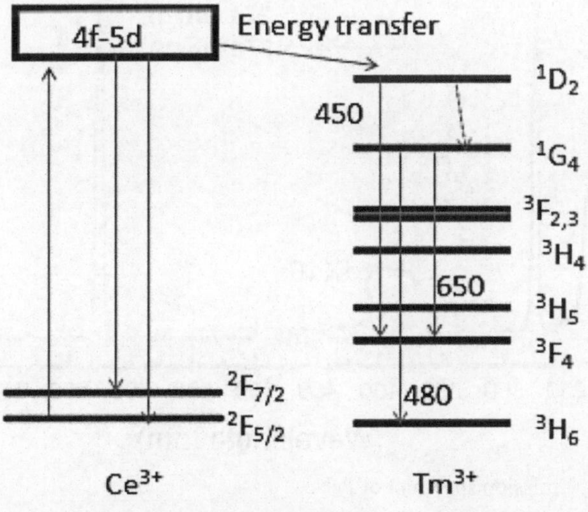

FIGURE 10.12 Schematic diagram of energy transfer Ce^{3+} to Tm^{3+}.

The energy transfer mechanism between Ce^{3+} and Eu^{3+} is investigated in many literatures and some of them are listed in Table 10.5.

TABLE 10.5 Phosphors Giving Tm^{3+} Emission via Ce^{3+} to Tm^{3+} Energy Transfer.

Sl. no.	Name of host	Crystal structure	Excitation wavelength (nm)	References
1	$NaYF_4$: Ce^{3+}, Tm^{3+}	Cubic	280	[90]
2	$CsCdBr_3$: Ce^{3+}, Tm^{3+}	Hexagonal	357	[91]

10.1.5.6 ENERGY TRANSFER FROM Ce^{3+} to Pr^{3+}

Pr^{3+} is well known for its red emission often used for compensating the scarcity of red components in YAG: Ce^{3+} white phosphors.[92] The emission spectrum of Pr^{3+}-doped tellurite glasses is given in Figure 10.13. The red spectral lines assigned to the $^1D_2$–$^3H_4$-level transitions are observed at 597 and 610 nm and those attributed to the $^3P_0$ – $^3H_6$ and $^3P_0$–$^3F_2$ transitions are observed at 620, 627, and 656 nm, respectively.[92] Pr^{3+} is also popular for blue–green emission peaks which are observed at 487, 498, 530, 543, and 556 nm assigned to the $^3P_0$–^3H_J transitions.[93]

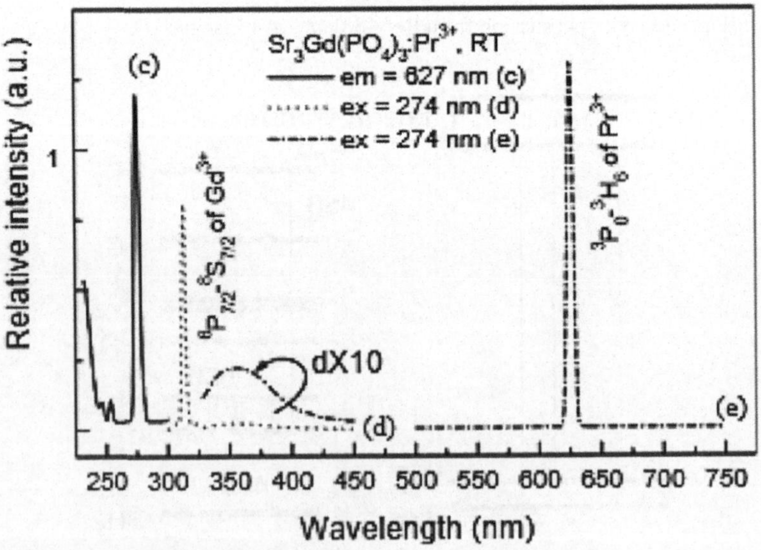

FIGURE 10.13 Emission spectrum of Pr^{3+}.

Source: Reprinted with permission from Ref. [94] @ 2004 Elsevier.

The energy transfer from Ce^{3+} to Pr^{3+} is explained as follows with the help of aschematic energy level diagram in Figure 10.14. The appearance of the red emission line of Pr^{3+} is the result of ET from the lowest 5d of Ce^{3+} to the $^1D_2$ and ^3P_J levels of Pr^{3+}. From these levels, emission occurs radiatively to the lower crystal field levels.

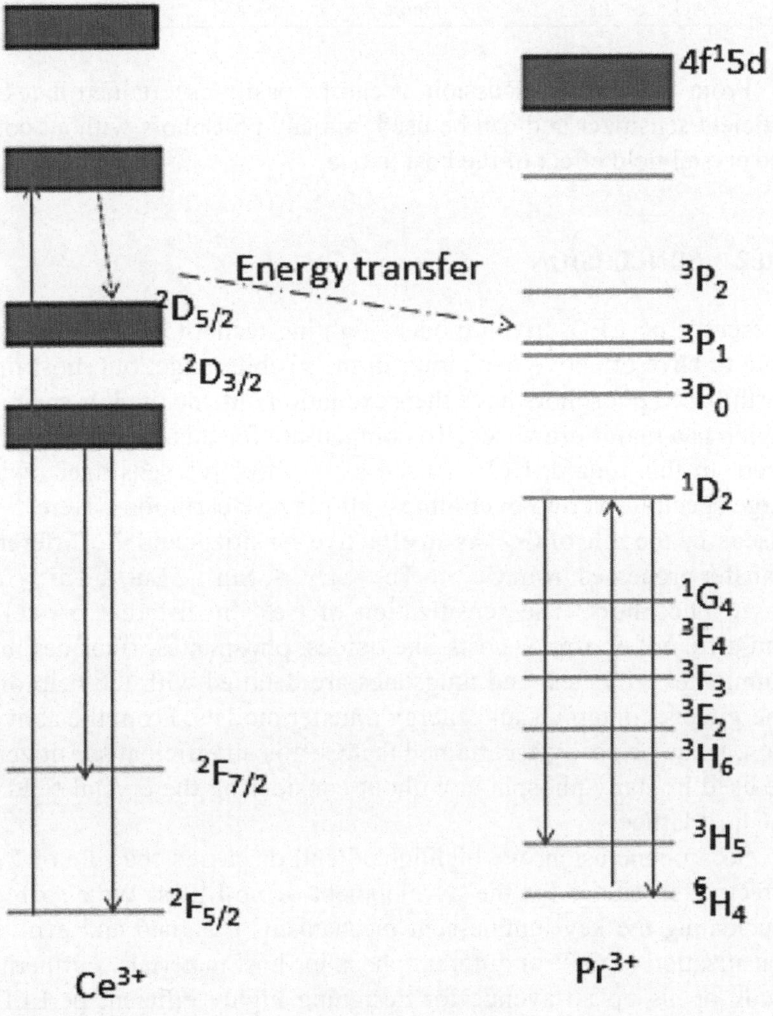

FIGURE 10.14 Schematics of the energy transfer mechanism between Ce^{3+} and Pr^{3+}.

The energy transfer mechanism between Ce^{3+} and Pr^{3+} is investigated in many literatures and some of them are listed in Table 10.6.

TABLE 10.6 Phosphors Giving Pr^{3+} Emission via Ce^{3+} to Pr^{3+} Energy Transfer.

Sl. No.	Name of host	Crystal structure	Excitation wavelength (nm)	References
1	YAG: Ce^{3+},Pr^{3+}	–	340	[92]
2	Li_2SiSrO_4: Ce^{3+},Pr^{3+}	Hexagonal	360	[93]
3	YAl_5O_{12}: Ce^{3+},Pr^{3+}	Monoclinic	284	[95]
4	YAG: Ce^{3+},Pr^{3+}	Cubic	434	[96]

From the above discussion, it can be easily ascertained that Ce^{3+} is an efficient sensitizer and can be used in many phosphors without considering the crystal field effect of the host lattice.

10.2 CONCLUSION

Presently, pc-LED-driven modern lighting technology is extremely desirable to have effective excitation in the visible range, but most of the rare earth-based phosphors have their excitation outside the UV spectral range, which is a major drawback. To compensate for this, sensitizers are widely used. In this regard, Ce^{3+} is used as an effective sensitizer owing to its large spectral overlap over almost all the rare earth ions. Here, we briefly discussed the role of Ce^{3+} as an effective sensitizer and the different energy transfer processes from Ce^{3+} to Tb^{3+}-, Dy^{3+}-, Sm^{3+}-, Eu^{3+}-, Tm^{3+}-, and Pr^{3+}-based phosphors. The sensitization of Ce^{3+} in different crystal systems and different phosphor hosts like oxides, phosphates, fluorides, bromides, aluminates, silicates, and tungstates are detailed with the help of suitable energy level diagrams and energy transfer models. From the above discussion, it can be easily ascertained that Ce^{3+} is an efficient sensitizer and can be used in many phosphors without considering the crystal field effect of the host lattice.

So, in conclusion, we highlighted and discussed the role of Ce^{3+} as an effective sensitizer for the development of pc-LEDs. We also focused on elucidating the key luminescent mechanisms designed and evolved in the sensitization of Ce^{3+} in different phosphor-host materials. Furthermore, this study opens up an avenue for designing highly efficient pc-LED for the modern lighting industry in the future.

KEYWORDS

- **energy transfer**
- **cerium**
- **phosphor**
- **solid-state lighting**
- **rare earths**

REFERENCES

1. Franz, J; Jain, V. K. Optical Communications: Components and Systems: Analysis Design Optimization Application, 2000. https://books.google.com.my/books?id=iUld6xblvo0C
2. Matvoz, D.; Maksic, M. Comparison of LED and CFL Lamps and Their Impact on Electric Power Network. In: *Proceedings of International Conference on Harmonics and Quality of Power*; ICHQP, 2012; pp 320–27. https://doi.org/10.1109/ICHQP.2012.6381219
3. Schmid, S.; Corbellini, G.; Mangold, S.; Gross, T. R. An LED-to-LED Visible Light Communication System with Software-Based Synchronization. In: *2012 IEEE Globecom Workshops* **2012,** 2012, 1264–1268. DOI: 10.1109/GLOCOMW.2012.6477763
4. Braun, D. Semiconducting Polymer LEDs. *Mater. Today* **2002,** *5* (6), 32–39. https://doi.org/10.1016/S1369-7021(02)00637-5
5. Yang, Z.; Gao, M.; Wu, W.; Yang, X.; Sun, X. W.; Zhang, J.; Wang, H. C.; et al., Recent Advances in Quantum Dot-Based Light-Emitting Devices: Challenges and Possible Solutions. *Mater. Today* **2019,** *24*, 69–93. https://doi.org/10.1016/j.mattod.2018.09.002.
6. Khanna, V. K. *Fundamentals of Solid-State Lighting: LEDs, OLEDs and Their Applications in Illumination and Displays*, CRC Press, Taylor &Francis, Boca Raton, 2018; 604, 249
7. Ozgur, Ü.; Liu, H.; Li, X.; Ni, X.; Morkoç, H. GaN-Based Light-Emitting Diodes: Efficiency at High Injection Levels. *Proc. IEEE* **2010,** *98* (7), 1180–1196. DOI: 10.1109/JPROC.2010.2043210
8. Ma, L.; Yu, P.; Wang, W.; Kuo, H. C.; Govorov, A. O.; Sun, S.; Wang, Z. Nanoantenna-Enhanced Light-Emitting Diodes: Fundamental and Recent Progress. *Laser Photon. Rev.* **2021,** *15* (5), 2000367. DOI: 10.1002/lpor.202000367
9. Menéndez-Velázquez, A.; Morales, D.; BelénGarcía-Delgado, A. Sunlike White Light-Emitting Diodes Based on Rare-Earth-Free Luminescent Materials. *Materials* **2022,** *15* (5), 1680. DOI: 10.3390/ma15051680.
10. Guner, T.; Demir, M. M. A Review on Halide Perovskites as Color Conversion Layers in White Light Emitting Diode Applications. *Phys. Status Solidi(A) App. Mater. Sci.* **2018,** *215* (13). DOI: 10.1002/pssa.201800120
11. Phillips, J. M., Coltrin, M. E.; Crawford, M. H.; Fischer, A. J.; Krames, M. R.; Mueller-Mach, R.; Mueller, G. O.; et al. Research Challenges to Ultra-Efficient Inorganic Solid-State Lighting. *Laser Photon. Rev.* **2007,** *1* (4), 307–333. DOI: 10.1002/lpor.200710019

12. Yuhua, W.; Ge, Z. H. U.; Shuangyu, X. I. N.; Qian, W. A. N. G.; Yanyan, L. I.; Quansheng, W. U. et al. Recent Development in Rare Earth Doped Phosphors for White Light Emitting Diodes. *J. Rare Earths* **2015,** *33* (1), 1–12.
13. Shang, M.; Li, C.; Lin, J. How to Produce White Light in aSingle-Phase Host? *Chem. Soc. Rev.* **2014,** *43* (5), 1372–1386. DOI: 10.1039/c3cs60314h
14. Murthy, K. V. R.; Virk, H. S. Luminescence Phenomena: An Introduction. *Defect Diffusion Forum* **2014,** *347*, 1–34. DOI: 10.4028/www.scientific.net/DDF.347.1
15. Ye, S.; Xiao, F.; Pan, Y. X.; Ma, Y. Y.; Zhang, Q. Y. Phosphors in Phosphor-Converted White Light-Emitting Diodes: Recent Advances in Materials, Techniques and Properties. *Mater. Sci. Eng. R: Rep.* **2010,** *71* (1), 1–34. DOI: 10.1016/j.mser.2010.07.001
16. Cesaria, M.; Di Bartolo, B. Luminescence Spectroscopy of Nanophosphors. In: *NATO Science for Peace and Security Series B: Physics and Biophysics*; Di Bartolo, B., Collins, J., Silvestri, L., Eds.; Springer Netherlands: Dordrecht, 2017; pp 15–42. DOI: 10.1007/978-94-024-0850-8_2
17. Chepyga, L. Development of New Phosphors for High Temperature Thermometry. Friedrich-Alexander-Universität Erlangen-Nürnberg (FAU), 2018.
18. Yen, W. M.; Shionoya, S.; Yamamoto, H. *Handbook of* Phosphors, 2nd edn. *Optic. Mater*. **2003,** *23*. http://link.aip.org/link/OPEGAR/v36/i6/p1584/s1&Agg=doi
19. Palaspagar, R. S. Chapter Alumino-Borate Phosphors for Luminescence Application. *Recent Res. Intersection Sci. Technol.* **2022,** *39*.
20. Tucureanu, V.; Matei, A.; Avram, A. M. Synthesis and Characterization of YAG: Ce Phosphors for White LEDs. *Opto-Electron. Rev.* **2015,** *23* (4), 239–251.
21. Parauha, Y. R.; Dhoble, S. J. Enhancement of Photoluminescence and Tunable Properties for Ce^{3+}, Eu^{2+} Activated Na_2CaSiO_4Downconversion Phosphor: A Novel Approach Towards Spectral Conversion. *J. Luminescence* **2022,** *251*, 119173. DOI: 10.1016/j.jlumin.2022.119173
22. Gao, P.; Dong, P.; Zhou, Z.; Li, Q.; Li, H.; Zhou, Z.; Xia, M.; Zhang, P. Enhanced Luminescence and Energy Transfer Performance of Double Perovskite Structure Gd_2MgTiO_6:Bi^{3+}, Mn^{4+} Phosphor for Indoor Plant Growth LED Lighting. *Ceram. Int.* **2021,** *47* (12), 16588–16596. DOI: 10.1016/j.ceramint.2021.02.230
23. Ronda, C. Energy Transfer Processes and Their Applications in Luminescent Materials. In *Advances in Energy Transfer Processes*; 2001; pp 377–409. DOI: 10.1142/9789812810960_0015
24. Raukas, M.; Kelso, J.; Zheng, Y.; Bergenek, K.; Eisert, D.; Linkov, A.; Jermann, F. Ceramic Phosphors for Light Conversion in LEDs. *ECS J. Solid State Sci. Technol.* **2012,** *2* (2), R3168.
25. Clapp, A. R.; Medintz, I. L.; Mauro, J. M.; Fisher, B. R.; Bawendi, M. G.; Mattoussi, H. Fluorescence Resonance Energy Transfer Between Quantum Dot Donors and Dye-Labeled Protein Acceptors. *J. Am. Chem. Soc.* **2004,** *126* (1), 301–310. https://doi.org/10.1021/ja037088b
26. Hussain, S. A. An Introduction to Fluorescence Resonance Energy Transfer (FRET), 2009. http://arxiv.org/abs/0908.1815
27. Sisira, S.; Alexander, D.; Thomas, K.; Biju, P. R.; Unnikrishnan, N. V.; Joseph, C. Effect of Annealing Temperature on the Luminescence of $Ce_{1-x}PO_4$:Tb_x^{3+} Nanocrystals: A Novel Theoretical Model and Experimental Verification. *J. Mater. Sci.* **2018,** *53* (2), 1380–1394. DOI: 10.1007/s10853-017-1600-x.

28. Vimal, G.; Mani, K. P.; Biju, P. R.; Joseph, C.; Unnikrishnan, N. V.; Ittyachen, M. A. Synthesis, Structural and Spectroscopic Investigations of Nanostructured Samarium Oxalate Crystals. *Spectrochim. Acta – Part A: Mol. Biomol. Spectroscopy* **2014,** *122*, 624–630. https://doi.org/10.1016/j.saa.2013.11.080
29. Kamal, P. M.; Vimal, G.; Biju, P. R.; Joseph, C.; Unnikrishnan, N. V.; Ittyachen, M. A. Spectroscopic Investigation on Tunable Luminescence by EnergyTransfer in $Tb_{2-X}Sm_x(MoO_4)_3$ Nanophosphors. *Optic. Mater.* **2015,** *42*, 237–244. https://doi.org/10.1016/j.optmat.2014.12.038
30. Mani, K. P.; Vimal, G.; Biju, P. R.; Joseph, C.; Unnikrishnan, N. V.; Ittyachen, M. A. Structural and Spectral Investigation of Terbium Molybdate Nanophosphor. *Spectrochim. Acta – Part A: Mol. Biomol. Spectroscopy* **2015,** *148*, 412–419. https://doi.org/10.1016/j.saa.2015.04.016
31. Vimal, G.; Mani, K. P.; Biju, P. R.; Joseph, C.; Unnikrishnan, N. V.; Ittyachen, M. A. Structural Studies and Luminescence Properties of $CeO_2:Eu^{3+}$Nanophosphors Synthesized by Oxalate Precursor Method. *Appl. Nanosci.* **2015,** *5* (7), 837–846. https://doi.org/10.1007/s13204-014-0375-5
32. Jacob, L. A.; Sisira, S.; Thomas, K.; Alexander, D.; Biju, P. R.; Unnikrishnan, N. V.; Joseph, C. A Reddish-Orange Emitting Samarium Doped α-$Na_3Y(VO_4)_2$ Nanocrystals for Single Phased UV Excitable White Light Applications. *J. Solid State Chem.* **2019,** *280*. DOI: 10.1016/j.jssc.2019.120998
33. Jacob, L. A., Sisira, S.; Mani, K. P.; Thomas, K.; Alexander, D.; Biju, P. R.; Unnikrishnan, N. V.; Joseph, C. High Purity Blue Photoluminescence in Thulium Activated α-$Na_3Y(VO_4)_2$ Nanocrystals via Host Sensitization. *J. Luminescence* **2020,** *223* (July). DOI: 10.1016/j.jlumin.2020.117169.
34. Rajendran, V., Chang, H.; R. S. Liu. Recent Progress on Broadband Near-Infrared Phosphors-Converted Light Emitting Diodes for Future Miniature Spectrometers. *Optic. Mater: X* **2019,** *1*, 100011. https://doi.org/10.1016/j.omx.2019.100011
35. Sreeja, E.; Gopi, S.; Vidyadharan, V.; Mohan, P. R.; Joseph, C.; Unnikrishnan, N. V.; Biju, P. R. Luminescence Properties and Charge Transfer Mechanism of Host Sensitized $Ba_2CaWO_6:Eu^{3+}$ Phosphor. *Powder Technol.* **2018,** *323*, 445–453. https://doi.org/10.1016/j.powtec.2017.09.036
36. Smets, B. M. J. Phosphors Based on Rare-Earths, a New Era in Fluorescent Lighting. *Mater. Chem. Phys.* **1987,** *16* (3–4), 283–299. DOI: 10.1016/0254-0584(87)90103-9
37. Zhao, B., Wang, J.; Li, H.; Jia, X.; Dong, L.; den Engelsen, D. Rare Earth – Activated Y_2O_3 Phosphors with Novel Morphology for Dye-Sensitized Solar Cells. *ChemistrySelect* **2016,** *1* (6), 1136–1139. DOI: 10.1002/slct.201600137
38. Mikhailik, V. B., Kraus, H.; Dorenbos, P. Efficient VUV Sensitization of Eu^{3+}Emission by Tb^{3+} in Potassium Rare-Earth Double Phosphate. *Phys. Status Solidi—Rapid Res. Lett.* **2009,** *3* (1), 13–15. https://doi.org/10.1002/pssr.200802211
39. Xie, R. J., Mitomo, M.; Uheda, K.; Xu, F. F.; Akimune, Y. Preparation and Luminescence Spectra of Calcium- and Rare-Earth (R = Eu, Tb, andPr)-Codoped α-SiAlON Ceramics. *J. Am. Ceram. Soc.* **2002,** *85* (5), 1229–1234. https://doi.org/10.1111/j.1151-2916.2002.tb00250.x
40. Ege, A.; Ayvacikli, M.; Dinçer, O.; Satilmiş, S. U. Spectral Emission of Rare Earth (Tb, Eu, Dy) Doped $Y_2Sn_2O_7$ Phosphors. *J. Luminescence* **2013,** *143*, 653–656. DOI: 10.1016/j.jlumin.2013.05.027

41. Deshazer, L. G.; Dieke, G. H. Spectra and Energy Levels of Eu^{3+} in $LaCl_3$. *J. Chem. Phys.* **1963,** *38* (9), 2190–2199. DOI: 10.1063/1.1733949
42. Xu, G. X.; Lian, J. B.; Wu, N. C.; Zhang, X.; He, J. Co-Precipitation Synthesis of $La_2O_2SO_4$:Tb^{3+}Phosphor and Its Conversion to La_2O_2S:Tb^{3+}Ceramic Scintillator via Pressure Less Sintering in Hydrogen. *J. Ceram. Sci. Technol.* **2018,** *9* (3), 345–352. https://doi.org/10.4416/JCST2018-00018
43. Wang, Y., Guo, X.; Endo, T.; Murakami, Y.; Ushirozawa, M. Identification of Charge Transfer (CT) Transition in (Gd, Y)BO 3:Eu Phosphor Under100–300 Nm. *J. Solid State Chem.* **2004,** *177* (7), 2242–2248. https://doi.org/10.1016/j.jssc.2004.03.003
44. Ding, S.; Zhang, W.; Xu, B.; Wang, J. [Spectra of Ce^{3+}, Tb^{3+} and Gd^{3+} Ions in Ln(BO_3,PO_4) [Ln = La, Y]]. *GuangPuXue Yu GuangPu Fen Xi.* **2001,** *21*(3),:275–278. Chinese. PMID: 12947644.
45. Xia, Z.; Liu, R. S. Tunable Blue-Green Color Emission and Energy Transfer of $Ca_2Al_3O_6F$:Ce^{3+},Tb^{3+} Phosphors for Near-UV White LEDs. *J. Phys. Chem. C* **2012,** *116* (29), 15604–15609. https://doi.org/10.1021/jp304722z
46. Bhonsule, S. U.; Wankhede, S. P.; Moharil, S. V.; Synthesis and Characterization of $KCe(PO_3)_4$ Doped with Some Lanthanide Activators. *Luminescence* **2018,** *33* (2), 356–363. https://doi.org/10.1002/bio.3421
47. Lima, J. F.; de Sousa Filho, P. C.; Serra, O. A. Single Crystalline Rhabdophane-Type $CePO_4$ Nanoparticles as Efficient UV Filters. *Ceram. Int.* **2016,** *42* (6), 7422–7431.
48. Bobade, D. S.; Parauha, Y. R.; Dhoble, S. J.; Undre, P. B. Structural and Luminescence Study of Ce^{3+} and Eu^{3+} Doped $ZnAl_{12}O_{19}$ Nano-Structured Novel Phosphors. *Optik* **2021,** *227*, 166119. https://doi.org/10.1016/j.ijleo.2020.166119
49. Sumaletha, N.; Rajesh, K.; Mukundan, P.; Warrier, K. G. K. Environmentally Benign Sol-Gel Derived Nanocrystalline Rod Shaped Calcium Doped Cerium Phosphate Yellow-Green Pigment. *J. Sol-Gel Sci. Technol.* **2009,** *52* (2), 242–250. https://doi.org/10.1007/s10971-009-2020-4
50. Lima, J. F.; De Sousa Filho, P. C.; Serra, O. A. Single Crystalline Rhabdophane-Type $CePO_4$ Nanoparticles as Efficient UV Filters. *Ceram. Int.* **2016,** *42* (6), 7422–7431. https://doi.org/10.1016/j.ceramint.2016.01.146
51. Yang, D.; Kang, X.; Shang, M.; Li, G.; peng, C.; Li, C.; Lin, J. Size and Shape Controllable Synthesis and Luminescent Properties of$BaGdF_5$:Ce^{3+}/Ln^{3+} (Ln = Sm, Dy, Eu, Tb) Nano/Submicrocrystals by a Facile Hydrothermal Process. *Nanoscale* **2011,** *3* (6), 2589–2595. https://doi.org/10.1039/c1nr10203f
52. Rawat, K.; Vishwakarma, A. K.; Jha, K. Thermally Stable $Ca_2Ga_2SiO_7$:Tb^{3+} Green Emitting Phosphor for Tricolor w-LEDs Application. *Mater. Res. Bull.* **2020,** *124*, 110750. https://doi.org/10.1016/j.materresbull.2019.110750
53. Berdowski, P. A. M.; Lammers, M. J. J.; Blasse, G. $^5D_3$–$^5D_4$ Cross Relaxation inTb^{3+} Pairs in $CsCdBr_3$ Crystals. *J. Chem. Phys.* **1985,** *83* (2), 476–479. https://doi.org/10.1063/1.449562
54. Hoaksey, A.; Woods, J.; Taylor, K. N. R. Luminescence of Tb^{3+} Ions in Silicate Glasses. *J. Luminescence* **1978,** *17* (4), 385–400. https://doi.org/10.1016/0022-2313(78)90048-0
55. Yi, Z.; Lu, W.; Zeng, S.; Wang, H.; Rao, L.; Li, Z. Tunable Multicolor and White Luminescence in Tb^{3+}/Dy^{3+}/Mn^{2+} Doped $CePO_4$ via Energy Transfer. *J. Alloys Compd* **2015,** *637*, 489–496.
56. Sharma, R. K.; Chouryal, Y. N.; Nema, S.; Nigam, S.; Bera, S. P.; Bhargava, Y.; Ghosh, P. Green EmittingCe^{3+}/ Tb^{3+}-Doped BaF_2 Nanocrystals and Their Impact on Skeletal

Muscle of Developing Zebrafish Larvae. *Chemistry Select* **2020,** *5* (29), 9105–9110. https://doi.org/10.1002/slct.202001268

57. Sun, S., Wu, L.; Yi, H.; Wu, L.; Ji, J.; Zhang, C.; Zhang, Y.; Kong, Y.; Xu, J. Energy Transfer between Ce^{3+} and Tb^{3+} and the Enhanced Luminescence of a Green Phosphor SrB_2O_4:Ce^{3+}, Tb^{3+}, Na^+. *Optic. Mater. Express* **2016,** *6* (4), 1172–1185. https://doi.org/10.1364/ome.6.001172
58. Sahu, N. K., Singh, N. S.; Pradhan, L.; Bahadur, D. Ce^{3+} Sensitized $GdPO_4$:Tb^{3+} with Iron Oxide Nanoparticles: A Potential Biphasic System for Cancer Theranostics. *Dalton Trans.* **2014,** *43* (30), 11728–11738. https://doi.org/10.1039/c4dt00792a
59. Zhang, W., Hua, R.; Qi, X.; Zhao, J.; Qin, L.; Liu, T. Photoluminescence Properties and Energy Transfer of Ce^{3+}-Tb^{3+} Co-Doped $SrAlF_5$Nanorods by a Hydrothermal Method. *Cryst. Eng. Comm.* **2017,** *19* (35), 5214–5222. https://doi.org/10.1039/c7ce01054k
60. Sun, J.; Zhang, X.; Xia, Z.; Du, H. Luminescent Propertiesand Energy Transfer of Ce^{3+},Tb^{3+} Co-Doped $NaBaPO_4$ Phosphor. *J. Electrochem. Soc.* **2011,** *158* (11), J368. https://doi.org/10.1149/2.058111jes
61. Li, J.; Li, J. G.; Liu, S.; Li, X.; Sun, X.; Sakka, Y. Greatly Enhanced Dy^{3+} Emission via Efficient Energy Transfer in Gadolinium Aluminate Garnet ($Gd_3Al_5O_{12}$) Stabilized with Lu^{3+}. *J. Mater. Chem. C* **2013,** *1* (45), 7614–7622. https://doi.org/10.1039/c3tc31413h
62. Tariwong, Y.; Chanthima, N.; Kaewkhao, J.; Sangwaranatee, N. Optical Properties and Judd-Ofelt Analysis of Dy^{3+} Doped MgOBaO-P_2O_5 Glass Systems. In: *Proceedings—2015 4th International Conference on Instrumentation, Communications, Information Technology and Biomedical Engineering, ICICI-BME* 2015, 2016 pp 264–267. https://doi.org/10.1109/ICICI-BME.2015.7401375
63. Vinodkumar, P.; Panda, S.; Jaiganesh, G.; Padhi, R. K.; Madhusoodanan, U.; Panigrahi, B. S. $SrBPO_5$: Ce^{3+}, Dy^{3+}—A Cold White-Light Emitting Phosphor. *Spectrochim. Acta Part A: Mol. Biomol. Spectroscopy* **2021,** *253*, 119560. DOI: 10.1016/j.saa.2021.119560
64. Kumar, A.; Dhoble, S. J.; Peshwe, D. R.; Bhatt, J.; Terblans, J. J.; Swart, H. C. Crystal Structure, Energy Transfer Mechanism and Tunable Luminescence in Ce^{3+}/Dy^{3+} Co-Activated Ca20Mg3Al26Si3O68 Nanophosphors. *Ceram. Int.* **2016,** *42* (9), 10854–10865. DOI: 10.1016/j.ceramint.2016.03.216.
65. Nair, G. B.; Dhoble, S. J. White Light Emission Through Efficient Energy Transfer from Ce^{3+} to Dy^{3+} Ions in Ca3$Mg_3(PO_4)_4$ Matrix Aided by Li^+Charge Compensator. *J. Luminescence* **2017,** *192*, 1157–1166. DOI: 10.1016/j.jlumin.2017.08.047.
66. Li, J. Y.; Hou, D.; Zhang, Y.; Li, H.; Lin, H.; Lin, Z.; Zhou, W.; Huang, R. Luminescence, Energy Transfer and Temperature Sensing Property of Ce^{3+}, Dy^{3+} Doped $LiY_9(SiO_4)_6O_2$ Phosphors. *J. Luminescence* **2019,** *213*, 184–190. DOI: 10.1016/j.jlumin.2019.05.027.
67. Gong, Y.; Wang, Y.; Li, Y.; Xu, X. Ce^{3+},Dy^{3+} Co-Doped White-Light Long-Lasting Phosphor: Sr_2AlSub$_2$SiOSub$_7$Through Energy Transfer. *J. Electrochem. Soc.* **2010,** *157* (6), J208. DOI: 10.1149/1.3371488.
68. Stojadinović, S.; Ćirić, A. Dy^{3+} and Dy^{3+}/Ce^{3+} Doped Al_2O_3Coatings Obtained by Plasma Electrolytic Oxidation: Photoluminescence and Energy Transfer from Ce^{3+} to Dy^{3+}. *J. Luminescence* **2020,** 226, 117403.DOI: 10.1016/j.jlumin.2020.117403.
69. Krishna, K. G.; Krishna Bharat, L.; Yu, J. S. Energy Transfer and Luminescence Properties of Ce^{3+}/Dy^{3+} Co-Doped Sr_3SiO_5 Phosphors for WLEDs. *Ceram. Int.* **2017,** *43* (2), 2586–2591. DOI: 10.1016/j.ceramint.2016.11.065.
70. Xia, Z.; Chen, D. Synthesis and Luminescence Properties of $BaMoO_4$:Sm^{3+} Phosphors. *J. Am. Ceram. Soc.* **2010,** *93* (5), 1397–1401.DOI: 10.1111/j.1551-2916.2009.03574.x.

71. SambasivaRao, T.; Krishna Reddy, D. V.; Taherunnisa, S. K.; Rudramamba, K. S.; Sesha Reddy, A. S.; Veeraiah, N.; Reddy, M. R. Energy Transfer (Ce^{3+}→ Sm^{3+}) Influence on PL Emission ofCe^{3+}/Sm^{3+} Co-Doped Barium Gallium Borosilicate Glasses. *Phys. B: Condensed Matter* **2019**. DOI: 10.1016/j.physb.2019.01.053
72. Yang, M.; Lu, L.; Zhao, Q.; Du, S.; Zhang, X.; Li, G.; Yifan, D. Effect of Concentration of Ce^{3+} and Sm^{3+} on Luminescence Performance of YLuAG Phosphors. *Mater. Res. Express* **2019,** 6 (11). DOI: 10.1088/2053-1591/ab492c.
73. Sisira, S.; Jacob, L. A.; Thomas, K.; Alexander, D.; Biju, P. R..; Unnikrishnan, N. V.; Joseph, C. Effect of Doping Concentration and Annealing Temperature on the Luminescence of Novel Orange Red Emitting $CePO_4$:Sm^{3+}Nanocrystals. *J. Luminescence* **2019,** *209*, 197–207. DOI: 10.1016/j.jlumin.2019.01.043.
74. Shi, R.; Xu, J.; Liu, G.; Zhang, X.; Zhou, W.; Pan, F.; Huang, Y.; Tao, Y.; Liang, H. Spectroscopy and Luminescence Dynamics of $Ce^{3+}$and Sm^{3+} in $LiYSiO_4$. *J. Phys. Chem. C* **2016,** *120* (8), 4529–4537. DOI: 10.1021/acs.jpcc.5b12501.
75. Du, F.; Kuang, J.; Chen, L.; Xiao, Y.; Liu, L.; Peng, J. Study on Energy Transfer in Ce^{3+}/Sm^{3+} Activated $LaSi_3N_5$ Phosphor for Warm White-Light LED. *Phys. B: Condensed Matter 202, 624*, 413393. DOI: 10.1016/j.physb.2021.413393.
76. Tang, H.; Zhang, X.; Cheng, L.; Xie, J.; Yu, X.; Mi, X.; Liu, Q. Broadband Emission of $Lu_3Mg_2GaSi_2O_{12}$: Ce^{3+}, Sm^{3+}Phosphors and Their Potential Application for w-LEDs. *Ceram. Int.* **2021,** *47* (18), 26410–26420. DOI: 10.1016/j.ceramint.2021.06.052.
77. Krumpel, A. H.; Bos, A. J. J.; Bessière, A.; Van Der Kolk, E.; Dorenbos, P. Controlled Electron and Hole Trapping in YPO_4: Ce^{3+}, Ln^{3+} and$LuPO_4$: Ce^{3+}, Ln^{3+} (Ln=Sm, Dy, Ho, Er, Tm). *Phys. Rev. B—Condensed Matter Mater. Phys.* **2009,** *80* (8). DOI: 10.1103/PhysRevB.80.085103.
78. Ramakrishna, G.; Nagabhushana, H.; Basavaraj, R. B.; Prashantha, S. C.; Sharma, S. C.; Naik, R.; Anantharaju, K. S. Green Synthesis, Structural Characterization and Photoluminescence Properties of Sm^{3+} Co-Doped Y_2SiO_5:Ce^{3+}Nanophosphors for WLEDs. *Optik* **2016,** *127* (13), 5310–5315. DOI: 10.1016/j.ijleo.2016.03.034.
79. Li, M.; Wu, J.; Jia, H.; Wang, M.; Liu, Z. Luminescence Properties and Energy Transfers of $NaLa(WO_4)2$:Sm^{3+}:Ce^{3+} Phosphor. *J. Mater. Sci.* **2019,** *30* (11), 10465–10474. DOI: 10.1007/s10854-019-01389-8.
80. de Almeida, R.; da Silva, D. M.; Kassab, L. R. P.; de Araújo, C. B. Eu^{3+} Luminescence in Tellurite Glasses with Gold Nanostructures. *Optic. Commun.* **2008,** *281* (1), 108–112. DOI: 10.1016/j.optcom.2007.08.072.
81. Kitagawa, Y.; Ueda, J.; Fujii, K.; Yashima, M.; Funahashi, S.; Nakanishi, T.; Takeda, T. et al. Site-Selective Eu^{3+}Luminescence in the Monoclinic Phase of YSiO2N.*Chem. Mater.* **2021,** *33* (22), 8873–8885. DOI: 10.1021/acs.chemmater.1c03139.
82. Van Do, P.; Ngoc, T.; Ca, N. X.; Thanh, L. D.; ThanhNga, P. T.; Chung Thuy, T. T.; Van Nghia, N. Study of Spectroscopy of Eu^{3+} and Energy Transfer from Ce^{3+} to Eu^{3+} in Sodium-Zinc-Lead-Borate Glass. *J. Luminescence* **2020,** 117660. DOI: 10.1016/j.jlumin.2020.11766
83. Meejitpaisan, P.; Doddoji, R.; Kothan, S.; Jayasankar, C. K.; Kaewkhao, J. Intense Red Emission via Energy Transfer from (Ce^{3+}/Eu^{3+}):P_2O_5+NaF+CaF_2+AlF_3Glasses for Warm Light Sources. *Ceram. Int.* **2021,** *47* (2), 1962–1969. DOI: 10.1016/j.ceramint.2020.09.026.

84. Nazarov, M.; Noh, D. Y.; Sohn, J.; Yoon, C. Influence of Additional Eu^{3+}Coactivator on the Luminescence Properties of $Tb_3Al_5O_{12}$:Ce^{3+}, Eu^{3+}.*Optic. Mater.* **2008,** *30* (9), 1387–1392. DOI: 10.1016/j.optmat.2007.07.005.
85. Deng, S.; Qiu, Z.; Zhang, M.; Zhou, W.; Zhang, J.; Li, C.; Rong, C.; Yu, L.; Lian, S. Tricolor Emitting and Energy Transfer in the Phosphor $Ba_2ZnSi_2O_7$:Ce^{3+},Eu^{3+},Eu^{2+} for White-LED Based Near-UVChips. *J. Rare Earths* **2015,** *33* (5), 463–468. DOI: 10.1016/S1002-0721(14)60441-5.
86. Shinde, K. N.; Nagpure, I. M.; Dhoble, S. J. Energy Transfer Between Ce^{3+} and Eu^{2+} in Doped $Sr_5(PO_4)_3F$ Phosphor. *Synth. React. Inorg. Metal-Org. Nano-Metal Chem.* **2011,** *41* (1), 107–113. DOI: 10.1080/15533174.2010.538021.
87. Cai, L.; Ying, L.; Zheng, Z; Fan, B.; Chen, R.; Chen, C. Luminescent Properties of $Sr_2B_2O_5$: Tm^{3+}, Na^+ Blue Phosphor. *Ceram. Int.* **2014,** *40* (5), 6913–6918. DOI: 10.1016/j.ceramint.2013.12.012.
88. Li, Y.; Zheng, J.; Li, Z.; Yang, X.; Chen, J.; Chen, C. Luminescence Properties of $NaSrPO_4$: Tm^{3+} as Novel Blue Emitting Phosphors with High Color Purity. *Optik* **2018,** *169*, 257–263. DOI: 10.1016/j.ijleo.2018.05.065.
89. Adusumalli, V.; Heramba, N, K. B.; Koppisetti, V. S. R. M.; Mahalingam, V. Ce^{3+}-Sensitized Tm^{3+}/Mn^{2+}-Doped $NaYF^4$ Colloidal Nanocrystals:Intense Cool White Light from a Phosphor-Coated UV LED. *Chem.— Eur. J.* **2017,** *23* (72), 18134–18139. DOI: 10.1002/chem.201704156.
90. McPherson, A. M.; McPherson, G. L. Spectroscopic Properties of aCe^{3+}-Tm^{3+} Center in $CsCdBr_3$ Crystals: Absorption, Emission, and Energy Transfer. *Solid State Commun.* **1981,** *37* (6), 501–505. DOI: 10.1016/0038-1098(81)90487-7.
91. Guan, R.; Cao, L.; You, Y.; Cao, Y. The Luminescence Properties and Energy Transfer from Ce^{3+} to Pr^{3+} for YAG:Ce^{3+},Pr^{3+} Phosphors. *J. Nanomater.* **2015,** *2015*, 1–9. DOI: 10.1155/2015/549208.
92. Chen, J.; Guo, C.; Yang, Z.; Li, T.; Zhao, J. Li_2SrSiO_4:Ce^{3+}, Pr^{3+} Phosphor with Blue, Red, and Near-Infrared Emissions Used for Plant Growth LED. *J. Am. Ceram. Soc.* **2016,** *99* (1), 218–225. DOI: 10.1111/jace.13952.
93. Liang, H.; Tao, Y.; Xu, J.; He, H.; Wu, H.; Chen, W.; Su, Q. Photoluminescence of Ce^{3+}, Pr^{3+} and Tb^{3+} Activated $Sr_3Ln(PO_4)_3$ Under VUV–UV Excitation. *J. Solid State Chem.* **2004,** *177* (3), 901–908. DOI: 10.1016/j.jssc.2003.09.037
94. Wang, L.; Zhang, X.; Hao, Z.; Luo, Y.; Zhang, J.; Wang, X. J. Interionic Energy Transfer in $Y_3Al_5O_{12}$: Ce^{3+}, Pr^{3+} Phosphor. *J. Appl. Phys.* **2010,** *108* (9), 93515.
95. Marin, R.; Sponchia, G.; Riello, P.; Sulcis, R.; Enrichi, F. Photoluminescence Properties of YAG:Ce^{3+},Pr^{3+} Phosphors Synthesized via the Pechini Method for White LEDs. *J. Nanoparticle Res.* **2012,** *14* (6), 886. DOI: 10.1007/s11051-012-0886-5

Index

A

AC electrical conductivity, 68–69
Activated carbon units, 226–228
Advanced nanomaterials, 10
  agglomeration/state of aggregation, 15
  application, 19–20
  BET approach, 12
  carbon-based nanomaterials, 14, 17–18
  challenges with, 22–23
  chemical products, 11
  classification, 16–19
  crystallinity and defect structure, 16
  dendrimer nanomaterials, 14
  dendrimers, 18
  distribution of sizes, 15
  feasibility, 20–22
  gamma irradiation procedures, 12
  green chemistry principles (GCPs), 11
  metal-based nanomaterials, 14, 18
  nanocomposite nanomaterials, 14
  nanomaterials, 18–19
  physical characteristics, 13
  properties, 13–16
  solubility, 16
  specific surface area (SSA), 13
  surface topography and morphology, 15–16
Antenna Solar Energy Conversion (ASEC), 107

B

Bacteria testing, 263–264
Bacterial endotoxin, 176–177
Bacterial endotoxin testing, 264–265
Bacteriostatic water for injection, 175
Biomedical imaging, 93–94
Boltzmann constant, 61
British Pharmacopeia (BP), 171

C

Carbon, 106
  advantages and disadvantages, 114–115
  carbon nanotubes (CNTS), 115–116
  mechanism, 117–119
  optical applications
    information encryption, 114
    sensors, 112–113
  optical limiting, 116–117
  optical properties
    absorption behavior, 108
    fluorescence, 110–112
    intersystem crossing (ISC), 110
    phosphorescence, 109–110
    photoluminescence (PL), 108–109
  saturable absorber
    mode-locked lasers, 119–120
Carbon nanotubes (CNTs), 106
Cartridge filtration, 235–236
Certificate of Operation, 258
Chemical injection systems, 220–221
Chemical lasers, 84–85
Chemical testing, 265
Chemical vapor deposition (CVD), 132
Clean energy devices, 126
Color center lasers, 87
Continuous electro-deionization (CEDI), 188–189
Crystalline size, 60

D

DC electrical, 66–68
DC electrical conductivity, 57
DC electrical conductivity studies, 61
Defect level conduction, 58
Design qualification (DQ), 244–245
Dye-sensitized solar cells (DSSC), 32

E

Electrical conductivity studies
  AC electrical conductivity, 68–69
  DC electrical, 66–68
Environmental, social, and governance (ESG), 5
Excimer lasers, 83

F

Factory Acceptance Test (FAT), 204

G

Gamma irradiation procedures, 12
Gas lasers, 78
Graphene, 121–122
  applications, 122
  limitations, 129
  nanomaterials, 130
    chemical vapor deposition (CVD), 132
    mechanical exfoliation, 131
    SiC thermal deposition, 132
  optical devices
    clean energy devices, 126
    light-emitting diodes (LEDs), 123–124
    liquid crystal displays, 123
    photodetectors and phototransistors, 127–128
    polymer nanocomposites, 126–127
    sensors, 125
    solar cells, 124–125
    touch screen, 122–123
    transparent conductive films (TCFs), 125–126
    ultrafast laser, 128–129
  optical properties, 129–130
Graphene oxide (GO)
  DC electrical conductivity, 57
  defect level conduction, 58
  electrical conductivity studies
    AC electrical conductivity, 68–69
    DC electrical, 66–68
  experimental studies, 59–60
  result and discussion
    structural studies, 64
    surface morphological studies, 63–64
  theory
    Boltzmann constant, 61
    crystalline size, 60
    DC electrical conductivity studies, 61
    Joncher's universal dynamic response (UDR), 62
    Jonscher's Power Law (JPL), 62
    Mott characteristic, 61
  x-ray energy, 59
Green chemistry principles (GCPs), 11
Green energy
  advanced nanomaterials
    agglomeration/state of aggregation, 15
    application, 19–20
    BET approach, 12
    carbon-based nanomaterials, 14, 17–18
    challenges with, 22–23
    chemical products, 11
    classification, 16–19
    crystallinity and defect structure, 16
    dendrimer nanomaterials, 14
    dendrimers, 18
    distribution of sizes, 15
    feasibility, 20–22
    gamma irradiation procedures, 12
    green chemistry principles (GCPs), 11
    metal-based nanomaterials, 14, 18
    nanocomposite nanomaterials, 14
    nanomaterials, 18–19
    physical characteristics, 13
    properties, 13–16
    solubility, 16
    specific surface area (SSA), 13
    surface topography and morphology, 15–16
  defined, 10
  large-scale research, 11
  production and usage, 10

H

Heat exchangers, 223–224
Holography, 94–95

I

Injection storage tank
  operational qualification protocol, 257
    Certificate of Operation, 258
    conclusion, 258–259
    critical components, 259
    critical parameters, 259–261
    data sheets, 259–261
    equipment description and list of components, 259
    equipment justification, 258
    protocol format, 258
  performance qualification protocol, 261
    acceptance of protocol format, 262

bacteria testing, 263–264
bacterial endotoxin testing, 264–265
chemical testing, 265
log sheet, 266–267
logbook page, 266–267
sample bottle label, 266–267
sample locations, 267
sample schedule, 267
sampling techniques, 265–266
summary, 262
testing for chemicals, 262
Inline ultraviolet unit, 232–234
Installation qualification (IQ), 245–246, 252
accessory inspection, 255
Certificate of Installation, 253
component examination, 254–255
components and accessories, 254
control system inspection, 255–256
documentation verification, 256–257
equipment rationale, 253
instrumentation inspection, 256
IQ document's approval, 254
Mconclusion, 254
purpose, 253
utilities inspection section, 256
Intersystem crossing (ISC), 110
Ion lasers, 79–80
Ion removal techniques, 185
continuous electro-deionization (CEDI), 188–189
design considerations, 191–192
electrodialysis, 189
and electrodialysis reversal, 189, 190
pharmaceutical applications, 188
physical elements, 187
resins, 186

J

Japanese Pharmacopeia (JP), 171
Joncher's universal dynamic response (UDR), 62
Jonscher's Power Law (JPL), 62

L

Large-scale research, 11
Laser cutting, 100
Laser device, 73
applications
biomedical imaging, 93–94
communications, 90–91
holography, 94–95
laser cutting, 100
laser drilling, 98–99
laser hardening, 100–101
laser micro/nanomachining, 99–100
laser range finder, 95–96
laser-guided antitank missile, 97–98
lidar (laser radar), 95
medical applications, 92–93
ring laser gyroscope, 98
super resolution, 93–94
underwater laser, 96–97
classification
chemical lasers, 84–85
color center lasers, 87
excimer lasers, 83
gas lasers, 78
ion lasers, 79–80
liquid lasers, 89–90
metal vapor lasers, 80
molecular lasers, 81–82
neutral atom lasers, 78–79
rare-earth ion lasers, 86–87
semiconductor lasers, 88–89
solid-state lasers, 85
transition metal ion lasers, 85–86
laser working principles, 75
light amplification, 76–77
Laser drilling, 98–99
Laser hardening, 100–101
Laser micro/nanomachining, 99–100
Laser range finder, 95–96
Laser working principles, 75
Laser-guided antitank missile, 97–98
Lidar (laser radar), 95
Light amplification, 76–77
Light Amplification by Stimulated Emission of Radiation (LASER), 74
Light-emitting diodes (LEDs), 123–124, 274
Liquid crystal displays, 123
Liquid lasers, 89–90

M

Magnetic cellulose nanocrystals (MCNs), 164
Metal vapor lasers, 80

Molecular lasers, 81–82
Multiwalled carbon nanotube (MWCNT), 107

N

Nanocellulose, 162
eradication, 164
literature survey, 165
exceptional cellulosic, 163
magnetic cellulose nanocrystals (MCNs), 164
Nanocelluloses (NCs), 144
Nanomaterials, 10, 130
chemical vapor deposition (CVD), 132
mechanical exfoliation, 131
SiC thermal deposition, 132
Neutral atom lasers, 78–79

O

Operational qualification (OQ), 246
Operational qualification protocol, 257
Certificate of Operation, 258
conclusion, 258–259
critical components, 259
critical parameters, 259–261
data sheets, 259–261
equipment description and list of components, 259
equipment justification, 258
protocol format, 258
Optical applications
information encryption, 114
sensors, 112–113
Optical devices
clean energy devices, 126
light-emitting diodes (LEDs), 123–124
liquid crystal displays, 123
photodetectors and phototransistors, 127–128
polymer nanocomposites, 126–127
sensors, 125
solar cells, 124–125
touch screen, 122–123
transparent conductive films (TCFs), 125–126
ultrafast laser, 128–129
Optical limiting, 116–117
Optical properties
absorption behavior, 108
fluorescence, 110–112
intersystem crossing (ISC), 110
phosphorescence, 109–110
photoluminescence (PL), 108–109
Organic scavenging resins, 231–232

P

Particulate removal filters, 224–226
Performance qualification (PQ), 246–248
Performance qualification protocol, 261
acceptance of protocol format, 262
bacteria testing, 263–264
bacterial endotoxin testing, 264–265
chemical testing, 265
log sheet, 266–267
logbook page, 266–267
sample bottle label, 266–267
sample locations, 267
sample schedule, 267
sampling techniques, 265–266
summary, 262
testing for chemicals, 262
Pharmaceutical waters, 170
British Pharmacopeia (BP), 171
chemical specifications
added substance, 177–178
alert limits and bacteria action, 175–176
bacterial endotoxin, 176–177
bacteriostatic water for injection, 175
drinking water, 172–173
purified water, 173–174
sterile water for inhalation, 175
sterile water for injection, 174
sterile water for irrigation, 175
USP purified water, 174
water for injection, 174
European Pharmacopeia (EP), 171
ion removal techniques, 185
continuous electro-deionization (CEDI), 188–189
design considerations, 191–192
electrodialysis, 189
and electrodialysis reversal, 189, 190
pharmaceutical applications, 188
physical elements, 187
resins, 186

Japanese Pharmacopeia (JP), 171
operation, 204
and commissioning, 205–207
facility considerations, 207–211
water system startup, 205–207
pretreatment techniques
activated carbon units, 226–228
cartridge filtration, 235–236
chemical injection systems, 220–221
heat exchangers, 223–224
inline ultraviolet unit, 232–234
organic scavenging resins, 231–232
particulate removal filters, 224–226
recirculation and repressurization pumps, 234–235
residual disinfectant agent, 221
sodium hydroxide injection, 221–222
thermal blending systems, 222–223
water-softening units, 228–230
reverse osmosis (RO)
ion removal techniques, 178–181
membranes, 183–184
polishing system design (PSD), 181
pressure vessels, 184–185
sampling provisions for, 182–183
thin-film composite membranes, 182
unit integral cleaning system, 185
specification of, 171–172
system design
commercial perspective, 200–204
control system requirements, 198
documentation requirements, 198–199
drawing requirements, 197–198
Factory Acceptance Test (FAT), 204
quality control/assurance, 199–200
requirements, 194–196
theory and operation, 192–194
system installation
constant system observation, 218–219
credential, 217–218
maintenance of water systems, 219
system evaluations, 219–220
technical considerations, 214–216
utility considerations, 212–214
water systems, validation, 236–237
alert and action, levels, 248–249
change control, 250
concurrent validation, 251–252
construction materials, 242–243
design qualification (DQ), 244–245
installation, 241–242
installation qualification (IQ), 245–246
lifecycle of, 240
maintenance and control, 239–240
monitoring programme, 249
operation procedure, 249
operational qualification (OQ), 246
performance qualification (PQ), 246–248
pharmaceutical, 239
pharmaceuticals, 238
preventative maintenance, 250
prospective validation, 251
protocol for, 240–241
and qualification, 243–244
retrospective validation, 250
revalidation, 248, 252
sanitation procedures, 248
sanitization, 249
treating water, methods, 240
water quality, 238–239
Phosphor-converted light-emitting diode (pc-LED), 273
Phosphorconverted WLEDs (pc-WLED)
activators, 277
energy transfer mechanism, 276–277
phosphor, 275–276
photoluminescence, 275
Photoluminescence (PL), 108–109
Photovoltaic technology, 32
Polishing system design (PSD), 181
Pretreatment techniques
activated carbon units, 226–228
cartridge filtration, 235–236
chemical injection systems, 220–221
heat exchangers, 223–224
inline ultraviolet unit, 232–234
organic scavenging resins, 231–232
particulate removal filters, 224–226
recirculation and repressurization pumps, 234–235
residual disinfectant agent, 221
sodium hydroxide injection, 221–222
thermal blending systems, 222–223
water-softening units, 228–230

Q

Quality control/assurance, 199–200
Quantum dots (QDs), 107
Quasi-zero-dimentional (0-D), 107
Quaternary Cu_2ZnSnS_4 (CZTS) solar cells, 32
copper zinc tin sulfide, 35–36
fluorine-doped tin oxide (FTO), 53
indium-doped tin oxide (ITO), 53
nanoparticles and nanocrystals, 49–53
nanostructures in, 41–48
quantum dot solar cells (QDSC), 37–41
successive ionic layer adsorption and reaction (SILAR), 53
nanotechnology
fabrication cost, 33–34
installation cost, 34

R

Rare-earth ion lasers, 86–87
Recirculation and repressurization pumps, 234–235
Residual disinfectant agent, 221
Retrospective validation, 250
Reverse osmosis (RO)
ion removal techniques, 178–181
membranes, 183–184
polishing system design (PSD), 181
pressure vessels, 184–185
sampling provisions for, 182–183
thin-film composite membranes, 182
unit integral cleaning system, 185
Ring laser gyroscope, 98

S

Saturable absorber
mode-locked lasers, 119–120
Science, technology, engineering, and mathematics (STEM), 75
Semiconductor lasers, 88–89
Sensitizers
Cerium (III) ions, role, 278–279
dysprosium (III) ions, 281–282
energy transfer, 279–281
Eu^{3+}, 285–286
Pr^{3+}, 288–290
rare earth elements, 277–278
samarium (III) ions, 283
Tm^{3+} ions, 286–288
Sensors, 125
Single-walled carbon nanotubes (SWCNT), 107
Sodium hydroxide injection, 221–222
Solar cells, 124–125
Solid-state lasers, 85
Solid-state lighting (SSL), 273
phosphorconverted WLEDs (pc-WLED)
activators, 277
energy transfer mechanism, 276–277
phosphor, 275–276
photoluminescence, 275
sensitizers
Cerium (III) ions, role, 278–279
dysprosium (III) ions, 281–282
energy transfer, 279–281
Eu^{3+}, 285–286
Pr^{3+}, 288–290
rare earth elements, 277–278
samarium (III) ions, 283
Tm^{3+} ions, 286–288
Specific surface area (SSA), 13
Sterile water for inhalation, 175
Sterile water for injection, 174
Sterile water for irrigation, 175
Super resolution, 93–94
Surface morphological studies, 63–64
Surface topography and morphology, 15–16
Sustainability, 2
Sustainable Development Goals (SDGs), 2, 3
System installation
constant system observation, 218–219
credential, 217–218
maintenance of water systems, 219
system evaluations, 219–220
technical considerations, 214–216
utility considerations, 212–214

T

Theories
Boltzmann constant, 61
crystalline size, 60
DC electrical conductivity studies, 61
Joncher's universal dynamic response (UDR), 62
Jonscher's Power Law (JPL), 62
Mott characteristic, 61
Thermal blending systems, 222–223

Touch screen, 122–123
Transition metal ion lasers, 85–86
Transparent conductive films (TCFs), 125–126

U

Ultrafast laser, 128–129
Underwater laser, 96–97
Unit integral cleaning system, 185
United States Pharmacopeial Convention, 171

W

Water purification
  nanocellulose-based, 146
    desalination, 154–155
    heavy metal contamination, 147–148
    oil spills, 148–149
    organic pollutants, 152–154
    toxic dyes, 149–152
  nanocomposites
    characteristics of, 145–146
Water storage tank
  installation qualification (IQ), 252
    accessory inspection, 255
    Certificate of Installation, 253
    component examination, 254–255
    components and accessories, 254
    control system inspection, 255–256
    documentation verification, 256–257
    equipment rationale, 253
    instrumentation inspection, 256
    IQ document's approval, 254
    Mconclusion, 254
    purpose, 253
    utilities inspection section, 256
Water systems
  validation, 236–237
    alert and action, levels, 248–249
    change control, 250
    concurrent validation, 251–252
    construction materials, 242–243
    design qualification (DQ), 244–245
    installation, 241–242
    installation qualification (IQ), 245–246
    lifecycle of, 240
    maintenance and control, 239–240
    monitoring programme, 249
    operation procedure, 249
    operational qualification (OQ), 246
    performance qualification (PQ), 246–248
    pharmaceutical, 239
    pharmaceuticals, 238
    preventative maintenance, 250
    prospective validation, 251
    protocol for, 240–241
    and qualification, 243–244
    retrospective validation, 250
    revalidation, 248, 252
    sanitation procedures, 248
    sanitization, 249
    treating water, methods, 240
    water quality, 238–239
Water-softening units, 228–230

X

X-ray energy, 59